SOCIAL WORK AND GERIATRIC SERVICES

SOCIAL WORK AND GERIATRIC SERVICES

Sharon Duca Palmer, CSW, LMSW

School Social Worker, ACLD Kramer Learning Center,
Bay Shore, New York; Certified Field Instructor,
Adelphi University School of Social Work,
Garden City, New York, U.S.A.

Apple Academic Press

TORONTO   NEW JERSEY

Social Work and Geriatric Services

© Copyright 2011*
Apple Academic Press Inc.

First Published in the Canada, 2011
Apple Academic Press Inc.
3333 Mistwell Crescent
Oakville, ON L6L 0A2
Tel. : (888) 241-2035
Fax: (866) 222-9549
E-mail: info@appleacademicpress.com
www.appleacademicpress.com

The full-color tables, figures, diagrams, and images in this book may be viewed at www.appleacademicpress.com

---

First issued in paperback 2021

ISBN 13: 978-1-77463-251-2 (pbk)
ISBN 13: 978-1-926692-87-6 (hbk)

Sharon Duca Palmer, CSW, LMSW

Cover Design: Psqua

Library and Archives Canada Cataloguing in Publication Data
CIP Data on file with the Library and Archives Canada

CONTENTS

Acknowledgments and How to Cite   8

Introduction   9

1. Chronic Disease Prevalence and Care Among the Elderly   13
    in Urban and Rural Beijing, China—A 10/66 Dementia
    Research Group Cross-Sectional Survey

    Zhaorui Liu, Emiliano Albanese, Shuran Li, Yueqin Huang,
    Cleusa P. Ferri, Fang Yan, Renata Sousa, Weimin Dang and
    Martin Prince

2. Challenges to Conducting Research with Older People   33
    Living in Nursing Homes

    Sue Hall, Susan Longhurst and Irene J. Higginson

3. Association between Subjective Memory Complaints and   45
    Health Care Utilisation: A Three-Year Follow Up

    Frans Boch Waldorff, Volkert Siersma and Gunhild Waldemar

4. Cognitive Function, Social Integration and Mortality in a   59
    U.S. National Cohort Study of Older Adults

    Thomas O. Obisesan and R. F. Gillum

5.   Magnitude of Potentially Inappropriate Prescribing in   75
     Germany Among Older Patients with Generalized Anxiety Disorder
     Ariel Berger, Marko Mychaskiw, Ellen Dukes, John Edelsberg and
     Gerry Oster

6.   Social Vulnerability, Frailty and Mortality in Elderly People   89
     Melissa K. Andrew, Arnold B. Mitnitski, and Kenneth Rockwood

7.   Impact of Exercise in Community-Dwelling Older Adults   106
     Ruth E. Hubbard, Nader Fallah, Samuel D. Searle, Arnold Mitnitski
     and Kenneth Rockwood

8.   Social Participation and Independence in Activities of Daily Living: 119
     A Cross Sectional Study
     Encarnación Rubio, Angelina Lázaro and Antonio Sánchez-Sánchez

9.   Age-Related Attenuation of Dominant Hand Superiority   139
     Tobias Kalisch, Claudia Wilimzig, Nadine Kleibel, Martin Tegenthoff
     and Hubert R. Dinse

10. Are Sedatives and Hypnotics Associated with Increased Risk of   162
     Suicide in the Elderly?
     Anders Carlsten and Margda Waern

11. Mental Rotation of Faces in Healthy Aging and Alzheimer's Disease   173
     Cassandra A. Adduri and Jonathan J. Marotta

12. Preventing Falls in Older Multifocal Glasses Wearers by   194
     Providing Single-Lens Distance Glasses: The Protocol for the
     VISIBLE Randomized Controlled Trial
     Mark J. Haran, Stephen R. Lord, Ian D. Cameron, Rebecca Q. Ivers,
     Judy M. Simpson, Bonsan B. Lee, Mamta Porwal, Marcella M. S. Kwan
     and Connie Severino

13. Emotional Stress as a Trigger of Falls Leading to Hip or   208
     Pelvic Fracture. Results from the Tofa Study—A Case-Crossover
     Study Among Elderly People in Stockholm, Sweden
     Jette Möller, Johan Hallqvist, Lucie Laflamme, Fredrik Mattsson,
     Sari Ponzer, Siv Sadigh and Karin Engström

14. Health Status Transitions in Community-Living Elderly with   223
     Complex Care Needs: A Latent Class Approach
     Louise Lafortune, François Béland, Howard Bergman and Joël Ankri

15. Effectiveness of a Mobile Smoking Cessation Service in Reaching   248
     Elderly Smokers and Predictors of Quitting

     Abu Saleh M. Abdullah, Tai-Hing Lam, Steve K. K. Chan,
     Gabriel M. Leung, Iris Chi, Winnie W. N. Ho and Sophia S. C. Chan

16. Factors Associated with Self-Rated Health in Older People Living   264
     in Institutions

     Javier Damián, Roberto Pastor-Barriuso and
     Emiliana Valderrama-Gama

17. Discomfort and Agitation in Older Adults with Dementia   276

     Isabelle Chantale Pelletier and Philippe Landreville

18. Factors Influencing Elderly Women's Mammography Screening   293
     Decisions: Implications for Counseling

     Mara A. Schonberg, Ellen P. McCarthy, Meghan York,
     Roger B. Davis and Edward R. Marcantonio

19. Do Social Networks Affect the Use of Residential Aged Care   307
     Among Older Australians?

     Lynne C. Giles, Gary F. V. Glonek, Mary A. Luszcz and
     Gary R. Andrews

     Index   329

ACKNOWLEDGMENTS AND HOW TO CITE

The chapters in this book were previously published in various places and in various formats. By bringing these chapters together in one place, we offer the reader a comprehensive perspective on recent investigations into this important field.

We wish to thank the authors who made their research available for this book, whether by granting permission individually or by releasing their research as open source articles or under a license that permits free use provided that attribution is made. When citing information contained within this book, please do the authors the courtesy of attributing them by name, referring back to their original articles, using the citations provided at the end of each chapter.

INTRODUCTION

Social work is a difficult field to operationally define, as it is practiced differently in many settings. It is a very diverse occupation and one that can be practiced in settings such as hospitals, clinics, welfare agencies, schools, and private practices.

The main goal of all social work practice is to assist the client to function at the best of their ability and assess what their needs are. Social workers help clients with problem-solving strategies, such as defining personal goals, focusing on what is necessary to make changes, and helping them through the process.

Social work is a demanding field and is often emotional draining. Many social workers have large caseloads, limited resources for their clients, and often work for relatively low salaries. But the personal rewards can be very satisfying.

The social work profession is committed to promoting social and economic policy though helping to improve people's lives. Research is conducted to improve social services, community development, program evaluation, and public administration. The importance of research in these areas is to examine variables that can be addressed in order to resolve issues. Research can lead to what is called "best practice". By utilizing "best practice", a social worker is engaging clients based on research that is intended to increase successful outcomes.

Social work is one of the most diverse careers available. Most social workers are employed by health care facilities and government agencies. These facilities can include hospitals, mental health clinics, nursing homes, rehabilitation centers, schools, child welfare agencies, and private practice.

Social work's interface with mental health promotion and the treatment of mental illness dates to the earliest roots of our profession. While many social workers provide mental health services in private practice settings, the majority of services are offered in community-based agencies, both public and private, and in hospitals and prisons. Social workers are the largest provider of mental health services, providing more services than all other mental health care providers combined. These workers also often provide services to those who are struggling with substance abuse.

Twenty-first century health issues are complex and multidimensional, requiring innovative responses across professions at all levels of society. Public health social workers work to promote health in hospitals, schools, government agencies and local community-based settings, making connections between prevention and intervention from the individual to the whole population.

In an ideal world, every family would be stable and supportive. Every child would be happy at home and at school. Every elderly person would have a carefree retirement. Yet in reality, many children and families face daunting challenges. For example, single parents struggle to raise kids while working. Teens may become parents before they are ready. Child social workers help kids get back on track so they can lead healthy, happy lives.

Rapid aging populations are expected worldwide. With the rapid growth of this population, social work education and training specializing in older adults and practitioners interested in working with older adults are increasingly in demand. Geriatric social workers typically provide counseling, direct services, care coordination, community planning, and advocacy in an array of organizations including in homes, neighborhoods, hospitals, senior congregate living and nursing facilities. They work with older people, their families and communities, as well as with aging-related policy, and aging research

In whatever subcategory they work, social workers help provide support services to individuals and communities by assessing their needs in order to improve the quality of life and overall well-being. This can lead to positive changes in people's environments, dignity, and self-worth. It can also lead to changes in social policy for those who are vulnerable and oppressed. Social workers change entire communities for the better.

There have been many changes emerging in the social work profession. The uses of the Internet and online counseling have been major trends. Some people are more likely to seek assistance and information first through the use of the Internet. There has also been a strong move for collaborating between professions when providing services in order to offer clients more options for success. Keeping up to date with best practice research, licensing requirements, continuing education, and professional ethics make this an exciting and challenging time to be a social worker!

— **Sharon Duca Palmer, CSW, LMSW**

Chronic Disease Prevalence and Care Among the Elderly in Urban and Rural Beijing, China — A 10/66 Dementia Research Group Cross-Sectional Survey

Zhaorui Liu, Emiliano Albanese, Shuran Li,
Yueqin Huang, Cleusa P. Ferri, Fang Yan,
Renata Sousa, Weimin Dang and Martin Prince

ABSTRACT

Background

Demographic ageing is occurring at an unprecedented rate in China. Chronic diseases and their disabling consequences will become much more common.

Public policy has a strong urban bias, and older people living in rural areas may be especially vulnerable due to limited access to good quality healthcare, and low pension coverage. We aim to compare the sociodemographic and health characteristics, health service utilization, needs for care and informal care arrangements of representative samples of older people in two Beijing communities, urban Xicheng and rural Daxing.

Methods

A one-phase cross-sectional survey of all those aged 65 years and over was conducted in urban and rural catchment areas in Beijing, China. Assessments included questionnaires, a clinical interview, physical examination, and an informant interview. Prevalence of chronic diseases, self-reported impairments and risk behaviors was calculated adjusting for household clustering. Poisson working models were used to estimate the independent effect of rural versus urban residence, and to explore the predictors of health services utilization.

Results

We interviewed 1002 participants in rural Daxing, and 1160 in urban Xicheng. Those in Daxing were more likely to be younger, widowed, less educated, not receiving a pension, and reliant on family transfers. Chronic diseases were more common in Xicheng, when based on self-report rather than clinical assessment. Risk exposures were more common in Daxing. Rural older people were much less likely to access health services, controlling for age and health. Community health services were ineffective, particularly in Daxing, where fewer than 3% of those with hypertension were adequately controlled. In Daxing, care was provided by family, who had often given up work to do so. In Xicheng, 45% of those needing care were supported by paid caregivers. Caregiver strain was higher in Xicheng. Dementia was strongly associated with care needs and caregiver strain, but not with medical helpseeking.

Conclusion

Apparent better health in Daxing might be explained by under-diagnosis, under-reporting or selective mortality. Far-reaching structural reforms may be needed to improve access and strengthen rural healthcare. The impact of social and economic change is already apparent in Xicheng, with important implications for future long-term care.

Background

China, with over 1.3 billion citizens is the world's most populous country [1]. Demographic ageing is occurring at a rate unprecedented for any world region;

the proportion of Chinese aged 65 and over will increase from 4% in 2000 to 14% by 2025, amounting to 200 million older people [2]. The prevention and control of chronic diseases is recognized as an urgent priority in China [3,4]. The long term care needs of dependent older people has, comparatively, received much less attention [5-7]. A little over half of China's population lives in rural settings, [8] the proportion among older people is likely to be higher.

The literature on rural/urban differences in health in China suggests three main areas for further study. First, there may be differences in chronic disease prevalence and mortality. Data from the China Health and Nutrition survey showed a 30% increased mortality for those aged 50 and over in rural compared with urban areas, partly mediated through the lack of amenities and lower wages in rural areas [9]. However, in the third Chinese National Health Service Survey (CNHSS), [10] the prevalence of self-reported physician diagnoses of chronic diseases was lower among rural than urban residents. In the 1991 Beijing Longitudinal Study on Aging (BLAS), [11] urban residents were 3.2 times more likely to report chronic disease diagnoses, 4.5 times more likely to report diabetes and 2.5 to three times more likely to report respiratory disease, heart disease and hypertension. On the other hand, levels of disability were similar in the CNHSS, [10] and considerably higher in rural areas in BLAS [11]. Hypertension, when measured directly as opposed to by self-report, is more prevalent in rural areas, [11] and awareness is higher in urban areas [12]. Hence, rural communities seem to be advantaged with respect to some health outcomes, and disadvantaged in others. More research is required to confirm these findings, and to clarify the reasons for the apparent discrepancies.

Second, in China, financing, coverage and access to healthcare depends, largely upon where you live. Only 61% of rural residents, compared with 82% of urban dwellers can access health services within one kilometre of their homes [13]. In urban China, there are two employee-based health insurance schemes, one for government and the other for public and private company employees. There is limited cover for dependents, based on a personal annual subscription. Discounts are available for poor people, those with mental disorders, and retirees. In rural China, the government contributes to a common fund which covers healthcare costs but only proportionate to the amount contributed. In 2003, 79% of rural and 45% of urban residents did not have meaningful health insurance [12]. Almost 50% of health care costs are covered by out-of-pockets payments [14] and more than 35% of urban and 45% of rural households cannot afford any health care [15]. The consequences of these disparities for rural and urban older people, in terms of their ability to access healthcare and to manage their chronic diseases, need to be determined.

Third, social protection (encompassing the range of formal and informal mechanisms to provide safety nets and support to poor and disadvantaged members of society) is under threat for older people in China, as its population ages rapidly. In traditional societies social protection is provided by the family and community. With modernisation, these responsibilities are shared by wider society through intergenerational transfers legislated for and managed by the state [16]. However, in the future there will be fewer children available to provide support and care because of the one child policy [17], and social protection by the state, both in terms of pension coverage and insurance against catastrophic healthcare expenditure, remains very patchy [18]. It is important to understand how these processes may be playing out with respect to older people living in urban and rural communities.

In summary, the urban bias of public policy is particularly marked in China, and older people living in rural areas may be especially vulnerable. The Research Agenda on Ageing Project [19] has advocated more research on this group, including demographic and migration patterns; social transitions; family exchanges; health behaviors and use of and access to healthcare. In this paper we seek to pursue this agenda by

(a) comparing the socio-demographic and health characteristics of representative samples of older people in two Beijing communities, urban Xicheng and rural Daxing.

(b) describing the patterns of recent health service utilization among rural and urban elderly and estimating the independent effect of health conditions, socio-demographic and socioeconomic factors on access to and use of these services.

(c) comparing the levels of disability and needs for care, informal care arrangements and extent of carer strain with respect to dependent older people in urban Xicheng and rural Daxing.

Methods

Study Design and Catchment Areas

This study is part of the 10/66 Dementia Research Group's multi-site international program on dementia and ageing in low and middle income countries. The full study protocol and procedures are available in an open access online publication [20]. From January 2004 to April 2005 we carried out one-phase cross-sectional surveys of all older people (aged 65 years and over) residing in two defined catchment areas: the urban district of Xicheng in the center of Beijing, close to

Tiananmen Square and the 14 villages of Daxing, a rural district 40 kilometres away. In China, rural regions are divided into four classes mainly according to income level, with Class I being the richest and Class IV the poorest [21]—Daxing in common with other rural areas of Beijing is classified under Class I. Participants were identified by means of door-knocking and no exclusion criteria were applied. Sample sizes of 1,000 would allow an estimation of a typical 4.5% dementia prevalence with a standard error of 1.2%. Community doctors administered the interviews, lasting two to three hours, in participants' homes. Informed consent was obtained from informants and participants. The institutional review boards of the Institute of Psychiatry, King's College London and the Institute of Mental Health, Peking University approved the project.

Measures

The 10/66 protocol comprises questionnaires on participants' sociodemographic characteristics, health status, risk factor exposures and health service use. A physical examination was carried out and an informant interview administered for information on care arrangements, and caregiver practical, psychological and economic strain. All measures and assessments of the 10/66 protocol are described in detail elsewhere, [20] only those relevant for the purposes of this analysis are described here:

(1) Sociodemographic characteristics: gender, age, education, marital status, household living circumstances, sources of income (pension and family support), number of household assets.

(2) Health Status:

   a. Chronic disease diagnoses established through structured clinical interviewing and or physical examination. Dementia—all those who on the basis of cognitive testing, clinical and informant interview met either or both of the cross-culturally validated 10/66 dementia [22] or DSM IV dementia criteria [23]. Depression - ICD-10 depressive episodes, ascertained using the Geriatric Mental State structured clinical interview [24]. Self-reported stroke, ischemic heart disease and diabetes ("have you ever been told by a doctor that you had a stroke/heart attack/angina/diabetes?"). Chronic obstructive pulmonary disease, defined as reporting a chronic cough with production of sputum for three or more months. Hypertension—all those with self-reported hypertension ("have you ever been told by a doctor that you have high blood pressure?") and/or a blood pressure measurement meeting the European Society of Hypertension criteria [25]. (systolic blood pressure > = 140 mm Hg and/or diastolic blood pressure > = 95 mm Hg) were considered to have hypertension. Those with self-reported hypertension were considered to be detected. Those with self-reported hypertension but not meeting ESH criteria were considered to be controlled. Those with self-reported treatment ("Were you started on treatment?") were considered to be treated.

b. Physical impairments—self-reported: arthritis or rheumatism; eyesight problems; hearing difficulties or deafness; persistent cough; breathlessness, difficulty breathing or asthma; heart trouble or angina; stomach or intestine problems; faints or blackouts; paralysis, weakness or loss of one leg or arm; skin disorders [26]. Each impairment was rated if it interfered with activities 'a little' or 'a lot.'

c. Disability—measured using the 12 item World Health Organization Disability Assessment Schedule (WHODAS II., developed by WHO for cross-cultural research [27])

d. Risk Factors: Smoking ("ever smoked," "current smoker," "life-time smoking"), hazardous alcohol use ("ever been a heavy drinker" currently and before the age of 60), physical activity ("have you walked at least 0.5 Km in the last month?"), obesity (direct waist circumference in cm and meets waist circumference criteria for metabolic syndrome)

(3) Health Service Utilization: Any use of community services (primary care, hospital outpatient, private doctor, dentistry) and hospital admission during the three months preceding the interview.

(4) Dependency (needs for care): we used a series of open-ended questions addressed to a key informant, to define the family network, to establish if the older person needed and received care from family members or others and to identify who was responsible for organising and providing 'hands on' care. On the basis of these questions, the interviewer coded whether the older person required no care, care for some of the time or care for much of the time.

(5) Care arrangements and impact of providing care

a. Informal care arrangements (time spent with the participant, and time spent assisting to communicate, to use transport, to dress, to eat, and for personal hygiene), paid day and night care

b. The gender of the main carer and their relationship to the cared for older person

c. Carer strain. Carer perceived strain was assessed using the Zarit Burden Interview (ZBI) [28] with 22 items that assess the carer's appraisal of the impact their involvement has had on their lives. The ZBI has been widely used in the USA and Europe, but also in Nigeria and Taiwan [29,30] and in Japan, where it was formally validated [31]. When used in the 10/66 Dementia Research Group pilot studies in 24 centers in Latin America, India, China and Africa the ZBI was found to be practical, culturally relevant, and to have robust psychometric properties [32]. Carer psychological strain was assessed using the 20 item Self-Reporting Questionnaire [33]. Economic strain was assessed according to whether the carer had had to cut back on work to care.

Analysis

We describe and compare the characteristics of the urban and rural samples according to sociodemographic circumstances (χ2 and t-test, as appropriate), health status (Poisson regression with robust prevalence ratios controlling for age and gender, and for WHODAS II scores negative binomial regression) and health service utilisation (Poisson regression controlling for age, gender and physical impairments).

(1) We compared (χ^2 tests) the proportion of all hypertension cases that are detected, the proportion of detected cases that are a) treated and b) controlled and the proportion of all cases that are controlled.

(2) We estimated the independent predictors of health service utilization (accessing any community health service in the last three months) separately for urban and rural samples. Crude and adjusted prevalence ratios (PRs) with 95% confidence intervals were calculated using Poisson regression adjusting for household clustering, because sometimes two participants of a spouse pair were recruited to the survey. These two persons could not be regarded as statistically independent [34]. Variables included in the model were: dementia diagnosis, one or more physical impairments, age, gender, education, marital status, assets, health insurance and pension.

(3) Dementia has previously been shown to be an important determinant of the level and type of informal care, and the extent of carer strain [35,36]. Therefore, among those needing care we assessed the effect of dementia diagnosis, and the effect of rural or urban residence (adjusting for age, gender and independently of dementia status) upon: disability (WHODAS II) and dependency, time spent assisting with activities of daily living (ADL), carer strain (ZBI scores) and mental health (SRQ-20), and care arrangements (daytime or night time paid carer, cutting back on work to care, additional informal care). To contrast rural and urban samples, we calculated prevalence ratios (PRs) using Poisson regression adjusting for household clustering for dichotomous outcomes, rate ratios using ordinal

regression adjusting for household clustering for ordered categorical variables, and mean differences using General Linear Modeling adjusting for household clustering for continuous variables, all with 95% confidence intervals controlling for age, gender and dementia status.

All analyses were carried out on release 1_5 of the 10/66 dataset using STATA 9.2 (Stata 9.1; Stata Corp., College Station, Texas)

Results

In all, 2162 (1160 urban and 1002 rural) participants completed the survey, with 95.7% responding in rural Daxing and 74.3% in urban Beijing, where more eligible people refused to participate or could not be contacted after at least four attempts. The urban elderly were better educated and older, and less likely to be widowed than their rural counterparts (Table 1). Living alone was unusual in either setting, but urban residents were more likely to be living with their spouse only, and less likely to be living with children. Pension coverage was much lower in the rural (3.8%) than the urban sample (90.5%). Conversely, family transfers and rental income were much more common in the rural sample. Only nine urban participants and one rural participant reported receiving a disability pension.

Table 1. Social-demographic characteristics in Daxin (rural) and Beijing (urban)

	Urban (n = 1160)	Rural (n = 1002)	χ^2 /Z	df	P
Age group (MV)	0	0			
65-69 years	316(27.2%)	383(38.2%)	40.4	3	< 0.001
70-74 years	362(31.2%)	296(29.5%)			
75-79 years	254(21.9%)	202(20.2%)			
80+ years	228(19.7%)	121(12.1%)			
Gender (MV)	0	0			
Female	661(57.0%)	556(55.5%)	0.5	1	0.49
Education level (MV)	0	0			
No education	232(20.0%)	579(57.8%)	491.9	3	< 0.001
Primary education only	456(39.3%)	373(37.2%)			
Completed secondary	335(28.9%)	45(4.5%)			
Completed tertiary	137(11.8%)	5(0.5%)			
Marital status (MV)	0	0			
Never married	3(0.3%)	22(2.2%)	52.0	3	< 0.001
Married or cohabiting	829(71.5%)	585(58.4%)			
Widowed	326(28.1%)	394(39.3%)			
Divorced or separated	2(0.2%)	1(0.1%)			
Previous occupation[1] (MV)	1	1			
Professional/Managerial/Clerical	518(45.0%)	37(3.7%)	1888.6	3	< 0.001
Skilled or semi-skilled	421(36.5%)	14(1.4%)			
Unskilled	207(18.0%)	13(1.3%)			
Agricultural worker	7(0.6%)	938(93.6%)			
Living arrangements (MV)	0	0			
Alone	54(4.7%)	49(4.9%)	202.0	3	< 0.001
With spouse only	415(35.8%)	194(19.4%)			
With children	446(38.4%)	679(67.8%)			
With others	245(21.1%)	80(8.0%)			
With children under 16	217(18.7%)	462(46.1%)	187.4	1	< 0.001
Source of income (MV)	0	0			
Government or occupational pension	1050 (90.5%)	38(3.8%)	1617.5	1	< 0.001
Family transfers	54(4.7%)	305(36.4%)	347.3	1	< 0.001
Disability pension	9(0.8%)	1(0.1%)	5.3	1	0.02
Rent	0	122(12.2%)	149.7	1	< 0.001
Paid work	0	6(0.6%)	7.0	1	0.008
Have health insurance plan	14(1.2%)	769(76.9%)	1327.2	1	< 0.001
Number of household assets[2] (MV)	0	0			
0-3 assets	6(0.5%)	108(10.8%)	113.3	1	< 0.001
More than 3 assets	1154(99.5%)	894(89.2%)			

MV = Missing values
[1] Participants were asked 'What was your best (highest level) job?'
[2] household assets: television, fridge, water and electricity utilities, telephone, plumbed toilet and plumbed bathroom

All of the self-reported diagnoses and impairments, except hearing problem and limb impairment, were much less common in the rural sample (Table 2). Older people in the rural sample were four times less likely to report three or more limiting impairments and nearly five times more likely to rate their health positively. The picture was different for diagnoses made on the basis of clinical interview and examination. The prevalence of dementia was similar, while that of hypertension was just 20% lower, and that of uncontrolled hypertension 30% higher in the rural sample. Significantly more rural (22.2%) than urban (13.6%) elderly reported chronic pain. Rural residents were more likely to have smoked, and to continue to smoke, to have engaged in hazardous drinking, and to be sedentary, but were less likely to be obese.

Table 2. Health status in Xicheng (urban) and Daxing (rural).

Health condition	Urban, Xicheng (n = 1160)	Rural, Daxing (n = 1002)	Prevalence ratio (rural vs urban) adjusted for age and gender
Diagnosed diseases (MV)	0	0	
Dementia	84(7.2%)	56(5.6%)	0.96(0.70-1.32)
History of hypertension, and/or meets ESH criteria	726(62.6%)	500(49.9%)	0.80(0.74-0.87)
Uncontrolled hypertension (ESH criteria)	471(40.6%)	487(48.6%)	1.29(1.10-1.33)
ICD-10 Depression	3(0.3%)	7(0.7%)	3.05(0.83-11.2)
Self-reported diagnoses (MV)	1	0	
Diabetes	195(16.8%)	9(0.9%)	0.05(0.03-0.10)
Ischaemic heart disease (myocardial infarction or angina)	115(9.9%)	12(1.2%)	0.12(0.07-0.23)
Stroke	109(9.4%)	18(1.8%)	0.20(0.12-0.34)
Chronic obstructive pulmonary disease	36(3.1%)	16(1.6%)	0.54(0.30-0.96)
Self-reported impairments (MV)	0	0	
Arthritis	165(14.2%)	20(2.0%)	0.14(0.09-0.23)
Eye problem	194(16.7%)	65(6.5%)	0.41(0.31-0.55)
Hearing problem	142(12.2%)	86(8.6%)	0.83(0.64-1.08)
Cough problem	33(2.8%)	14(1.4%)	0.51(0.27-0.96)
Breathing problem	52(4.5%)	19(1.9%)	0.45(0.27-0.74)
Heart problem	329(28.4%)	31(3.1%)	0.11(0.08-0.16)
Gastrointestinal problem	67(5.8%)	12(1.2%)	0.22(0.12-0.40)
Fainting problem	62(5.3%)	10(1.0%)	0.19(0.10-0.37)
Limb problem	72(6.2%)	44(4.4%)	0.77(0.53-1.12)
Skin problem	12(1.0%)	2(0.2%)	0.20(0.04-0.94)
Three or more physical impairments	208(17.9%)	39(3.9%)	0.23(0.17-0.33)
Pain that interferes with life	158(13.6%)	222(22.2%)	1.60(1.31-1.94)
Locomotion (observed) (MV)	0	0	
Obvious abnormality of walking	53(4.6%)	27(2.7%)	0.64(0.41-1.04)
WHODAS II disability score (MV)	10	2	
Mean	8.1 ± 20.1	8.0 ± 14.6	1.20(0.96-.49)[1] 0.59(0.51-0.67)[2]
Self-rated health (MV)	0	0	
'Good' or 'very good'	176(15.2%)	690(68.9%)	4.59(3.94-5.35)
Dependency	0	0	
Needs any care	183(15.8%)	54(5.4%)	0.41(0.30-0.54)
Needs much care	119(10.3%)	30(3.0%)	0.34(0.23-0.51)
Chronic disease risk factors (MV)	12	4	
Ever smoked	284(24.5%)	336(33.5%)	1.28(1.12-1.46)
Current smoker	193(16.6%)	305(30.4%)	1.68(1.43-1.97)
20 or more pack years of smoking	180(15.5%)	274(27.3%)	1.62(1.38-1.91)
Hazardous drinker in early life	26(2.2%)	73(7.3%)	2.89(1.83-4.52)
Current hazardous drinker	17(1.5%)	42(4.2%)	2.66(1.55-4.56)
No walks of > 0.5 km in last month	209(18.0%)	384(38.3%)	1.39(1.32-1.47)
Obesity (meets waist circumference criterion for metabolic syndrome)	530(45.7%)	158(15.8%)	0.35(0.30-0.41)

[1]Negative binomial regression
[2]Zero inflated negative binomial regression
MV = Missing values

Rural residents (6.1%) were strikingly less likely than urban residents (38.6%) to have used any health services over the three months prior to the survey. Underutilisation of services by rural residents was apparent even after controlling for age, gender and number of limiting physical impairments (Prevalence ratio 0.24, 95% CI 0.19 to 0.32). Underutilisation by rural elderly was equally apparent for primary care services (3.8% versus 20.9%, adjusted PR 0.32, 95% CI 0.23 to 0.45), hospital doctor services (2.2% versus 23.4%, adjusted PR 0.14, 95% CI 0.09 to 0.22) and hospital admission (0.5% versus 2.4%, adjusted PR 0.43, 95% CI 0.15-1.20). Hypertension was less likely to be detected among rural compared with urban residents, and detected cases were much less likely to be controlled (Table 3). The net result was that only 2.6% of all cases of hypertension were controlled in rural Daxing compared with 35.1% in urban Xicheng.

Table 3. Detection and control of hypertension, by site

	Urban, Xicheng (n = 1160)	Rural, Daxing (n = 1002)	χ^2	v	P
Detection and control of hypertension (MV)	0	0			
The proportion of all hypertension cases that are detected	78.5%(570/726)	50.8%(254/500)	103.2	1	< 0.001
The proportion of detected cases that are treated	96.8%(552/570)	99.6%(253/254)	6.0	1	0.015
The proportion of detected cases that are controlled	44.7%(255/570)	5.1%(13/254)	125.7	1	< 0.001
The proportion of all cases that are controlled	35.1% (255/726)	2.6% (13/500)	181.5	1	< 0.001

MV = Missing values

In both urban and rural sites, numbers of physical impairments were the strongest independent predictors of health service utilisation, after controlling for age, gender, education, assets, pension availability and health insurance (Table 4). Dementia was associated with health service utilization only in rural Daxing, but the association was no longer apparent after controlling for covariates. Economic factors (household assets, receipt of pension and health insurance) predicted health service utilization only in urban Xicheng.

Table 4. Predictors of health service utilization (crude and adjusted robust Prevalence Ratios [PRs] with 95% confidence intervals [CI])

	Crude PRs (95% CI)		Adjusted PRs* (95% CI)	
	Urban	Rural	Urban	Rural
Dementia	1.12(0.86-1.45)	2.19(1.05-4.57)	0.94(0.73-1.20)	1.54(0.82-3.06)
Number of limiting physical illnesses				
None	1 (ref)	1 (ref)	1(ref)	1 (ref)
1-2	2.32(1.76-2.84)	4.02(2.31-7.00)	2.26(1.79-2.87)	3.82(2.12-6.85)
3 or more	3.78(2.98-4.81)	8.91(4.52-17.6)	3.74(2.94-4.75)	8.31(4.06-17.0)
Age (per 5 year increment)	1.03(0.96-1.10)	0.93(0.73-1.17)	1.00(0.96-1.08)	0.80(0.61-1.04)
Gender (male vs. female)	0.91(0.79-1.04)	0.87(0.53-1.41)	0.89(0.77-1.03)	0.99(0.58-1.70)
Education (per level)	1.04(0.99-1.09)	0.90(0.77-1.07)	1.02(0.96-1.08)	0.86(0.63-1.19)
Assets (per quarter)	1.28(1.14-1.44)	0.90(0.77-1.05)	1.23(1.09-1.38)	0.84(0.70-1.02)
Any government or occupational pension	1.51(1.10-2.08)	1.31(0.43-3.98)	1.46(1.07-1.99)	1.17(0.39-3.57)
Have health insurance	1.87(1.33-2.62)	1.75(0.90-3.41)	1.94(1.28-2.95)	1.58(0.82-3.06)

* Mutually adjusted for all other exposures in the crude model

Among the 237 participants who were rated as needing care we described levels of disability and dependency, informal care arrangements and carer strain by site (Table 5). In both settings, people with dementia were more disabled than other needing care (mean WHODAS II score 61.1, SD 30.6 versus 33.1, SD 25.7 in Xicheng; 65.5, SD 24.3, p < 0.001 versus 33.6, SD 22.7, p < 0.001 in Daxing) and more likely to be rated as needing care 'much of the time' (77% versus 57%, p = 0.005 in Xicheng; 64% versus 46%, p = 0.18 in Daxing). Carers of people with dementia spent more time assisting with basic activities of daily living (tests for trend $\chi^2 = 14.1$, P = 0.001 in Xicheng, $\chi^2 = 9.9$, P = 0.007 in Daxing), particularly communication, dressing, eating, grooming and toileting. Caregiver strain, measured using the Zarit Burden Interview was also higher among those caring for people with dementia (mean ZBI score 26.4, SD 20.6, p < 0.001 versus 12.1, SD 12.6, p < 0.001 in Xicheng; 17.1, SD 14.9 versus 5.3, SD 7.7 in Daxing, p < 0.001). It was therefore important to control for dementia diagnosis, as well as the age and gender of the participant when comparing care-related variables between rural and urban settings (Table 5). Adjusted analyses suggested no differences in levels of disability or dependency between rural and urban older people needing

Table 5. Levels of disability and dependency, informal care arrangements and carer strain (among those identified as needing care), by site

	Xicheng (urban) Total (n = 183)	Daxing (rural) Total (n = 54)	Effect size for Rural vs. urban contrast, adjusting for age, gender and dementia status
Disability and dependency (in the care recipient) (MV)	0	2	
WHODAS 12 Disability score (mean [SD])	44.2 (30.9)	50.2 (28.4)	1.7 (-6.8,10.3)[1]
Needs care 'much of the time'	119 (65.0%)	30 (55.6%)	0.81 (0.63, 1.05)[2]
Time spent by the carer assisting with ADL (MV)	0	0	
0 hours	28 (15.3%)	7 (13.0%)	0.52 (0.43-0.63)[3]
1-4 hours	57 (31.1%)	26 (48.1%)	
5 hours +	98 (53.6%)	21 (38.9%)	
Assistance provided for specific ADL (> one hour/day) (MV)	0	0	
Supervision	16 (8.7%)	8 (14.8%)	1.64 (0.70-3.85)[2]
Communication	65 (35.5%)	22 (40.7%)	1.13 (0.78, 1.64)[2]
Using transport	8 (4.4%)	2 (3.7%)	0.93 (0.23, 3.75)[2]
Dressing	55 (30.1%)	10 (18.5%)	0.59 (0.33, 1.05)[2]
Eating	57 (31.1%)	15 (27.8%)	0.78 (0.50, 1.22)[2]
Grooming	57 (31.1%)	10 (18.5%)	0.55 (0.31, 1.00)[2]
Toileting	73 (39.9%)	14 (25.9%)	0.61 (0.38, 0.96)[2]
Bathing	57 (31.1%)	10 (18.5%)	0.55 (0.31, 1.00)[2]
Caregiver Strain	3	0	
Zarit Burden Interview Score (mean [SD])	17.9 (17.7)	11.4 (13.3)	-8.7 (-3.9, -13.5)[1]
Caregiver mental health SRQ-20 Score			
Mean (SD)	0.9(2.3)	1.3(2.6)	0.1(-0.6,0.8)[1]
Median (IQR)	0(0,1.0)	0 (0,1.0)	
Characteristics of main carer (MV)	1	2	
Relationship to older person			
Spouse	71 (38.8%)	21 (38.9%)	-
Child	69 (37.7%)	21 (38.9%)	
Daughter-/son-in-law or other relative	13 (7.1%)	11 (20.4%)	
Non-relative	30 (16.4%)	1 (1.9%)	
Gender			
Female	123 (67.2%)	27 (50.0%)	1.15 (1.05, 1.27)[2]
Care arrangements (MV)	0	1	
Daytime paid carer	83 (45.4%)	1 (1.9%)	0.05 (0.01, 0.33)[2]
Night time paid carer	81 (44.3%)	0	-
Carer cut back on work to care	7 (3.8%)	26 (48.1%)	11.7 (5.20, 26.4)[2]
Additional informal care	13 (7.1%)	12 (22.2%)	2.78 (1.37, 5.63)[2]

Abbreviations used in the table: MV+ missing values; WHO-DAS 2.0 = World Health Organization Disability Assessment Schedule (12-item version); ADL = activities of daily living; SRQ-20 = Self Reporting Questionnaire (20 items).
[1] Mean difference from a general linear model, with 95% confidence intervals
[2] Prevalence ratios (PRs) from Poisson regression (robust 95% confidence intervals)
[3] Prevalence ratios (PRs) from ordinal regression,(robust 95% confidence intervals)

care. However, rural carers spent less time assisting with core activities of daily living, and reported lower levels of strain. Paid care was a common option in urban Beijing; one half of dependent people with dementia and slightly less than one half of all urban dependent people paid for daytime care, with a similar proportion using night time care. Only one rural family used paid daytime care. Instead, rural carers were nearly 12 times more likely to give up or cut back on work to care, and nearly three times more likely to benefit from additional informal care from friends or family.

Discussion

We carried out a comprehensive one phase survey of two catchment areas in Beijing province; Daxing's rural villages and Xicheng in the heart of Beijing city. There were relatively few non-responders, but the higher proportion in urban Xicheng (25.7%) compared with rural Daxing (4.3%) creates some potential for response bias. We applied the same catchment area sampling techniques and research protocol in both settings, and the same research group supervised the implementation of the research. Given the proximity, shared language and culture of the two sites, we believe that the comparison was apt and likely to be informative regarding the impact of contrasting infrastructure, policies, lifestyles and family structures on health outcomes and chronic disease care. However, clearly, findings from this comparison cannot be generalised to urban and rural settings in China as a whole. In particular, Daxing is less remote, and better resourced than the majority of rural locations in China. We set out to compare rural and urban samples with respect to the health status of older people, their use of health services, and their needs for informal care. For older people, these three elements are very much inter-related. Other studies that have addressed just one or other of these elements in isolation have not provided a comprehensive overview of chronic diseases, their consequences and their management, and how these might differ in urban and rural populations. However, the broad agenda for this paper has meant that we have not been able to address each topic in detail, for which more in-depth dedicated studies will be required.

Self-reported chronic disease diagnoses (diabetes, heart disease and stroke) were more prevalent in urban Xicheng than in rural Daxing. These findings are consistent with reports from previous Chinese surveys, [37-39] but need to be interpreted with caution. There may be systematic under-ascertainment in rural sites because of low levels of awareness and help-seeking, under-detection and under-treatment. Of note, hypertension and dementia, ascertained from clinical assessments in the survey, were similarly prevalent in both sites. Low levels of education may have contributed to ignorance of chronic diseases and under-reporting

[11] On the other hand, the prevalence of self-reported impairments was also much lower among older people in rural Daxing, consistent with their better self-rated overall health. Also, when zero inflation was accounted for the disability score count in the rural site was 40% lower. The lesser needs for care among rural elderly, based upon global assessment by the interviewer, is again consistent with a lower prevalence of chronic disease in Daxing. However, in interpreting these differences in health perception we should bear in mind Amartya Sen's allusion to the substantial evidence that "people in states that provide more education and better medical and health facilities are in a better position to diagnose and perceive their own morbidities than the people in less advantaged states, where there is less awareness of treatable conditions (to be distinguished from "natural" states of being)" [40]. Selective mortality may be an additional explanation for the differences in health outcomes. For rural residents a 30% excess mortality is consistently observed across several data sets, from midlife onwards. The younger age and higher proportion of widows and widowers in Daxing compared with Xicheng is consistent with a difference in midlife mortality between the two populations. Unhealthier lifestyles among the rural elderly may have contributed. Consistent with our findings, a survey in Hubei Province showed higher levels of smoking and alcohol use, and much lower levels of physical activity among older people in rural compared with urban districts [41]. While our data suggests a decline in the prevalence of current smoking among older people compared with the Beijing Longitudinal Ageing Study conducted in 1991; [11] this decline was more pronounced in urban (from 48.2% to 16.6%) than in rural districts (from 43.5% to 30.4%). In summary, our data, considered in the context of other Chinese surveys, is in no way reassuring regarding the underlying health status of the Chinese rural elderly population.

The differences in our survey in the accessibility and effectiveness of the urban and rural health services were striking. In the Third Chinese National Health Services Survey, rural and urban residents with an illness in the past two weeks were equally likely to seek help from a physician; hospitalizations were less frequent among rural residents, but only among those aged 65 and over [37,42]. However, fewer than 7% of our rural sample as opposed to nearly 40% of the urban used any health service in the three months preceding the interview. In both sites, physical health was the strongest predictor of the use of health services. Our findings were not explained by the younger age and better health of rural residents. The limited availability of local health services, [38,43] rural poverty, [37,44] the lack of effective insurance cover after the collapse of the rural Cooperative Medical System, and sharp increases in charges under the new fee-for-service system [42] are all likely to be implicated. Economic factors (household assets, receipt of pension and possessing health insurance) were all independently associated with accessing healthcare in urban Xicheng, and may have explained some of the differences in help-seeking between the two sites; limited variance of these factors probably accounts for the lack of association in rural Daxing. Detection and control of hypertension is an important index of the effectiveness of community healthcare. The control of blood-pressure-related disease is a global health priority [45]. The prevalence of hypertension among older people in China has risen sharply over the period 1991-2006, [12,46,47] and prevention and control are also clear national priorities. Parameters for awareness and control in urban Xicheng were similar to those recently reported for older people in urban Chengdu, [48] while those for rural Daxing were a little worse than those from the national InterASIA survey of 2000-2001, described at that time as 'unacceptably low' [49].

Underutilisation of health services, and lack of routine medical checks may explain the low detection rates [13,22,50]. Lack of control among those who were detected and treated was a particular problem in rural Daxing. In Chengdu, [4] lack of control of hypertension was associated with infrequent blood pressure checks, under-treatment, poor treatment adherence, and ignorance of risk factors and potential complications. Hypertension in mid-life is a recognized risk factor for dementia [51-53] Therefore, the extent to which prevention and control of hypertension can be established early in the coming epidemic in China and other LAMIC may have important implications for the size of the predicted increase in numbers of people with dementia in those regions [54]. As others have noted, there is an urgent need to promote access to healthcare in China [42]. Adequate insurance or subsidy to cover health care costs, need to be extended to those outside of the urban cadres, particularly rural residents, those without formal employment and older people [18]. Community healthcare services need to be strengthened. However, attention needs also to be given to increasing the demand for healthcare; health promotion and education to encourage healthy behaviors and help-seeking [55]. Older people need to be targeted [41].

In Daxing, the burden of support and care, where it was required, fell mainly on family members who had often given up work to care. In Xicheng, family members rarely gave up work to care, paid caregivers being employed instead. These stark differences are understandable in the context of China's rapid economic development. Urban Beijing is experiencing a boom, while development in rural areas stagnates. Widening differentials in salary levels between the city and the country drive the trend towards the employment of women from less developed provinces to care for dependent older people in the city. Residential care is costlier, and associated with considerable stigma. Some caution is indicated in interpreting the higher levels of carer strain among urban compared with rural carers, since measurement bias between urban and rural settings may have been implicated; nevertheless, the finding seems plausible. Although the literature is inconsistent on this point, [56] juggling work roles with those of parent, organisational and 'hands-on' caregiver for an older relative can be stressful. In Daxing, traditional extended family living arrangements are still the norm, with neighbours and relatives available to provide additional informal care. In China, as in the Dominican Republic [35] and the USA [36] dementia is consistently associated with greater needs for care, more time spent caregiving and greater caregiver strain. Non-communicable diseases are already leading causes of mortality in China [55] and the pace of demographic ageing in China is such that predicted increases in numbers of dependent people [57], and numbers of people with dementia [54] will be greater in absolute and relative terms than for almost any other world region. Developing policies and investing in long-term care should be key priorities, alongside health sector reform.

Conclusion

Self-reported diabetes, heart disease and stoke were more prevalent in urban Xicheng than in rural Daxing, conversely hypertension and dementia, ascertained from clinical assessments in the survey, were similarly prevalent in both sites. Apparent better health in rural Daxing might be explained by under-diagnosis (and limited access to health care facilities), under-reporting or selective mortality. Care need was common in both sites but whilst informal care was the norm in rural Daxing, paid caregivers were very common in urban Xicheng. The health reform in China should ensure access and long term care in rural settings and at the same time meet the important implications of the socio-cultural and economic changes already apparent in urban China.

Competing Interests

The 10/66 Dementia Research Group works closely with Alzheimer's Disease International (ADI), the non-profit federation of 77 Alzheimer associations around the world. ADI is committed to strengthening Alzheimer associations worldwide, raising awareness regarding dementia and Alzheimer's Disease and advocating for more and better services for people with dementia and their caregivers. ADI is supported in part by grants from GlaxoSmithKline, Novartis, Lundbeck, Pfizer and Eisai.

Authors' Contributions

MP leads the 10/66 Dementia Research Group study and CF acts as research coordinator assisted by EA and RS. SL and YH are the principal investigators in China, and ZL is the study coordinator in China. They were assisted in the conduction of the study by FY and WD. ZL and EA wrote the first draft of the paper and carried out the analyses with the assistance of MP and YH. All other authors reviewed the report and provided further contributions and suggestions. All authors approved the final manuscript.

Acknowledgements

The 10/66 Dementia Research Group study in China has been funded by the World Health Organization (baseline survey) and the Wellcome Trust (GR08002). The authors thank staff in Beijing Xicheng Ping An Hospital and Beijing Daxing Institute of Mental Health Care for assistance with fieldwork and data entry.

References

1. United Nations Department of Economic and Social Affairs: World Population Prospects: 2006 revision. [http://esa.un.org/unpp/, 2006.

2. The International Institute for Applied Systems Analysis: China's Population by Age and Sex, 1950–2050. [http:/ / www.iiasa.ac.at/ collections/ IIASA_Research/ SRD/ ChinaFood/ data/ anim/ pop_ani.htm], 2008.

3. Yang GH: The Transition of Health Mode and the Control Strategy on Chronic Diseases in China. Chinese Journal of Prevention and Control of Chronic Non-communicable diseases 2001, 9:145–148.

4. Zhang PH, Jiao SF, Zhou Y, Li G, Shi Y, Li H, Ren ZY, Wu F, Jiang Y, Guo XH, et al.: Studies on prevalence and control of several common chronic diseases among Beijing adults in 2005. Chinese Journal of Epidemiology 2007, 28:625–630.

5. World Health Organization: Towards an International Consensus on Policy for Long-Term Care of the Ageing. Geneva. 2000.

6. Prince M, Acosta D, Albanese E, Arizaga R, Ferri C, Guerra M, Huang Y, Jacob K, Jimenez-Velazquez IZ, Rodriguez JL, et al.: Ageing and dementia in low and middle income countries - using research to engage with public and policymakers. Int Rev Psychiatry 2008, 20:332–343.

7. Wang L, Yang G, Wang S, Zhou Y: Health service need and utility, and community nursing for the elderly person in urban area. Chinese Journal of Gerontology 2007, 1947–1948.

8. The Ministry of Health PRC: Abstract of Health Statistic Report. [http:/ / www.moh.gov.cn/ publicfiles/ business/ htmlfiles/ zwgkzt/ ptjty/ 200805/ 35671.htm], 2008.

9. Zimmer Z, Kaneda T, Spess L: An examination of urban versus rural mortality in China using community and individual data. J Gerontol B Psychol Sci Soc Sci 2007, 62:S349–S357.

10. Shi J, Liu M, Zhang Q, Lu M, Quan H: Male and female adult population health status in China: a cross-sectional national survey. BMC Public Health 2008, 8:277.

11. Woo J, Zhang XH, Ho S, Sham A, Tang Z, Fang XH: Influence of different health-care systems on health of older adults: a comparison of Hong Kong, Beijing urban and rural cohorts aged 70 years and older. Australas J Ageing 2008, 27:83–88.

12. Xu L, Wang S, Wang YX, Wang YS, Jonas JB: Prevalence of arterial hypertension in the adult population in rural and urban China: the Beijing eye study. Am J Hypertens 2008, 21:1117–1123.

13. Center for Health Statistics and Information-Ministry of Health: An Analysis report of national health services survey in 2003. Beijing - China: Chinese Academy Science & Peking Union Medical College; 2004.

14. China National Health Economics Institute: China national health accounts report 2005. Beijing - China: China National Health Economics Institute; 2005.

15. Liu Y, Rao K, Hu SL: People's Republic of China: Toward establishing a rural health protection system. [http:/ / www.adb.org/ Documents/ Reports/ PRC_ Rural_Health_Protection_System/ default.asp], Manila, Asian Development Bank 2002.

16. Diao L, Tang Z, Sun F: A survey on care need of elderly in Beijing. Chinese Journal of Gerontology 2005, 25:985–986.

17. Li JX: Fertility policy and population ageing in China. Population Research 2000, 24:9–15.

18. Wagstaff A, Lindelow M, Gao J, Xu L, Qian J: Extending health insurance to the rural population: an impact evaluation of China new cooperative scheme, policy research working paper 4150. Washington, DC: World Bank; 2007.

19. Andrews G, Clark MJ: The International Year of Older Persons: putting aging and research onto the political agenda. J Gerontol B Psychol Sci Soc Sci 1999, 54:7–10.

20. Prince M, Ferri CP, Acosta D, Albanese E, Arizaga R, Dewey M, Gavrilova SI, Guerra M, Huang Y, Jacob KS, et al.: The protocols for the 10/66 dementia research group population-based research program. BMC Public Health 2007, 7:165.

21. Liu Y, Rao K, Wu J, Gakidou E: China's health system performance. Lancet 2008, 372:1914–1923.

22. Prince M, Acosta D, Chiu H, Scazufca M, Varghese M: Dementia diagnosis in developing countries: a cross-cultural validation study. Lancet 2003, 361:909–917.

23. American Psychiatric Association: Diagnostic and Statistical Manual of Mental Disorders, (DSM-IV). Fourth edition. Washington DC: American Psychiatric Association; 1994.

24. Copeland JR, Dewey ME, Griffiths-Jones HM: A computerized psychiatric diagnostic system and case nomenclature for elderly subjects: GMS and AGE-CAT. Psychol Med 1986, 16:89–99.

25. O'Brien E, Asmar R, Beilin L, Imai Y, Mancia G, Mengden T, Myers M, Padfield P, Palatini P, Parati G, et al.: Practice guidelines of the European Society of Hypertension for clinic, ambulatory and self blood pressure measurement. J Hypertens 2005, 23:697–701.

26. George LK, Fillenbaum GG: OARS methodology. A decade of experience in geriatric assessment. J Am Geriatr Soc 1985, 33:607–615.

27. Rehm J, UTSS: On the development and psychometric testing of the WHO screening instrument to assess disablement in the general population. International Journal of Methods in Psychiatric Research 2000, 8:110–122.

28. Zarit SH, Reever KE, Bach-Peterson J: Relatives of the impaired elderly: correlates of feelings of burden. Gerontologist 1980, 20:649–655.

29. Uwakwe R, Modebe I: Disability and care-giving in old age in a Nigerian community. Niger J Clin Pract 2007, 10:58–65.

30. Chou KR, LaMontagne LL, Hepworth JT: Burden experienced by caregivers of relatives with dementia in Taiwan. Nurs Res 1999, 48:206–214.

31. Arai Y, Kudo K, Hosokawa T, Washio M, Miura H, Hisamichi S: Reliability and validity of the Japanese version of the Zarit Caregiver Burden interview. Psychiatry Clin Neurosci 1997, 51:281–287.

32. 10/66 Dementia Research Group: Care arrangements for people with dementia in developing countries. Int J Geriatr Psychiatry 2004, 19:170–177.

33. Mari JJ, Williams P: A comparison of the validity of two psychiatric screening questionnaires (GHQ-12 and SRQ-20) in Brazil, using Relative Operating Characteristic (ROC) analysis. Psychol Med 1985, 15:651–659.

34. Donner A: A regression approach to the analysis of data arising from cluster randomization. Int J Epidemiol 1985, 14:322–326.

35. Acosta D, Rottbeck R, Rodriguez G, Ferri CP, Prince MJ: The epidemiology of dependency among urban-dwelling older people in the Dominican Republic; a cross-sectional survey. BMC Public Health 2008, 8:285.

36. Ory MG, Hoffman RR III, Yee JL, Tennstedt S, Schulz R: Prevalence and impact of caregiving: a detailed comparison between dementia and nondementia caregivers. Gerontologist 1999, 39:177–185.

37. Feng X, Wang D: Medical health service need of Chinese elderly person. Chinese Journal of Health Statistics 1999, 16:287–289.

38. Tang Z, Fang X, Xiang M, Wu X, Diao L, Lin H, Sun F: Research on the health care needs of the elderly in Beijing. Journal of Chinese Hospital Management 2004, 20:464–469.

39. Zhang T, Yang H, Feng W, Zhang X, Guo H, Gong G, Zhang S, Tang L, Xi X: Survey on physical health and social support among elderly in two communities in Beijing. Chinese Journal of Epidemiology 2002, 23:240.

40. Sen A: Health: perception versus observation. BMJ 2002, 324:860–861.

41. Mao Z, Wu B: Urban-rural, age and gender differences in health behaviors in the Chinese population: findings from a survey in Hubei, China. Public Health 2007, 121:761–764.

42. Liu M, Zhang Q, Lu M, Kwon CS, Quan H: Rural and urban disparity in health services utilization in China. Med Care 2007, 45:767–774.

43. The Ministry of Health PRC: The Health Statistics Almanac in 2007. [http:/ / www.moh.gov.cn/ publicfiles/ business/ htmlfiles/ zwgkzt/ ptjnj/ 200807/ 37168.htm], 2008.

44. Li Z, Qin H, Guo Q, Cen M: Survey on health status and health care need among urban and rural elderly in Guangxi. Chinese Journal of Gerontology 2007, 27:374–375.

45. MacMahon S, Alderman MH, Lindholm LH, Liu L, Sanchez RA, Seedat YK: Blood-pressure-related disease is a global health priority. Lancet 2008, 371:1480–1482.

46. Wang LD: Comprehensive Report, Chinese nutrition and health survey in 2002. Beijing, China: People's Medical Publishing House; 2005.

47. Department of Disease Control and Prevention Chinese Center for Disease Control: Report on Chronic Diseases in China, Beijing - China. 2006.

48. Zhang X, Zhu M, Dib HH, Hu J, Tang S, Zhong T, Ming X: Knowledge, awareness, behavior (KAB) and control of hypertension among urban elderly in Western China. Int J Cardiol 2009, 137:9–15.

49. Gu D, Reynolds K, Wu X, Chen J, Duan X, Muntner P, Huang G, Reynolds RF, Su S, Whelton PK, et al.: Prevalence, awareness, treatment, and control of hypertension in china. Hypertension 2002, 40:920–927.

50. Collaborative research group on the practice guideline on detection evaluation treatment and prevention of hypertension for primary health care doctors: Comparison of the status in control of hypertension between rural and urban community health service centers in Beijing. Chinese Journal of Cardiology 2004, 32:1021–1025.

51. Skoog I, Lernfelt B, Landahl S, Palmertz B, Andreasson LA, Nilsson L, Persson G, Oden A, Svanborg A: 15-year longitudinal study of blood pressure and dementia. Lancet 1996, 347:1141–1145.

52. Launer LJ, Masaki K, Petrovitch H, Foley D, Havlik RJ: The association between midlife blood pressure levels and late-life cognitive function. The Honolulu-Asia Aging Study. JAMA 1995, 274:1846–1851.

53. Launer LJ, Ross GW, Petrovitch H, Masaki K, Foley D, White LR, Havlik RJ: Midlife blood pressure and dementia: the Honolulu-Asia aging study. Neurobiol Aging 2000, 21:49–55.

54. Ferri CP, Prince M, Brayne C, Brodaty H, Fratiglioni L, Ganguli M, Hall K, Hasegawa K, Hendrie H, Huang Y, et al.: Global prevalence of dementia: a Delphi consensus study. Lancet 2005, 366:2112–2117.

55. Yang G, Kong L, Zhao W, Wan X, Zhai Y, Chen LC, Koplan JP: Emergence of chronic non-communicable diseases in China. Lancet 2008, 372:1697–1705.

56. Edwards AB, Zarit SH, Stephens MA, Townsend A: Employed family caregivers of cognitively impaired elderly: an examination of role strain and depressive symptoms. Aging Ment Health 2002, 6:55–61.

57. Harwood RH, Sayer AA, Hirschfeld M: Current and future worldwide prevalence of dependency, its relationship to total population, and dependency ratios. Bull World Health Organ 2004, 82:251–258.

CITATION

Originally published under the Creative Commons Attribution License. Liu Z, Albanese E, Li S, Huang Y, Ferri CP, Yan F, Sousa R, Dang W, Prince M. "Chronic disease prevalence and care among the elderly in urban and rural Beijing, China - a 10/66 Dementia Research Group cross-sectional survey," in BMC Public Health 2009, 9:394. © 2009 Liu et al; licensee BioMed Central Ltd. doi:10.1186/1471-2458-9-394.

Challenges to Conducting Research with Older People Living in Nursing Homes

Sue Hall, Susan Longhurst and Irene J. Higginson

ABSTRACT

Background

Although older people are increasingly cared for in nursing homes towards the end of life, there is a dearth of research exploring the views of residents. There are however, a number of challenges and methodological issues involved in doing this. The aim of this paper is to discuss some of these, along with residents' views on taking part in a study of the perceptions of dignity of older people in care homes and make recommendations for future research in these settings.

Methods

Qualitative interviews were used to obtain the views on maintaining dignity of 18 people aged 75 years and over, living in two private nursing homes in

South East London. Detailed field notes on experiences of recruiting and interviewing participants were kept.

Results

Challenges included taking informed consent (completing reply slips and having a 'reasonable' understanding of their participation); finding opportunities to conduct interviews; involvement of care home staff and residents' families and trying to maintain privacy during the interviews. Most residents were positive about their participation in the study, however, five had concerns either before or during their interviews. Although 15 residents seemed to feel free to air their views, three seemed reluctant to express their opinions on their care in the home.

Conclusion

Although we experienced many challenges to conducting this study, they were not insurmountable, and once overcome, allowed this often unheard vulnerable group to express their views, with potential long-term benefits for future delivery of care.

Background

In many countries older people are increasingly cared for in nursing homes or other long term care facilities towards the end of life. Three systematic reviews of research conducted in these settings have highlighted the need for empirical research in this area [1-3]. In particular, the need to represent the views of residents and their families has been highlighted[2]. There are, however, a number of challenges and methodological issues involved in conducting research in care homes for older people, which can restrict the conduct of research and prevent the views of residents from being heard.

Older people often experience a range of symptoms, including pain, fatigue and hearing or visual problems, which can severely impact all aspects of the research process, including participant recruitment, data collection, quality and analysis[4]. The increasing likelihood of cognitive impairment and dementia amongst older people is a particular challenge when it comes to taking informed consent. Whilst the importance of informed consent is widely acknowledged, the circumstances under which it is obtained amongst older people remains contentious[5]. Although it has been shown that older people may readily agree to participate in research studies, either to increase their human contact, or for the benefit of diversion, the practical aspects of conducting research in this environment can present a unique set of challenges[6]. There have also been concerns over institutionalised participants feeling an overwhelming reluctance to criticise health care professionals or feeling coerced to participate in research as a 'captive audience'[7]. In general however, it has been shown that older people at the end of life regard their participation in research as a valuable contribution to the future lives of others and that such participation can have substantial therapeutic benefits[8,9].

Poor staff compliance with research protocols, inflexibility of established routines, policies and practices in the nursing home environment, together with the potential 'gate-keeping' role of family members could create substantial obstacles to the research process. To fully represent the views of older people in care home settings, research design must be sufficiently robust to meet the strict ethical standards for vulnerable groups which govern some of the following specialist challenges; equity of participant selection, informed consent, confidentiality, risk/benefit ratio and the special protection of resident's rights[6,7].

One area, which is likely to be of great concern to residents of care homes, is preserving dignity[10]. The aim of this paper is to highlight some of the methodological challenges we experienced whilst conducting a study of the perceptions of dignity of older people living in nursing homes with a view to describing the lessons learned, responses and strategies developed and recommendations for future research study design and delivery.

Methods

We used qualitative research methods to obtain the views on dignity of people aged 75 years and over, living in two private care homes in South East London [11]. Following local ethical and research and development approval of our protocol (King's College Hospital Research Ethics Committee: ref 07/Q0703/22), two private nursing homes in South East London were approached to participate in the study. These cared for between 40–44 residents each. The inclusion criteria for residents was aged 75 years and over. The exclusion criteria were, being unable to speak English or to provide informed consent, or too ill or distressed to take part in the study. Since we did not want to exclude the views of the oldest residents, we placed no upper age limit in eligibility.

Care home managers identified residents who they felt were eligible for the study. Since we planned to interview 20 residents, we randomly selected 15 eligible residents from each nursing home. The managers gave the selected residents our information sheets 'expression of interest' forms (all printed in large font). As some residents were later found to be unable to provide informed consent and a substantial number declined to take part, it was necessary to repeat this process several times to try to achieve the desired sample size.

Managers gave the completed expression of interest slips to the researcher (SL), who visited each home a week later to obtain written informed consent from those residents interested in taking part. Potential participants were required to have a 'reasonable' understanding of their participation in the study before informed consent was taken. They needed to (i) recall receiving the patient information sheet, (ii) give a brief account of the study and (iii) describe their involvement in the study.

Of the 86 residents in the two homes 23 (27%) were excluded by the managers, 39 (45%) did not return expression of interest forms, and six (7%) were unable to understand their participation in the study. Participants' ages ranged from 78 to 98. All but one was female, all but one was white-British, and all had multiple co-morbidities. Barthel scores[12] (ability to perform activities of daily living) ranged from total dependence (0) to nearly maximum independence (90).

Eighteen of the planned 20 interviews were eventually conducted. These were conducted by SL, who already had considerable interviewing experience, including interviews with older people and patients with serious illnesses. Interviews were in-depth and semi-structured, exploring a variety of issues including factors which either supported or undermined the participants' sense of dignity. Interview topics were closely based on recent research conducted with Canadian cancer patients[13] and lasted 45–75 minutes. At the end of each interview participants were asked how they felt about participating in the study. Two interviews were conducted with a family member present and two residents were interviewed together. To record the challenges to conducting this research SL kept detailed field notes on her experiences of recruiting and interviewing participants. These were discussed regularly at team meetings.

Results

The Challenges

Challenges included taking informed consent (completing reply slips and having a 'reasonable' understanding of their participation); finding opportunities to conduct interviews; involvement of care home staff and residents' families and trying to maintain privacy during the interviews.

Taking Informed Consent

Although we had intended that residents complete their 'expression of interest' forms, five residents approached by the researcher said they knew nothing about the study and that they had not completed these forms. In these instances, the researcher left residents information about the study and promised to return the following week. Of these, only one consented to take part and was eventually interviewed. This, and the lower than expected response rate, meant that we needed to invite more residents into the study, which took another two weeks.

When the researcher returned to obtain written consent, 10 of the residents who were eventually interviewed could not initially remember completing the expression of interest form or reading the information about the study, which meant she had to spend some time explaining the study. All 10 residents then had a reasonable understanding of the study, in that they were able to recall the main objectives of the study and the extent of their involvement.

Six potential participants were unable to provide informed consent to participate in the study, despite having produced completed reply slips. In these cases it was clear to the researcher that they were unable to understand their involvement in the study. So that these residents did not feel 'rejected,' instead of conducting an interview, she engaged in a short neutral conversation with them and thanked them for their time. None appeared to be concerned about this or asked why they had not been included.

Seizing Opportunities

Finding time to conduct the interviews was sometimes difficult. It was necessary to avoid busy times of the day such as mealtimes or regular visits by GPs, hairdressers, chiropodists, etc and to seize opportunities to approach residents, preferably allowing them time to recover from previous activities before commencing the interview. For example, we found it best not to interview residents after lunch as they were often tired and lethargic at this time. The researcher spent a great deal of time waiting for residents to finish activities and interviews were often postponed at a moment's notice if the resident did not feel well, had an unexpected visitor, or simply did not "feel" like participating at the moment.

Staff Involvement

We were grateful for the help of care home staff, who were often eager to help, however, on three occasions this involved waking a resident and immediately sitting them upright. This resulted in residents feeling tired, disorientated and less likely to want to discuss the study or to be interviewed. The way in which care home staff introduced the researcher could also be problematic, as they would sometimes emphasise the name of the institution responsible for the research, resulting in some residents being confused as to who wanted to speak to them and why. Some initially thought that she was a hospital doctor visiting to discuss their health. These misunderstandings occasionally worried residents before they could be reminded of the purpose of the visit. However, once these initial problems were resolved the interviews generally progressed smoothly with the majority of residents saying that they had enjoyed taking part.

Privacy

Since interviews covered issues regarding the resident's care in the home, privacy was important. However in the majority of cases, the resident's door was left open or staff would enter the resident's room during the course of an interview. On two occasions staff had moved residents to a hallway and dining room where there was little or no privacy. Since most residents had mobility problems, moving them to places where they would have more privacy was time consuming and usually involved enlisting the help of busy nursing home staff.

On two occasions a member of the resident's family asked to be present for the interview. This may have had an impact on both the resident's privacy and the quality of responses given. In both cases, family members were anxious about the 'burden' the interview would place on the resident and the types of questions that would be asked. Both residents seemed to gain immense support and comfort from their relatives being present. For the relatives, it seemed also that the interview presented a chance for them to air problems that they felt would not otherwise have been raised:

"That was my main problem. I mean we used to come in, every time we'd say "Oh has she been put on the toilet?" and they'd say "Oh yeah" you know, but you know that they hadn't, and the other people (...) (whispers). But I mean you're happy though mum, aren't you?" (Daughter of Betty, a 94 year old woman who had heart failure and mild cognitive impairment)

They also often 'prompted' the participant to respond, or reminded them of events which they had forgotten. These residents then seemed to delegate all responsibility for answering certain questions to their relative, which could affect the objectivity and validity of some responses.

Residents' Feelings about Taking Part in the Research

Of the 18 residents interviewed, 13 commented positively about their experience. Comments included enjoying having some company, being able to express their opinions freely, and feeling that they had contributed something that might benefit others in the future:

"...it makes me feel that at least somebody's interested in me." (Anne, an 84 year old woman who had a stroke)

"Well it's quite nice being talked to and expressing my opinion of what I feel and how I don't feel and you taking part in it." (Ellie, an 88 year old woman who had a stroke)

However, although they appeared to enjoy the interview, two residents said that they had initially been uncertain about how the interview would be conducted and what the experience would be like:

"Oh I've enjoyed this conversation. But I thought, the way they asked me about a week ago, that they were sending these interviewers round and I thought there'd be about three or four here, so why have they picked me out?" (Ellie)

"Well I wondered what they were going to do to me. (smiles) I hope I answered the right questions." (Jack, an 85 year old man who died the day after the interview)

Three residents were a little concerned about their 'performance' during the interview, commenting that they hoped they had answered correctly and not "talked too much." In all cases, the researcher reassured the resident several times during the course of the interview that there were no right or wrong answers to the questions, and that all views were useful and valid. Some participants took pleasure in answering the questions as they felt that this showed that they were not suffering from cognitive problems:

"I'm glad I've got the brain to answer you really." (Ellie)

"Perhaps it's because I can talk better and converse with, better than some of the people here because some of them have Alzheimer's disease and Parkinson's disease, and perhaps it's because I'm more...brainy (laughs)." (Anne)

Feeling Free to Criticise

The majority of residents (15/18) seemed to feel free to comment on their day-to-day routines and the care they received in the home. Although residents were satisfied with much of the care they received in the home, most described situations which they felt could have been handled differently by staff and suggested ways in which their care could be improved. However, three residents felt uncomfortable about voicing such criticisms:

"I found it difficult when I first came in to co-operate with the night staff. They didn't have a lot of patience. The day girls, they've been wonderful. Perhaps I shouldn't say this, should I?" (Sara, an 81 year old woman with chronic obstructive lung disease)

This could indicate concerns about reprisal from care home staff and the desire to maintain the status quo of their 'home' environment.

Discussion

One of the most notable observations in conducting this research was the desire of residents to discuss a wide range of issues relating to dignity, which could be seen as particularly sensitive and/or emotionally challenging for older people living in a nursing home environment. These issues included the multiple losses many of them had already experienced (their homes, family and friends and their independence), as well as considering their future decline in health and death[11]. Nevertheless, most appreciated the opportunity to be heard and to make a useful contribution and were positive about taking part in the study. Their views have added to our understanding of the concerns of older people in care homes on maintaining dignity, and have led to the trial of an intervention which could help residents maintain a sense of dignity[14].

The main challenges we needed to overcome to achieve this involved obtaining informed consent, finding suitable opportunities to meet with residents, staff involvement and ensuring privacy during interviews. Although, to some extent, some of these were anticipated, we soon realised that we had underestimated the time it would take to conduct this study, and, since we had a time limit to complete it, this resulted in us conducting only 18 of the planned 20 interviews.

Taking Informed Consent

Obtaining informed consent from residents is both extremely important and time consuming. Difficulty in recalling details of events from previous weeks was a problem for many residents, and one requiring some flexibility with procedures. A great deal of time and patience was needed to ensure that residents recalled and understood the study, and their role in it, before they consented to take part. Making this extra effort meant that it was possible to hear the voices of people who may otherwise have 'failed' the eligibility criteria. Conversely, we also found we had to exclude some residents who we felt did not have the capacity to provide informed consent. In some ways, the fact that managers excluded relatively few residents was reassuring as it suggests that 'gate keeping' on their part was not too much of a problem. We found ourselves, however, faced with residents who had apparently expressed an interest in taking part in the study, but we felt could not be interviewed. Taking the time to chat with them about fairly neutral topics such as the weather seemed to solve this problem. We felt that they had forgotten about the study and enjoyed the company of an extra visitor.

For us, the lesson learned was that it is important to take as much time as necessary to check the understanding of participation in a gentle, non-threatening way, and to be very tactful when excluding people who cannot provide informed consent. Researchers have to be particularly patient, and the extra time and training for this needs to be built into the design of the research. It may be especially helpful for the research team to develop set protocols for researchers on how to respond and handle a variety of potential responses, to ensure uniformity and consistency. It is also important to ensure that all information given to potential study participants is clearly written and in an appropriate format for those with visual impairment. We found that residents were often suspicious of strangers and found disruptions to their expected daily routine a little unsettling. In future studies we plan to leave residents a card with the date, time and length of the next visit with a photograph of the researcher(s). This will also have clear information about which institution they come from.

Seizing Opportunities and Staff Involvement

Although we were well aware that there would be times when residents would not be free to meet with the researcher, we were surprised at how few and how unpredictable opportunities would be. We needed to be sensitive to the needs of residents and respect their decision not to be interviewed at the agreed time, without asking them to justify their decision. Although staff were generally helpful, they were usually very busy, and sometimes the researcher needed to wait some time to be introduced to a resident or for them to be moved to a place of privacy. We found that a great deal of flexibility and reciprocity was needed to conduct research in this setting. The time spent developing relationships with staff and on discovering established routines and practices in the homes was time well spent.

Ensuring Privacy and Reassuring Residents

It was unusual for residents to be concerned about being critical of the care they received in the home, despite the fact that it was sometimes difficult to maintain privacy during the interviews. It is possible that residents who seemed uncomfortable about expressing negative views had never liked to complain, or it could be that they have had negative experiences after complaining in the past. The number of unexpected interruptions certainly didn't help them feel at ease to express such views. We found that one of the factors that eroded a residents' sense of dignity was loss of privacy[11] and concerns about loss of privacy have been raised in other studies of older people living in care homes[15]. Ensuring confidentiality and privacy is usually outlined in study protocols and scrutinised by ethics committees. However, it is important to communicate the importance of maintaining privacy to care home staff and to continually remind residents that their interviews are confidential and would not be shared with care home staff. However, should they prefer to have someone else present during the interview, the interviewer needs to be skilled in ensuring that the resident still has the opportunity to be heard. Older people can become used to not being heard and loose confidence in voicing their opinions. Most have never taken part in research or been interviewed before. It is perhaps not surprising that some find this a little daunting. We found that regularly emphasising the 'informality' of the interviews process and a friendly and patient interviewer who gave them encouragement and reassurance throughout the interview helped them to adjust to and enjoy this new experience.

The main limitations of this study relate to recruitment (outlined previously) and the fact that only one researcher conducted the interviews. The interviewer cannot completely avoid bias, and the interpretation of participants' experience and consideration of the data by only one researcher is limited. Had more than one interviewer been used, and inter-rater reliability measures included, the interpretation would have been more robust.

Conclusion

Although the challenges experienced throughout this study were numerous, they were not insurmountable. The key lessons we learned from our experiences were to have patience and allow plenty of time. The extra time and costs are a small price to pay to hear the views of this under-represented section of society.

Competing Interests

The authors declare that they have no competing interests.

Authors' Contributions

SH and IJH designed the study. SL conducted the interviews with residents. SH and SL co-wrote the manuscript, and IJH critically revised it. All authors read and approved the final manuscript.

Acknowledgements

This study was funded by Cicely Saunders International through a grant from the Dunhill Medical Trust. We are very grateful to the residents who took part in this study and to both care homes for their help and support.

References

1. Cartwright JC: Nursing homes and assisted living facilities as places for dying. In Annual Review of Nursing Research: Focus on Geriatric Nursing. Volume 8. Springer; 2002:231–263.

2. Froggatt KA, Wilson D, Justice C, MacAdam M, Leibovici K, Kinch J, et al.: End-of-life care in long term care settings for older people: a literature review. International Journal of Older People Nursing 2006, 1:45–50.

3. Parker-Oliver D, Porock D, Zweig S: End-of-Life Care in U.S. Nursing Homes: A Review of the Evidence. Journal of the American Medical Directors Association 2005, 5:147–155.

4. Uman GC, Urman HN: The challenges of conducting clinical nursing research with elderly populations. Association of Perioperative Registered Nurses Journal 1990, 52:400–406.

5. Harris R, Dyson E: Recruitment of frail older people to research: lessons learned through experience. Journal of Advanced Nursing 2001, 36:643–651.

6. Maas ML, Kelley LS, Park M, Specht JP: Issues in conducting research in nursing homes. West J Nurs Res 2002, 24(4):373–389.

7. Addington-Hall J: Research sensitivities to palliative care patients. European Journal of Cancer Care 2002, 11(3):220–224.

8. Chouliara Z, Kearney N, Worth A, Stott D: Challenges in conducting research with hospitalized older people with cancer: drawing from the experience of an ongoing interview-based project. Eur J Cancer Care (Engl) 2004, 13(5):409–415.

9. Hutchinson SA, Wilson ME, Wilson HS: Benefits of participating in research interviewing. Image: Journal of Nursing Scholarship 1994, 26:161–164.

10. Pleschberger S: Dignity and the challenges of dying in nursing homes: the residents' view. Age and Ageing 2007, 36:197–202.

11. Hall S, Longhurst S, Higginson IJ: Living and dying with dignity: A qualitative study of the views of older people in care homes. Age and Ageing 2009, 38:411–416.

12. Mahoney FI, Barthel DW: Functional evaluation: The Barthel Index. Maryland State Medical Journal 1965, 14:61–65.

13. Chochinov HM, Hack T, McClement S, Kristjanson L, Harlos M: Dignity in the terminally ill: a developing empirical model. Soc Sci Med 2002, 54(3):433–443.

14. Hall S, Chochinov H, Harding R, Murray S, Richardson A, Higginson IJ: A phase II randomized controlled trial of the feasibility, acceptability and potential effectiveness of Dignity Therapy for older people in care homes: Study protocol. BMC Geriatrics 2009, 9(1):9.

15. Applegate M, Morse JM: Personal privacy and interactional patterns in a nursing home. Journal of Aging Studies 1994, 8:413–434.

CITATION

Association between Subjective Memory Complaints and Health Care Utilisation: A Three-Year Follow Up

Frans Boch Waldorff, Volkert Siersma and Gunhild Waldemar

ABSTRACT

Background

Subjective memory complaints (SMC) are common among elderly patients and little is know about the association between SMC and health care utilisation. Thus, the aim of this study was to investigate health care utilisation during a three-year follow-up among elderly patients consulting their general practitioner and reporting subjective memory complaints (SMC).

Methods

This study was conducted as a prospective cohort survey in general practice with three-year follow-up. Selected health care utilisation or costs relative to SMC adjusted for potential confounders were analyzed in a two-part model where the incidence of use of a selected health care service were analyzed separately from the quantity of use for those that use the service. The former analyzed in a Poisson regression approach, the latter in a generalized linear regression model.

Results

A total 758 non-nursing home residents aged 65 years and older consulted their GP in October and November 2002 and participated in the present study. The adjusted probability of nursing home placement was significantly increased in subjects with SMC relative to subjects without SMC (RR = 2.3). More generally, SMC was associated with an increase in the cost of selected health care utilisation of 60% over three years (p = 0.003).

Conclusion

The data of this study indicated that in an elderly primary care population the presence of SMC increased the cost of health care utilisation by 60% over three years. Thus, inquiry into SMC may contribute to a risk profile assessment of elderly patients and may identify patients with an increased use of health care services.

Background

In studies of older patients, the reported prevalence of subjective memory complaints (SMC) shows a huge variation with figures ranging from 10-56% [1,2]. The large variation may be explained by sample selection or by the methods applied for assessing SMC [1]. Studies have consistently associated SMC with depression [2-4], as well as personality traits [5], high age, low education and female gender [1]. A Danish study indicated that these patients rarely share their perception of SMC with their General Practitioner (GP) spontaneously [6], even though SMC may identify frail patients and inquiry into SMC may easily be implemented in a busy GP routine consultation.

In some studies, association has been found between memory complaints and cognitive impairment on testing, even after adjustment for depressive symptoms [7,8]. However, longitudinal studies assessing the value of SMC in predicting dementia or cognitive decline have shown varying results [9-16]. Thus, the nature of SMC is complex [17].

In a study from 1999 among 8775 non-institutionalized persons aged 65 or more, a single question about health strongly predicted subsequent health care utilisation after a year [18]. Other research suggests that patients with mental health conditions use general medical services at a higher rate than those without mental health conditions [19-21]. Furthermore, dementia has been associated with increased health care utilisation in several studies [22,23]. In our recent study, SMC was associated with an increased probability for nursing home placement over 4 years following the assessment [24]. However, we did not identify any other studies addressing the association between the presence of SMC and health care utilisation. Thus, the aim of the present prospective study was to investigate health care utilisation during a three-year follow-up among elderly patients with and without SMC consulting their general practitioner.

Methods

Study Population

All 17 practices comprising a total of 24 GPs in the central district of the municipality of Copenhagen, Denmark, participated in this study. A total of 40.865 patients were listed and 2.934 were 65 or older. Patients' aged 65 and older consulting their GP, regardless of reason for the encounter, were asked to participate in the study and received information both verbally and written. All participants signed an informed consent declaration and were not offered a refund. Patients not able to speak or read Danish, patients living in a nursing home, and patients with severe acute or terminal illness, or specialist-diagnosed patients with dementia were excluded. Non-participants were defined as those who were not excluded because of the exclusion criteria, but refused to participate. The participants were enrolled during October and November 2002.

Outcome

End-point variables were GP related contacts, out-of-hour services, hospitalization and nursing home placement within a three-year period from enrolment, and a cumulated value of these services.

Measurements

In brief, the examination contained:

1. A self-administered participant questionnaire concerning aspects of memory and sociodemographics. Information on SMC was obtained from the following item: "How would you judge your memory?" The response categories were: "excellent," "good," "less good," "poor," or "miserable." Patients rating their memory as "less good," "poor" or "miserable" were classified as patients with SMC, while patients rating their memory as "excellent" or "good" were defined as patients without SMC.

2. A self-administered quality of life assessment. The patients completed the Danish Validated Version of Euro-Qol-5D. Euro-Qol-5D is a standardised instrument for use as a measure of health outcome and measures five dimensions—mobility, self-care, usual activities, pain/discomfort, and anxiety/depression—each by three levels of severity [25]. The anxiety/depression dimension was used as a proxy for depression.

3. A GP- or nurse-administered Mini Mental State Exermination (MMSE). The MMSE, a widely distributed test recommended in GP guidelines as a cognitive screening test, was completed after the completion of the GP questionnaire [26]. The MMSE score ranges from 0-30; a score lower than 24 was taken as indicative of cognitive impairment.

Registry Data

In Denmark, much health information is collected in national registers based on a unique personal identification number allocated to each inhabitant [27]. Information concerning incident deaths, hospital contacts and GP consultations were retrieved from the central national databases by the statistical department of the Danish National Board of Health at the end of 2007. The municipality of Copenhagen provided information concerning nursing home placement at the end of 2006.

In this study the following outcomes were investigated in the three-year period from January 1st 2003 until 31st December 2005:

1. Practice consultations (number of consultations)

2. Home visit consultations by GP (number of visits)

3. GP out-of-hours contacts (number of contacts)

4. Hospital admission (days in hospital, not as out-patient)

5. Out-patient stay (days in outpatient clinic)

6. Emergency room consultations (number of visits)

7. Nursing home placement (days in institution).

Health care utilisation was defined as the sum of the number of services or time (days) of stay over the three-year follow-up period; or a valuation based on the prices in Table 1. For those, who had died (and thereby did not use health care services during all three years), the nominal outcome was multiplied with the inverse of the proportion of the three years the subject was alive. Annualized outcomes were constructed by dividing the three-year outcomes by three.

Table 1. Valuation of selected health care services

Service	Unit	Value[1]	Source
Practice consultations	1 consultation	€ 14,39	Danish health insurance register (SSR)
Visits by GP	1 visit	€ 23,81	Danish health insurance register (SSR)
Hospital stay (not as outpatient)	1 admission day	€ 470,84	Journal of the Danish Medical Association 2005; 167 (07): 807
Outpatient stay	1 admission day	€ 187,39	The National Board of Health (drg.dk)
Out-of-hours contacts	1 contact	€ 14,66	Danish health insurance register (SSR)
Emergency	1 visit	€ 105,74	The National Board of Health (drg.dk)
Nursing home	1 admission day	€ 127,80	Journal of the Danish Medical Association 2005; 167 (07): 807

[1]2004 prices in DKK converted to EUR using the july 1st 2004 spot rate DKK743.35 = EUR100 (source: Danish national bank http://www.nationalbanken.dk)

Statistical Analysis

Differences in characteristics and health care utilisation between participants with and without SMC were tested by chi-squared tests. A total cost for the health care utilisation was calculated using the valuation in Table 1; the difference in this cost between participants with and without SMC was analyzed with a Kruskal-Wallis non-parametric test. Differences in total cost between subgroups of the participants were tested by the F-test of the regression parameter(s) corresponding to the characteristic classifying the subgroups in a linear regression on total cost, additionally adjusted for SMC. These tests evaluated the effect of the characteristic on the total cost beyond the part of the effect that was mediated by SMC.

Multivariate analysis of health care utilisation followed a two-part model where the incidence of use (ever used) of a selected health care service was analyzed separately from the quantity of use for those that use the service [28]. The incidence was analyzed in a Poisson regression approach [29] so that the regression parameters were equivalent to the log of the relative risk (RR) of using the service ever in the study period. For the participants that use the service (or have cost>0) the quantity of use was analyzed in a generalized linear model using a Gamma distribution and a logarithmic link function; the parameters from this model were interpreted as the log of a (multiplicative) factor how much more the service was used compared to a baseline class. A combined (multiplicative) effect of having SMC compared to not having SMC was straightforwardly calculated by multiplying the RR from the first part and the factor from the second part.

Statistical significance was assessed at a 5% level. We adjusted for multiple testing by the method of Benjamini-Hochberg in the final multivariate analysis [30].

Ethics

The Scientific Ethical Committee for Copenhagen and Frederiksberg Municipalities evaluated the project. The Danish Data Protection Agency, the Danish College of General Practitioners Study Committee as well as The National Board of Health approved the project.

Results

The final cohort consisted of 775 non-nursing home residents of which 758 filled out the SMC item. Figure 1 shows the trial flow. The average age of participants at baseline was 74.8 of whom 38.6% were males; average MMSE was 28.2 (range: 16-30). According to our definition 177 (23%) had SMC at baseline. Non-participants were more likely to be males (OR = 1.4) and were, according to the GP, less likely to complain about memory problems, (OR = 1.8). All participants were followed up until the end of 2005 and none were lost to follow-up.

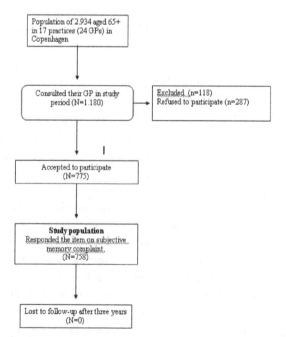

Figure 1. Flowchart of study population.

During the study period 88 (11.6%) died and 50 (6.6%) were admitted to nursing homes. A total of 701 (92.5%) had at least one GP consultation and 432 (60.0%) have at least one hospital admission during the study period. Furthermore, SMC is not seen to correlate with MMSE (Table 2). Valuations of selected health care services are shown in Table 1.

Table 2. Baseline characteristics and health care utilisation of the study participants (n = 758) by Subjective Memory Complaints (SMC)

		SMC					
		No (n = 581)		Yes (n = 177)		Sign.	Missing
		n	%	n	%		
Death	no	517	89,0	153	86,4		
	yes	64	11,0	24	13,6		
MMSE	≥ 24	555	95,5	165	93,2		
	< 24	26	4,5	12	6,8		
Age	60 - 74	318	54,7	86	48,6		
	75 - 84	207	35,6	68	38,4		
	85+	56	9,6	23	13,0		
Sex	male	233	40,1	61	34,5		
	female	348	59,9	116	65,5		
Living without partner	no	240	41,4	60	34,3		3
	yes	340	58,6	115	65,7		
Education	< 8 years	226	38,9	69	39,0		
	> 8 years	355	61,1	108	61,0		
Home care	no	473	81,7	126	72,0	***	4
	yes	106	18,3	49	28,0		
Mobility[1]	no problems	384	67,3	90	52,0	**	14
	some problems	187	32,7	83	48,0		
Self-care[1]	no problems	539	94,7	158	90,8		15
	some problems	30	5,3	16	9,2		
Usual activities[1]	no problems	412	72,5	84	48,6		
	some problems	145	25,5	84	48,6	***	17
	severe problems	11	1,9	5	2,9		
Pain/discomfort[1]	no	216	38,4	45	25,9		
	moderate	323	57,4	111	63,8	***	21
	extreme	24	4,3	18	10,3		
Anxiety/depression[1]	no	442	77,8	98	57,0		
	moderate	115	20,2	71	41,3	***	18
	extreme	11	1,9	3	1,7		
Health Care Utilization							
Practice consultations[2]	no	41	7,1	16	9,0		
	yes	540	92,9	161	91,0		
Visits by GP[2]	no	422	72,6	117	66,1		
	yes	159	27,4	60	33,9		
Hospital stay (no outpatient)[2]	no	259	44,6	67	37,9		
	yes	322	55,4	110	62,1		
Outpatient stay[2]	no	165	28,4	34	19,2	*	
	yes	416	71,6	143	80,8		
Out-of-hours contact[2]	no	548	94,3	170	96,0		
	yes	33	5,7	7	4,0		
Emergency[2]	no	345	59,4	82	46,3	**	
	yes	236	40,6	95	53,7		
Nursing home[2]	no	554	95,4	154	87,0	***	
	yes	27	4,6	23	13,0		

* significant at 5% level ** significant at 1% level *** significant at 0.1% level, [1]based on Euro-Qol-5D, for mobility and self-care the third category did not appear because of the method of data collection, [2]incidence in the period 2003-2005.

Annualized cost (in EUR) of health care utilisation by SMC and participant characteristics is shown in Table 3. Lower MMSE scores, increased age, lower education, home care and lower physical activity increased the cost of health care utilisation. The differences in health care utilisation and costs attributable to SMC, i.e. adjusted for the characteristics listed in Table 3, are shown in Table 4. The presence of SMC significantly increased the probability of nursing home placement (RR = 2.3). More generally, SMC was significantly associated with an increase in health care costs for the combined selected services over the three years of follow-up by 60%. When the cost of nursing home admission is omitted from the total cost analysis, SMC is associated only with a non-significant 23% increase

Table 3. Annualised cost (EUR) of health care utilisation by Subjective Memory Complaints (SMC) and participant characteristics

		SMC					
		No (n = 581)		Yes (n = 177)			
		Median	IQR		Median	IQR		Sign.[1]
Total cost (EUR)		838	192	3389	1577	597	9894	***2
MMSE	≥24	831	192	3209	1457	548	7620	**
	< 24	4572	183	13033	9888	2597	22082	
Age	60 - 74	566	178	2170	993	274	1659	
	75 - 84	1143	226	4438	3321	1076	14743	***
	85+	3277	494	24366	14609	1713	27545	
Sex	male	1036	202	3637	1190	322	3302	
	female	794	187	3360	2187	733	13738	
Living without partner	no	815	154	2601	1225	541	4135	
	yes	842	219	3931	2125	695	14770	
Education	< 8 years	944	185	4151	2998	528	17280	*
	> 8 years	831	197	2750	1383	612	6291	
Home care	no	660	182	2551	1069	307	2998	***
	yes	2642	682	9065	14609	3329	23801	
Mobility	no problems	613	163	2327	1177	280	4985	***
	some problems	1778	313	8509	2883	958	14770	
Self-care	no problems	832	192	3294	1431	548	6755	**
	some problems	2027	288	8896	13416	2788	20493	
Usual activities	no problems	605	163	2380	1194	301	4160	
	some problems	1891	433	6497	2556	767	13791	
	severe problems	1531	178	14788	3615	3248	23152	***
Pain/discomfort	no	594	133	2227	1811	695	14609	
	moderate	1063	222	4266	1510	543	6161	
	extreme	1877	324	6137	2669	1050	12867	
Anxiety/depression	no	794	187	3182	1494	548	8905	
	moderate	1036	226	3595	1577	682	9894	
	extreme	13606	887	17944	12512	7620	30547	

*significant at 5% level ** significant at 1% level *** significant at 0.1% level
[1]Significance of the regression parameter of the corresponding participant characteristic in a linear regression on total cost, adjusted for SMC
[2]Wilcoxon non-parametric test

Table 4. Selected health care utilisation and costs in subjects with Subjective Memory Complaints (SMC) relative to patients without SMC1

	The RR of any use of the corresponding service at all				Factor how much more people with SMC use the service				Combined effect
Service	RR	95% CI		p-value[2]	Factor	95% CI		p-value[2]	
GP contacts									
Practice consultations	0.976	0.922	1.032	0.3924	0.988	0.866	1.126	0.8559	0.964
Visits by GP	1.116	0.876	1.421	0.3863	0.967	0.717	1.304	0.8255	1.079
GP contacts (cost)	0.970	0.920	1.023	0.2610	1.001	0.883	1.135	0.9830	0.972
Hospital stay									
Hospital stay (not as outpatient, days)	1.052	0.911	1.216	0.4953	1.189	0.895	1.582	0.2282	1.252
Outpatient stay (days)	1.111	1.010	1.221	0.0344	1.082	0.875	1.338	0.4663	1.202
Hospital stay (cost)	1.061	0.978	1.151	0.1611	1.178	0.920	1.509	0.1896	1.250
Out-of-hours services									
Out-of-hours GP contacts	0.575	0.237	1.398	0.1686	1.437	1.092	1.891	0.0116	0.827
Emergency (visits)	1.209	1.007	1.452	0.0512	1.073	0.916	1.256	0.3828	1.297
Out-of-hours services (cost)	1.121	0.939	1.340	0.2183	1.172	0.977	1.405	0.0858	1.314
Nursing home									
Nursing home (days)	2.296	1.357	3.886	0.0075	0.922	0.686	1.238	0.5900	2.117
Nursing home (cost)	2.296	1.357	3.886	0.0075	0.922	0.686	1.238	0.5900	2.117
The above combined (cost)	**0.990**	**0.961**	**1.020**	**0.5070**	**1.615**	**1.234**	**2.114**	**0.0003**	**1.599**

[1]All analyses adjusted for the participant characteristics presented in Table 3
[2]Due to multiple testing the level of significance is set to 0.0081

Discussion

To our knowledge, this is the first study to demonstrate that in elderly patients SMC was attributable to an increase in cost by 60% over three years for selected health care services. Specifically, SMC increased the probability of nursing home placement. Much of the excess cost in the SMC group seems to be explained by the higher frequency of nursing home admission.

SMC is a commonly reported symptom in the elderly [1,2]. In this study we adjusted for commonly known confounders e.g. depression and cognitive performance, and the result indicated that the increase in health care utilisation attributed to SMC was substantial. The tendency, that nursing home placement was increased has been reported previously using data from this study. The increased health care utilisation may not solely be explained by nursing home admission. Tendencies of increased use of out-patient clinic admissions and out-of-hour services can be observed. In contrast, the use of GP daytime consultations and acute hospital admittance were not increased.

The reported effect of SMC was beyond various other potential confounders. It is well-known that the presence of dementia in general is associated with an increased health care utilisation [31]. This is in accordance with this study, where our item indicating that significant cognitive impairment (defined as MMSE less than 24) was an independent predictor for nursing home placement. Also, depression in old age has also consistently been associated with an increased health care costs, even after controlling for chronic medical co-morbidity [32]. Our study found that age, but not depressive symptoms were associated with an increased health care utilisation. Furthermore, low education increased health care utilisation. The absence of correlation between SMC and cognitive functioning (MMSE) stresses their different psychometric properties. We assume that SMC measures a global functioning in elderly patients. In Table 2 it can be seen that there is no notable difference in mortality between the subjects with and without SMC. Hence, the difference in health care utilisation and costs cannot be attributed to the high end-of-life utilisation and costs that are generally observed.

The mechanism by which SMC leads to increased health care utilisation is, in our view, not a direct causative relation. However, we see a statistical association between SMC and health care utilisation as residual confounding, i.e. there are certain factors—possibly unknown or immeasurable—beyond the covariates that are used in the analyses, that cause the subject to have memory complaints and cause increased health care utilisation.

The sampling of the participants reflects the population in which the GP has an opportunity to ask questions about SMC. Thus, we deliberately designed the study to include a patient sample, which reflects daily clinical practice. The nation-wide databases used in order to evaluate our main outcomes are regarded as highly valid. Thus, we believe that our findings are valid.

The statistical analysis was done in a two-part model according to recommendations [28]. Data tend to conform to the analytic assumptions for these models, and the models can be used to gain insight in the process of health care utilisation. The decision to have any use at all of a certain service is most likely made by the person and so is related primarily to personal characteristics, while the cost and frequency per user may be more related to characteristics of the health care system.

Several limitations must be addressed. This study had some selection biases at baseline, which may decrease generalizability. Only elderly persons who consulted their GP for whatever reason were included, and they may be more vulnerable than elderly persons in the general population. We did not have access to databases regarding medication, which would have been relevant to evaluate. Likewise, we did not obtain information about medical diagnosis in the participants, as diagnostic criteria are not systematically implemented in general practice in Denmark, and we wanted the study to reflect current standards. Participants who had already been diagnosed with dementia by a specialist were excluded from the study, which is reflected by the high average MMSE in our study population. A MMSE score less than 24 has been widely used as an indication of the presence of cognitive impairment in population based studies [33]. However, epidemiological research has shown that MMSE scores are affected by age, education, and cultural background [33] and MMSE is not sufficient to diagnose dementia. In our study we used the depression item in the Euro-Qol-5D to identify patients with self reported anxiety and depression. These patients may not fulfill international criteria for anxiety and depression. However, this item may serve as indicator for affective symptoms.

There is a lack of consensus concerning the assessment of SMC. Some studies have assessed the presence of SMC by a single item, others by several items. In this study, a single item was used to assess SMC. This item did not allow us to know whether the patient was calibrating the response by comparing to former functioning or to the functioning of others. Notably, our SMC item did not distinguish between short-term and long-term memory loss. We recommend that future studies give more attention to this specific aspect and also include informant reports on memory.

Conclusion

The data suggest that in an elderly primary care population SMC is associated with an increased health care utilisation by 60%, primarily because of increased nursing home placement. Therefore, the result of this study indicates that GPs may identify elderly patients with an increased probability of subsequent health care utilisation by routinely inquiring about memory problems.

Competing Interests

The authors declare that they have no competing interests.

Authors' Contributions

FBW conceived the study concept, design, funding, data analysis, interpretation, and wrote the first draft of the manuscript. VS participated in the data analysis, interpretation, and manuscript preparation. GW participated in the study concept, design, data analysis, interpretation, and manuscript preparation. All authors read and approved the final manuscript.

Acknowledgements

This work was supported by a General Practitioners' Foundation for Education and Development Grant (Grant nr: R56-A369-B186). The sponsor did not contribute to any part of the study or the preparation of the manuscript.

References

1. Jonker C, Geerlings MI, Schmand B: Are memory complaints predictive for dementia? A review of clinical and population-based studies. Int J Geriatr Psychiatry 2000, 15:983–991.

2. Jungwirth S, Fischer P, Weissgram S, Kirchmeyr W, Bauer P, Tragl KH: Subjective memory complaints and objective memory impairment in the Vienna-Transdanube aging community. J Am Geriatr Soc 2004, 52:263–268.

3. Bassett SS, Folstein MF: Memory complaint, memory performance, and psychiatric diagnosis: a community study. J Geriatr Psychiatry Neurol 1993, 6:105–111.

4. Kahn RL, Zarit SH, Hilbert NM, Niederehe G: Memory complaint and impairment in the aged. The effect of depression and altered brain function. Arch Gen Psychiatry 1975, 32:1569–1573.

5. Hanninen T, Reinikainen KJ, Helkala EL, Koivisto K, Mykkanen L, Laakso M, Pyorala K, Riekkinen PJ: Subjective memory complaints and personality traits in normal elderly subjects. J Am Geriatr Soc 1994, 42:1–4.

6. Waldorff FB, Rishoj S, Waldemar G: If you don't ask (about memory), they probably won't tell. J Fam Pract 2008, 57:41–44.

7. Gagnon M, Dartigues JF, Mazaux JM, Dequae L, Letenneur L, Giroire JM, Barberger-Gateau P: Self-reported memory complaints and memory performance in elderly French community residents: results of the PAQUID Research Program. Neuroepidemiology 1994, 13:145–154.

8. Jonker C, Launer LJ, Hooijer C, Lindeboom J: Memory complaints and memory impairment in older individuals. J Am Geriatr Soc 1996, 44:44–49.

9. Schmand B, Jonker C, Geerlings MI, Lindeboom J: Subjective memory complaints in the elderly: depressive symptoms and future dementia. Br J Psychiatry 1997, 171:373–376.

10. Schofield PW, Marder K, Dooneief G, Jacobs DM, Sano M, Stern Y: Association of subjective memory complaints with subsequent cognitive decline in community-dwelling elderly individuals with baseline cognitive impairment. Am J Psychiatry 1997, 154:609–615.

11. Wang L, van BG, Crane PK, Kukull WA, Bowen JD, McCormick WC, Larson EB: Subjective memory deterioration and future dementia in people aged 65 and older. J Am Geriatr Soc 2004, 52:2045–2051.

12. Jorm AF, Christensen H, Korten AE, Jacomb PA, Henderson AS: Memory complaints as a precursor of memory impairment in older people: a longitudinal analysis over 7-8 years. Psychol Med 2001, 31:441–449.

13. Smith GE, Petersen RC, Ivnik RJ, Malec JF, Tangalos EG: Subjective memory complaints, psychological distress, and longitudinal change in objective memory performance. Psychol Aging 1996, 11:272–279.

14. Wang PN, Wang SJ, Fuh JL, Teng EL, Liu CY, Lin CH, Shyu HY, Lu SR, Chen CC, Liu HC: Subjective memory complaint in relation to cognitive performance and depression: a longitudinal study of a rural Chinese population. J Am Geriatr Soc 2000, 48:295–299.

15. Palmer K, Backman L, Winblad B, Fratiglioni L: Detection of Alzheimer's disease and dementia in the preclinical phase: population based cohort study. BMJ 2003, 326:245.

16. Glodzik-Sobanska L, Reisberg B, De SS, Babb JS, Pirraglia E, Rich KE, Brys M, de Leon MJ: Subjective memory complaints: presence, severity and future outcome in normal older subjects. Dement Geriatr Cogn Disord 2007, 24:177–184.

17. Reid LM, Maclullich AM: Subjective memory complaints and cognitive impairment in older people. Dement Geriatr Cogn Disord 2006, 22:471–485.

18. Bierman AS, Bubolz TA, Fisher ES, Wasson JH: How well does a single question about health predict the financial health of Medicare managed care plans? Eff Clin Pract 1999, 2:56–62.

19. Fogarty CT, Sharma S, Chetty VK, Culpepper L: Mental health conditions are associated with increased health care utilization among urban family medicine patients. J Am Board Fam Med 2008, 21:398–407.

20. Thomas MR, Waxmonsky JA, Gabow PA, Flanders-McGinnis G, Socherman R, Rost K: Prevalence of psychiatric disorders and costs of care among adult enrollees in a Medicaid HMO. Psychiatr Serv 2005, 56:1394–1401.

21. Salsberry PJ, Chipps E, Kennedy C: Use of general medical services among Medicaid patients with severe and persistent mental illness. Psychiatr Serv 2005, 56:458–462.

22. Hill JW, Futterman R, Duttagupta S, Mastey V, Lloyd JR, Fillit H: Alzheimer's disease and related dementias increase costs of comorbidities in managed Medicare. Neurology 2002, 58:62–70.

23. Fillit H, Hill JW, Futterman R: Health care utilization and costs of Alzheimer's disease: the role of co-morbid conditions, disease stage, and pharmacotherapy. Fam Med 2002, 34:528–535.

24. Waldorff FB, Siersma V, Waldemar G: Association between subjective memory complaints and nursing home placement: a four-year follow-up. Int J Geriatr Psychiatry 2009, 24:602–609.

25. Rabin R, de CF: EQ-5D: a measure of health status from the EuroQol Group. Ann Med 2001, 33:337–343.

26. Folstein MF, Folstein SE, McHugh PR: "Mini-mental state." A practical method for grading the cognitive state of patients for the clinician. J Psychiatr Res 1975, 12:189–198.

27. Olivarius NF, Hollnagel H, Krasnik A, Pedersen PA, Thorsen H: The Danish National Health Service Register. A tool for primary health care research. Dan Med Bull 1997, 44:449–453.

28. Diehr P, Yanez D, Ash A, Hornbrook M, Lin DY: Methods for analyzing health care utilization and costs. Annu Rev Public Health 1999, 20:125–144.

29. Zou G: A modified poisson regression approach to prospective studies with binary data. Am J Epidemiol 2004, 159:702–706.

30. Benjamini Y, Hochberg Y: Controlling for the false discovery rate: a practical and powerful approach to multiple testing. Journal of the Royal Statistical Society: Series B 1995, 57:289–300.

31. Kronborg AC, Sogaard J, Hansen E, Kragh-Sorensen A, Hastrup L, Andersen J, Andersen K, Lolk A, Nielsen H, Kragh-Sorensen P: The cost of dementia in Denmark: the Odense Study. Dement Geriatr Cogn Disord 1999, 10:295–304.

32. Luppa M, Heinrich S, Matschinger H, Sandholzer H, Angermeyer MC, Konig HH, Riedel-Heller SG: Direct costs associated with depression in old age in Germany. J Affect Disord 2008, 105:195–204.

33. Tombaugh TN, McIntyre NJ: The mini-mental state examination: a comprehensive review. J Am Geriatr Soc 1992, 40:922–935.

CITATION

Originally published under the Creative Commons Attribution License. Waldorff FB, Siersma V, Waldemar G. "Association between subjective memory complaints and health care utilisation: a three-year follow up," in BMC Geriatrics 2009, 9:43. © 2009 Waldorff et al; licensee BioMed Central Ltd. doi:10.1186/1471-2318-9-43.

Cognitive Function, Social Integration and Mortality in a U.S. National Cohort Study of Older Adults

Thomas O. Obisesan and R. F. Gillum

ABSTRACT

Background

Prior research suggests an interaction between social networks and Alzheimer's disease pathology and cognitive function, all predictors of survival in the elderly. We test the hypotheses that both social integration and cognitive function are independently associated with subsequent mortality and there is an interaction between social integration and cognitive function as related to mortality in a national cohort of older persons.

Methods

Data were analyzed from a longitudinal follow-up study of 5,908 American men and women aged 60 years and over examined in 1988–1994 followed an average 8.5 yr. Measurements at baseline included self-reported social integration, socio-demographics, health, body mass index, C-reactive protein and a short index of cognitive function (SICF).

Results

Death during follow-up occurred in 2,431. In bivariate analyses indicators of greater social integration were associated with higher cognitive function. Among persons with SICF score of 17, 22% died compared to 54% of those with SICF score of 0–11 (p < 0.0001). After adjusting for confounding by baseline socio-demographics and health status, the hazards ratio (HR) (95% confidence limits) for low SICF score was 1.43 (1.13–1.80, p < 0.001). After controlling for health behaviors, blood pressure and body mass, C-reactive protein and social integration, the HR was 1.36 (1.06–1.76, p = 0.02). Further low compared to high social integration was also independently associated with increased risk of mortality: HR 1.24 (1.02–1.52, p = 0.02).

Conclusion

In a cohort of older Americans, analyses demonstrated a higher risk of death independent of confounders among those with low cognitive function and low social integration with no significant interaction between them.

Background

Both impaired cognitive function and social isolation are prevalent concomitants of aging in industrialized nations [1,2]. Cognitive function has been found to be inversely associated with subsequent mortality in elderly adults in a number of previous studies [3-6]. Mechanisms remain obscure but may include diminished adherence to medical regimens, and self care including healthful diet and exercise. Likewise, social integration (close social relationships and ties to community) has been found to be inversely related to mortality [7-9]. Further, low social integration may be a risk factor for cognitive decline and dementia [2]. Mechanisms may include support received or provided, improved coping with stressful life events, reduced depression, and positive emotions leading to health-promoting physiological effects of decreased chronic sympatho-adrenal activation, improved immune function, and less chronic inflammation [10-13]. Greater social integration was associated with lower levels of C-reactive protein (CRP) [13]. One study suggests an interaction between social networks and Alzheimer's disease pathology, such that even at more severe levels of disease, individuals with larger network sizes had higher cognitive function [14]. However, further studies of interaction versus independence of cognitive function and social integration in prediction of mortality are needed to document such an important finding in the literature.

Therefore, we tested hypotheses of independent, inverse associations of score on a test of cognitive function and a social integration index with mortality in the United States population. We further test the hypothesis that the effect of cognitive function score on mortality is modified by social integration index, the effect being less in the well integrated than in the less well integrated. We report the analysis of data on mortality in a national health examination and follow-up survey conducted with scientific sampling and state-of-the-art interviewing, examination and laboratory methods.

Methods

Subjects

The Third National Health and Nutrition Examination Survey (NHANES III) was conducted in 1988–1994 on a nationwide multi-stage probability sample of 39,695 persons from the civilian, non-institutionalized population aged 2 months and over of the United States. Details of the plan, sampling, operation, response and institutional review board approval have been published as have procedures used to obtain informed consent and to maintain confidentiality of information obtained [15]. The personal interviews, physical and laboratory examinations of NHANES III subjects provided the baseline data for the study. This analysis was based on follow-up data collection through 2000. Of 33,994 persons with baseline interview data, 13,944 were under age 17 and 26 lacked data for matching leaving 20,024 eligible for mortality follow-up. Two deaths were excluded for missing data on cause of death. The NHANES III Linked Mortality File contains information based upon the results from a probabilistic match between NHANES III and the National Death Index records. The NHANES III Linked Mortality File provides mortality follow-up data from the date of NHANES III survey participation (1988–1994) through December 31, 2000.

Of the 20,022 interviewed persons with mortality follow-up, 6,588 were aged 60 years and over and eligible to have cognitive function testing performed, 6,339 of whom had valid cognitive function data. After excluding persons with missing data for marital status, education, self-reported health status, cigarette smoking status, history of stroke, history of heart attack, history of cancer (other than skin), social integration and mean systolic blood pressure at home visit, 5,908 persons aged 60 and over with complete data remained for this mortality analysis.

The length of follow-up of survivors ranged from 75 to 146 months, mean 108 months.

Cognitive Function and Social Integration

During a home interview, an interviewer collected the socio-demographic variables such as age, gender and level of education used in this analysis. Questions assessing cognition were asked of respondents aged 60 or older and not to proxy respondents. These questionnaires were designed for administration in a bilingual (English/Spanish) format so that respondents could be interviewed in their preferred language. The neuropsychological measures used in the NHANES III study were selected to assess cognitive functions typically affected in dementia. The cognitive items on NHANES III are subsets of different cognitive screening instruments. For example, naming 3 objects comes from the Mini Mental Status Examination and subtracting 3 from 20 comes from the Short Portable Mental Status Questionnaire[16]. The orientation questions are common to most cognitive screens. These items were administered both at home interview and again at a mobile examination center to assess orientation, recall and attention [17,18]. To minimize non-response in older persons, a home examination consisting of abbreviated set of measures similar to those performed, was administered to 493 (8.6 percent of the sample for this analysis) participants who were unable or unwilling to come to a mobile examination center for a complete examination. Both examinations assessed memory function using the SICF. The version of SICF used consisted of six orientation, six recall and five attention items. The six orientation items include general information such as the day of the week, the date, and participant's complete address including street, city/town, state and zip code[17]. Each correct reply was scored one, with zero for an incorrect reply. Six recall items were tested in the home by naming three objects to the participant "apple," "table" and "penny," all of which were repeated immediately up to maximum of six trials and number of trials required to learn the task are noted. Each correct response was scored as one or scored zero for an incorrect answer irrespective of the number of trials required to learn the objects. The subjects were asked to recall after 2 minutes of distracting tasks. Again, each object recalled correctly was scored one and zero for an incorrect answer. Attention was evaluated in the home by asking the participant to count backwards by 3's from 20 each time. The series of digits were selected from those used in the Weschler Adult Intelligence Scale[19] Each correct digit was scored as one for correct count or zero for wrong count. Thus, the overall scores on SICF calculated from the replies use the sum of orientation, recall and attention and ranged from 0 to 17 with median of 13, 25th percentile 12 and 75th percentile 16, i.e. skewed to the left. For analysis, four groups were formed using these cut-points; quartiles with equal numbers of subjects could not be formed due to the skewed distribution and discrete nature of the variable. Two subscales (orientation/recall, range 0–12; counting, range 0–5) were also formed. Cronbach's apha for SICF was 0.77.

Social integration was measured using a social network index (SNI) as described in a previous analysis of these data [13]. Briefly, variables for marital status (1 married, 0 other), frequency of contacts in domains of friends and relatives (1 > = 156 0 < 156 contacts/year), religious attendance (1 > = 4/year, 0 <4/year) and voluntary associations (1 any memberships, 0 other) were summed and the resulting total ranging from 0 to 4 used in the analysis, as previously described.

Confounding and Mediating Variables

A review of the literature identified potential confounding variables: age, gender, race/ethnicity, education, and health status at baseline. Health status was assessed as self-reported general health, presence or absence of any history of major morbidity by physician diagnosis (heart attack, heart failure, stroke, medication for hypertension, diabetes, chronic bronchitis, emphysema, or non-skin cancer) and limitation of mobility (self-reported difficulty in climbing one flight of stairs or walking 1/4 mile with survey physician impression of mobility used to impute missing data). Possible mediators of the effect of social integration were leisure-time physical activity, regular clinic or physician, smoking, alcohol use, body mass index, blood pressure and C-reactive protein. Measurement of blood pressure, height, weight and serum analytes is described elsewhere [15].

Outcome Variable

NCHS conducted a mortality linkage of the Third National Health and Nutrition Examination Survey (NHANES III) with the National Death Index. The current linkage of the NHANES III includes deaths for adult participants occurring from the date of NHANES III interview through December 31, 2000. Information regarding the date of death and age of death, was collected from matched death certificates. Variables used in the selection step of the matching process were social security number, components of name and date of birth. This process detected 2,431 deaths in those in the present analysis. Efforts to trace all NHANES III participants who died may have been unsuccessful in some cases. However, previous validation studies of tracing in the NHANES I Epidemiologic Follow-up Study showed that only about one percent of deaths were not successfully identified using these methods [20]. For details about NHANES III Linked Mortality Files see http:/ / www.cdc.gov/ nchs/ r&d/ nchs_datalinkage/ nhanes3_data_linkage_mortality_activities.htm.

Statistical Analysis

Detailed descriptive statistics and measures of association were computed using the SUDAAN system (Version 9.0, Research Triangle Institute, Research Triangle Park, NC), to take into account the complex survey design and design effect in producing point and variance estimates using Taylor series linearization for variance estimation [21]. Kaplan-Meier survival curves were computed using PROC KAPMEIER. Estimates of the risk of death for persons with lower SICF score relative to those scoring 17 derive from Cox proportional hazards regression models with time to event as the time scale computed using the SURVIVAL procedure in SUDAAN. Follow-up time for survivors ended (was "censored") at the date of the last follow-up interview. An interaction term was initially included for social integration score with SICF score. Two models were fit: one controlled for likely confounders only and the second for confounders and likely mediators of the association of social integration with mortality. Validity of the proportional hazards assumption was confirmed by inspection of unweighted log negative log survival curves [22].

Results

The mean SICF score for persons aged 60 years and over in 1988–1994 was 13.5, range 0–17, median 13, inter-quartile range (IQR) 12–16. The distribution was skewed to the left. Table 1 shows age-adjusted prevalence of selected characteristics by SICF category. SICF score was significantly associated with age, ethnicity, region, marital status, education, self-reported health status, mobility limitation, smoking, alcohol use, physical activity, regular source of care, systolic blood pressure and body mass index.

Table 2 shows the percentage dying over the follow-up period by SICF score at baseline. (Due to rounding percentages may not total exactly 100). Compared to the survivors, scores of decedents were clustered in the two lowest categories (p < 0.0001). Kaplan Meier survival curves for persons aged 60 and over at baseline showed the poorest survival in those scoring 0–11 and the best survival in those scoring 17 (Figure 1). (The curve for the third group—line with stars—became unstable after 110 months due to the small number of events.) Similarly, Kaplan Meier survival curves for persons aged 60 and over at baseline showed the poorest survival in those in the lowest social integration index category and the best survival in those in the highest category (Figure 2).

Table 1. Prevalence (%) of selected socio-demographic characteristics in persons aged 60+y by level of cognitive function: NHANES III.

		SICF score				
	Total	0–11	12–13	14–16	17	P*
N	5908	1867	1985	1177	988	
Total	100	20	36	20	23	
Female	100	22	36	20	22	0.15
Male	100	19	36	21	24	
Age 80+ y	100	37	27	24	12	0.00
70–79 y	100	22	35	20	24	
60–69 y	100	15	39	20	26	
Mexican American	100	37	25	26	12	0.00
Non-Mexican American	100	20	36	20	23	
African American	100	41	27	17	15	0.00
Non-African American	100	19	37	21	24	
South region	100	29	35	18	18	0.00
Other regions	100	17	36	21	25	
Metropolitan residence	100	17	40	20	22	0.02
Non-metro residence	100	23	33	21	24	
Unmarried	100	27	33	21	19	0.00
Married	100	16	38	20	26	
Educ < 12 y	100	32	32	19	19	0.00
Educ > = 12 y	100	12	39	20	29	

• *Chi-square test

Table 2. Prevalence (%) of selected biomedical characteristics in persons aged 60+y by level of cognitive function: NHANES III.

		SICF score				
	Total	0–11	12–13	14–16	17	P*
N	5908	1867	1985	1177	988	
Total	100	20	36	20	23	
Fair-poor health	100	30	32	19	19	0.00
Good health	100	16	38	21	25	
> = I chronic illness	100	21	35	21	23	0.21
No chronic illness	100	19	37	20	24	
Mobility limitation	100	30	38	17	15	0.00
No mobility limitation	100	16	35	22	27	
Current smoking	100	22	36	17	24	0.02
No smoking	100	17	38	21	25	
Alcohol in past month	100	13	40	17	30	0.00
No alcohol	100	24	34	22	20	
Low physical activity	100	26	34	20	21	0.00
Mod/hi activity	100	21	35	20	24	
No regular physician	100	23	39	19	19	0.08
Regular physician	100	20	35	21	24	
No religious attendance	100	22	37	20	21	0.22
Wkly relig attendance	100	19	36	20	24	
Low social support	100	27	36	20	17	0.01
High social support	100	19	36	21	24	
Systolic BP > = 140 mmHg	100	19	38	19	24	0.01
Systolic BP < 140 mmHg	100	20	36	20	24	
BMI > = 25 kg/m2	100	18	35	22	25	0.03
BMI < 25 kg/m2	100	23	37	18	22	
Death during follow-up	100	32	35	17	15	0.00
Alive during follow-up	100	14	36	22	27	

• *Chi-square test

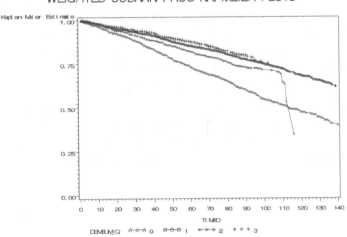

Figure 1. Weighted Kaplan-Meier plots of survival over the follow-up period by level of the score on the short cognitive function index (triangles, 0–11; circles 12–13; stars, 14–16; plusses, 17).

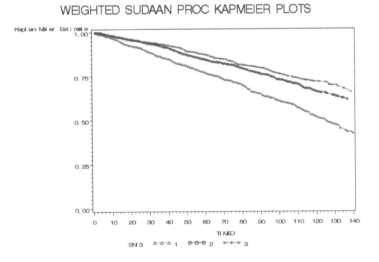

Figure 2. Weighted Kaplan-Meier plots of survival over the follow-up period by level of the score on the Social Network Index (triangles, 0–1; circles 2; stars, 3–4).

Proportional hazards regression analysis revealed a significant bivariate inverse association of SICF score category with mortality during follow-up: test for trend hazard ratio (HR) 0.79, 95% CL 0.73–0.85, p < 0.001. Compared to persons scoring 17, those scoring 0–11 had a hazards ratio of 2.2 (95% CL 1.7–2.7, p < 0.001). Regression models failed to find significant hypothesized interaction between SICF score and social integration index (P = 0.10).

Controlling for confounding by baseline demographic and socioeconomic variables and health status (Model 1), persons scoring 0–11 on the SICF had significantly higher risk of mortality than those scoring 17 (Table 3). A test for linear trend was significant (p = 0.03). Next, the effect was assessed after controlling for

Variable		Model I		Model II	
		Hazard ratio	95% CI	Hazard ratio	95% CI
	17	1.00		1.00	
SICF	14–16	1.10	0.82–1.80	1.11	0.82–1.49
score	12–13	1.07	0.85–1.35	1.06	0.84–1.35
	0–11	1.43*	1.13–1.80	1.36+	1.06–1.76
Age	Yr	1.09*	1.08–1.10	1.09*	1.08–1.10
Gender	Male	1.57*	1.42–1.73	1.57*	1.39–1.78
Ethnicity	AA	1.01	0.88–1.17	1.07	0.91–1.26
	MA	0.72	0.58–0.91	0.81	0.59–1.12
Education	< HS	1.01	0.88–1.16	0.96	0.84–1.09
Region	South	0.94	0.76–1.15	0.88	0.70–1.12
Urbanized	Yes	1.05	0.87–1.26	1.02	0.83–1.25
SR health	F/P	1.72*	1.49–1.99	1.47*	1.27–1.71
Morbidity	Yes	1.55*	1.35–1.79	1.48*	1.28–1.72
Mobility	Limited	1.25*	1.08–1.45	1.29*	1.13–1.48
SBP	Mm Hg			1.00	1.00–1.00
BMI	Kg/m2			0.96*	0.94–0.97
Smoking	Current			1.69*	1.39–2.06
	Former			1.34*	1.14–1.58
Alcohol	Yes			0.79*	0.67–0.94
Physical activity	Low			1.75*	1.52–2.02
	Average			1.26*	1.10–1.44
Reg. care	Yes			1.05	0.87–1.25
Social-network index	3–4			1.00	
	2			1.02	0.88–1.19
	0–1			1.24+	1.02–1.52

CI, confidence interval
SR self-reported; F/P fair-poor; Reg. care, regular care by personal physician.
* p < = 0.01, +P < 0.05

Table 3. Adjusted hazards ratios (95% confidence intervals) of SICF score for mortality from all causes among persons aged 60+ y in NHANES III

variables identified in the literature as potential mediators of a protective health effect of social integration (e.g. healthy behaviors) and social integration index itself. Table 3 shows that these variables explained little of the effect of SICF score (Model II). Further, a significant protective effect of high social integration index independent of SICF score and other variables was apparent (p = 0.02). Finally, to assess whether the beneficial effect of high social integration might be mediated by reduced inflammation, we added the log concentration of C-reactive protein to Model II (not shown). This had essentially no effect on the hazard ratio for low social integration (HR = 1.23) nor on the hazard ratio for low cognitive function (HR = 1.35), indicating that neither effect could be explained by inflammation.

Discussion

NHANES III data show that risk of mortality was higher among persons with low compared to high cognitive function and low compared to high social integration even after controlling for confounding by baseline health status and mobility, and health behaviors, supporting the hypothesis of an independent association. The association did not differ by gender, ethnicity or age. Cognitive function and social integration did not significantly interact.

Mechanisms by which low cognitive function or dementia adversely affect mortality remain obscure, for in most studies the effect cannot be fully explained by adjusting for comorbidity, functional status, or socio-demographic variables [3-6,23,24]. Low social integration has been linked to increased risk of cognitive decline [2,25]. Biopsychosocial mechanisms by which those who are well integrated in their society would be less likely to die when they experience cognitive impairment might include improved coping with the stressful life events of aging leading to decreased oxidative stress and free radical production leading to a slowing of the rate of progression of atherosclerosis in response to impaired cognitive function induced inactivity and adherence to therapy.

Recent work has explored the relationship of inflammation and cognitive function indicating an inverse association of levels of interleukin-6 and CRP with incident cognitive impairment, with mixed results regarding rate of change in cognitive function [26]. Another study in a sample with a mean age of about 45 years found that log IL-6 was significantly inversely correlated with performance on the Wisconsin Card Sorting Test, the Stroop color/word Test, the California Verbal Learning Test, and positively with the Trail Making Test (A & B); also, IL-1 alpha concentration was significantly inversely correlated with the California Verbal Learning Test [27]. After controlling for demographic variables and pro-inflammatory cytokines (linear regression), higher log IL-6 predicted lower scores on the color/word condition of the Stroop Test and higher scores on Trail Making

Test A. Additionally, after controlling for demographic variables, inflammatory illness status, and proinflammatory cytokine concentration (linear regression), Log IL-1-alpha predicted log perseverative errors on the Wisconsin Card Sorting Test. The directionality of the association is unclear. The present study suggests that excess mortality associated with low social integration or low cognitive function is not explained by inflammation as measured by CRP. However, CRP in NHANES III was not high sensitivity (hs-CRP) and may therefore not sensitive enough to detect a relationship with AD-related pathology.

Nor is there evidence that excess mortality is explained by adverse effects of cognitive dysfunction on social integration nor was there an interaction of the two variables. A small study of hospitalized or day patients with significant cognitive impairment found no significant association of living alone, support from relatives, day center care or home help with survival [28]. Receiving meals on wheels was associated with poorer survival. Mechanisms by which social integration may lessen risk of mortality include material aid from others, and emotional support to lessen the psychoneuroimmunologic effects of life stress on the organism [10-13].

NHANES III provides population-based data on the association of cognitive function, social integration and survival in a nationwide, representative sample of Americans. However, several unavoidable limitations of the present study include possible bias arising from survey non-response and from missing values for some variables and from possible changes in cognitive function and social integration or other variables over the follow-up period. A comparison of persons excluded with those included indicated that those excluded were significantly more likely to be over 80 and African American, but did not differ in gender, region or urbanization. Several special studies of NHANES III data have indicated little bias due to non-response [29]. Comparison of vital status and demographics of the analysis sample and those excluded for missing data but eligible for follow-up revealed those excluded were more likely to be female, Mexican American, in poor health, and more likely to die during the follow-up. Thus, selection bias cannot be excluded. It seems reasonable to suggest that the score of our index of cognitive function measures cognitive functioning, but the possible problems inherent in using a cognitive functioning screen that has not been used before, has no validity studies and cannot be directly compared to accepted cognitive screens used in the literature must be acknowledged [16]. We have explored two subscales (orientiation/recall and counting) as shown in the Appendix. The index of social integration used originates from work by Berkman and Syme [7]. However, their Social Network Index differs somewhat from what is described in this paper due to limitations of data in NHANES [13]. Further, it does not capture giving or receiving support, shown to be important for health [10]. Different results may have been obtained had other tests of cognitive function or social integration been used. Unfortunately we were unable to control for depressive symptoms due to lack of such data for persons over 60 years in the survey. CRP is a nonspecific inflammatory marker, which may not be sensitive to chronic brain tissue damage. The representativeness of the sample and the use of sample weights provide generalizability of the results to United States non-institutionalized population of the same ages. Data from longitudinal studies with multiple measures of multiple dimensions of cognition and integration would be helpful in delineating mechanisms involved.

Conclusion

In a nationwide cohort of Americans, analyses demonstrated a higher risk of death independent of confounders among those with low cognitive function or social integration compared to others. The two variables did not interact. Given the growing global burden of impaired cognitive function and social isolation among the elderly, further research is need on the mechanisms by which these factors influence mortality.

Competing Interests

The authors declare that they have no competing interests.

Authors' Contributions

RG participated in the conception, design, analysis and interpretation of the data. TO participated in the conception, design, analysis and interpretation of the data. All authors read and approved the final manuscript.

Appendix

Cognitive Function Subscale Analyses

The mean SICF orientation/recall score for persons aged 60 years and over in 1988–1994 was 8.7, range 0–12, median 8.0, inter-quartile range (IQR) 7–11. The distribution was skewed to the left. For analysis the score was categorized into quartiles. The mean SICF counting score for persons aged 60 years and over in 1988–1994 was 3.9, range 0–5, median 5, inter-quartile range (IQR) 1–5. Over

50% had a score of 5; therefore the score was dichotomized as 5 versus < 5. In unadjusted bivariate crosstabulation, mortality was 48% of those in the lowest quartile, 30% in the second, 31% in the third and 22% in the highest quartile of orientation/recall (p < 0.001) and mortality was 43% in those with counting score < 5 versus 29% in those scoring 5 (p < 0.001). In proportional hazards regression analysis.

There were no significant interactions of orientation/recall with age, gender or race/ethnicity. The baseline orientation/recall score was a significant predictor of future survival after adjusting for sociodemographic and health status variables (p = 0.03). Compared to the highest quartile, the hazard ratio for the bottom quartile of the score was 1.23 (95% CL 1.01–1.49, p = 0.04). Similarly, compared to those with counting score of 5, those with counting score of < 5 had an adjusted hazard ratio of 1.22 (95% CL 1.05–1.41, p = 0.01). Hence both subscales of the SICF were significantly associated with survival.

Disclaimer

The findings and conclusions in this report are those of the authors and do not necessarily represent the views of Howard University or the funding agency.

Acknowledgements

TOO was supported by career development award # AG00980 from the National Institute on Aging.

References

1. Hebert LE, Beckett LA, Scherr PA, Evans DA: Annual incidence of Alzheimer disease in the United States projected to the years 2000 through 2050. Alzheimer disease and associated disorders 2001, 15(4):169–173.

2. Bassuk SS, Glass TA, Berkman LF: Social disengagement and incident cognitive decline in community-dwelling elderly persons. Annals of internal medicine 1999, 131(3):165–173.

3. Dewey ME, Saz P: Dementia, cognitive impairment and mortality in persons aged 65 and over living in the community: a systematic review of the literature. International journal of geriatric psychiatry 2001, 16(8):751–761.

4. Smits CH, Deeg DJ, Kriegsman DM, Schmand B: Cognitive functioning and health as determinants of mortality in an older population. American journal of epidemiology 1999, 150(9):978–986.

5. Ostbye T, Hill G, Steenhuis R: Mortality in elderly Canadians with and without dementia: a 5-year follow-up. Neurology 1999, 53(3):521–526.

6. Ostbye T, Steenhuis R, Wolfson C, Walton R, Hill G: Predictors of five-year mortality in older Canadians: the Canadian Study of Health and Aging. Journal of the American Geriatrics Society 1999, 47(10):1249–1254.

7. Berkman LF, Syme SL: Social networks, host resistance, and mortality: a nine-year follow-up study of Alameda County residents. American journal of epidemiology 1979, 109(2):186–204.

8. Seeman TE, Kaplan GA, Knudsen L, Cohen R, Guralnik J: Social network ties and mortality among the elderly in the Alameda County Study. American journal of epidemiology 1987, 126(4):714–723.

9. Eng PM, Rimm EB, Fitzmaurice G, Kawachi I: Social ties and change in social ties in relation to subsequent total and cause-specific mortality and coronary heart disease incidence in men. American journal of epidemiology 2002, 155(8):700–709.

10. Krause N: Church-based social support and mortality. J Gerontol B Psychol Sci Soc Sci 2006, 61(3):S140–146.

11. Krause N: Exploring the stress-buffering effects of church-based and secular social support on self-rated health in late life. J Gerontol B Psychol Sci Soc Sci 2006, 61(1):S35–43.

12. Lutgendorf SK, Russell D, Ullrich P, Harris TB, Wallace R: Religious participation, interleukin-6, and mortality in older adults. Health Psychol 2004, 23(5):465–475.

13. Ford ES, Loucks EB, Berkman LF: Social integration and concentrations of C-reactive protein among US adults. Annals of epidemiology 2006, 16(2):78–84.

14. Bennett DA, Schneider JA, Tang Y, Arnold SE, Wilson RS: The effect of social networks on the relation between Alzheimer's disease pathology and level of cognitive function in old people: a longitudinal cohort study. Lancet neurology 2006, 5(5):406–412.

15. Plan and operation of the Third National Health and Nutrition Examination Survey, 1988–94. Series 1: programs and collection procedures Vital Health Stat 1 1994, (32):1–407.

16. Hooijer C: Short Screening-Tests for Dementia in the Elderly Population .1. A Comparison between AMTS, MMSE, MSQ and SPMSQ. International journal of geriatric psychiatry 1992, 7(8):559–571.

17. Folstein MF, Folstein SE, McHugh PR: "Mini-mental state." A practical method for grading the cognitive state of patients for the clinician. J Psychiatr Res 1975, 12(3):189–198.

18. Albert M, Smith LA, Scherr PA, Taylor JO, Evans DA, Funkenstein HH: Use of brief cognitive tests to identify individuals in the community with clinically diagnosed Alzheimer's disease. The International journal of neuroscience 1991, 57(3–4):167–178.

19. Weschler D: Weschler Adult Intelligence Scale-revised. New York: The Psychological Corporation; 1981.

20. Cox CS, Mussolino ME, Rothwell ST, Lane MA, Golden CD, Madans JH, Feldman JJ: Plan and operation of the NHANES I Epidemiologic Followup Study, 1992. Vital Health Stat 1 1997, (35):1–231.

21. Anon: SAS/STAT user's guide, version 8. Cary, NC: SAS Institute; 1999.

22. Kleinbaum DG: Survival Analysis: a self-learning text. New York: Springer-Verlag; 1995.

23. Pavlik VN, Hyman DJ, Doody R: Cardiovascular risk factors and cognitive function in adults 30–59 years of age (NHANES III). Neuroepidemiology 2005, 24(1–2):42–50.

24. Waring SC, Doody RS, Pavlik VN, Massman PJ, Chan W: Survival among patients with dementia from a large multi-ethnic population. Alzheimer disease and associated disorders 2005, 19(4):178–183.

25. Van Ness PH, Kasl SV: Religion and cognitive dysfunction in an elderly cohort. The journals of gerontology 2003, 58(1):S21–29.

26. Alley DE, Crimmins EM, Karlamangla A, Hu P, Seeman TE: Inflammation and rate of cognitive change in high-functioning older adults. J Gerontol A Biol Sci Med Sci 2008, 63(1):50–55.

27. Willis TD: Relationships among chronological age, proinflammatory cytokines, and neurocognitive abilities. In Ph.D. Dissertation. Washington, D.C.: Howard University; 2008.

28. Orrell M, Butler R, Bebbington P: Social factors and the outcome of dementia. International journal of geriatric psychiatry 2000, 15(6):515–520.

29. Mohadjer LBB, Waksberg J: National health and nutrition examination survey: III. Accounting for item nonresponse bias. Rockville, MD: WESTAT, Inc; 1996.

CITATION

Magnitude of Potentially Inappropriate Prescribing in Germany Among Older Patients with Generalized Anxiety Disorder

Ariel Berger, Marko Mychaskiw, Ellen Dukes,
John Edelsberg and Gerry Oster

ABSTRACT

Background

Several medications commonly used to treat generalized anxiety disorder (GAD) have been designated "potentially inappropriate" for use in patients aged ≥65 years because their risks may outweigh their potential benefits. The actual extent of use of these agents in clinical practice is unknown, however.

Methods

Using a database with information from encounters with general practitioners (GP) in Germany, we identified all patients, aged ≥65 years, with any GP office visits or dispensed prescriptions with a diagnosis of GAD (ICD-10 diagnosis code F41.1) between 10/1/2003 and 9/30/2004 ("GAD patients"). Among GAD-related medications (including benzodiazepines, tricyclic antidepressants [TCAs], selective serotonin reuptake inhibitors, venlafaxine, hydroxyzine, buspirone, pregabalin, and trifluoperazine), long-acting benzodiazepines, selected short-acting benzodiazepines at relatively high dosages, selected TCAs, and hydroxyzine were designated "potentially inappropriate" for use in patients aged ≥ 65 years, based on published criteria.

Results

A total of 975 elderly patients with GAD were identified. Mean age was 75 years, and 72% were women; 29% had diagnoses of comorbid depression. Forty percent of study subjects received potentially inappropriate agents— most commonly, bromazepam (10% of all subjects), diazepam (9%), doxepin (7%), amitriptyline (5%), and lorazepam (5%). Twenty-three percent of study subjects received long-acting benzodiazepines, 10% received short-acting benzodiazepines at relatively high doses, and 12% received TCAs designated as potentially inappropriate.

Conclusion

GPs in Germany often prescribe medications that have been designated as potentially inappropriate to their elderly patients with GAD—especially those with comorbid depressive disorders. Further research is needed to ascertain whether there are specific subgoups of elderly patients with GAD for whom the benefits of these medications outweigh their risks.

Background

Generalized anxiety disorder (GAD) is a chronic condition that is characterized by persistent worry or anxiety that occurs more days than not over a period of at least six months [1]. The condition is frequently difficult to diagnose because of the variety of clinical presentations and the common occurrence of comorbid medical or other psychiatric conditions. Lifetime prevalence has been estimated to be between 4% and 6% [2]; the disease is more common among women than men. GAD is the most common anxiety disorder among patients presenting to primary care physicians [3,4].

Several different types of medications are often used to treat GAD—specifically, benzodiazepines (e.g., flurazepam, diazepam, chlordiazepoxide), buspirone, tricyclic antidepressants (TCAs) (e.g., amitriptyline, imipramine, doxepin, opipramol), selective serotonin reuptake inhibitors (SSRIs) (e.g., paroxetine. escitalopram), and venlafaxine (a selective serotonin and norepinephrine reuptake inhibitor) [5-7]. Among these available therapies, benzodiazepines have long been the mainstay of pharmacologic treatment for GAD. While effective, benzodiazepines are associated with excessive sedation and motor impairment [8]; their long-term use is also associated with a risk of physical dependence as well as withdrawal when therapy is discontinued [6]. In one study comparing 4554 persons prescribed benzodiazepines with 13,662 persons receiving other (i.e., non-benzodiazepine) medications who were matched on age, sex, and calendar month in which therapy was initiated, Oster and colleagues found that patients in the former group had a 15% higher risk of an accident-related medical event; those who filled three or more prescriptions for benzodiazepines had a 30% higher risk compared with those who filled only one such prescription [9].

An expert panel convened by Beers in 1991 developed explicit criteria for identifying medication use among nursing home residents that was potentially inappropriate [10]. Recognizing that these criteria were developed specifically for a nursing home population, Beers convened another expert panel in 1997 to develop criteria applicable to the entire population of older persons ($\geq$65 years); the resulting criteria designated some of the drugs used to treat GAD (benzodiazepines, amitriptyline, doxepin) as potentially inappropriate for use in persons aged $\geq$ 65 years [11]. The panel compiled its list of potentially inappropriate medications without regard to diagnosis or place of residence, and sought to include only those agents whose ". . . potential for adverse outcomes is greater than the potential for benefit" [11].

While well-known and extensively cited, the Beers' criteria have been criticized as not providing a sufficient basis for identifying inappropriate prescribing, as they are not indication-specific [12]. A subsequent expert panel convened by Zhan et al. classified 33 medications on the Beers' list alternatively as always to be avoided, rarely appropriate, and appropriate for some indications [13]. Among drugs that are sometimes used to treat GAD, flurazepam was designated as "always to be avoided"; chlordiazepoxide and diazepam were designated as "rarely appropriate"; and amitriptyline and doxepin, "appropriate for some indications."

In their update of the Beers' criteria, Fick et al. designated flurazepam, amitriptyline, chlordiazepoxide, doxepin, and anything other than low doses of short-acting benzodiazepines (e.g., >3 mg lorazepam) as potentially inappropriate for use in older patients; adverse outcomes for all such medications were deemed by the authors to be of high (versus low) severity [14].

Despite their limitations, the 1997 Beers' criteria have been widely used by researchers to identify potential medication risks [13,15-19]. An epidemiologic study of noninstitutionalized persons who participated in the 1987 US National Medical Expenditure Survey reported that 23.5% of those aged ≥65 years received at least one of the 20 medications on the Beers' list [20]. Zhan et al. applied their revised list to persons participating in the 1996 US Medical Expenditure Panel Survey and reported that 21.3% received drugs that were potentially—albeit not necessarily—inappropriate, 2.6% received medications that should always be avoided, and 9.1% received drugs that were rarely appropriate; 3.4% of those aged >65 years had received amitriptyline [13].

Recently, an examination of 2707 older home-care patients from eight countries across Europe found that 19.8% received at least one medication designated as potentially inappropriate; in multivariate analysis, use of anxiolytics was associated with a twofold increase in the likelihood of receiving potentially inappropriate medications [21]. Some of the most commonly used, potentially inappropriate medications in this study were benzodiazepines (diazepam [3.1% of patients received such therapy] and chlordiazepoxide [0.6%]) and tricyclic antidepressants (amitriptyline [1.4%])—agents often used for the treatment of GAD.

Although older adults with GAD would appear to often receive potentially inappropriate medications, the actual extent of such use in clinical practice is unknown. Moreover, the generalizability of earlier findings—based largely on US data—to other countries is unknown. In this study, we examine the magnitude of exposure of patients aged ≥65 years with GAD to potentially inappropriate medications in Germany, a large European country in which observed patterns of treatment of GAD may possibly be reflective of those throughout Europe.

Methods

Data were Obtained from the IMS MediPlus

Disease Analyzer database; a detailed description of the database and its research capabilities may be found elsewhere [22]. The database is longitudinal in nature, and provides patient-level information on consultations, diagnoses, and treatments from over 900 GP practices throughout Germany, comprising more than 4.2 million patient records and 75 million prescriptions over a 10-year period. The database is compiled by sampling practices throughout Germany, and is designed to be representative of the general population in Germany. All patient identifiers in the database are fully encrypted.

Information in the IMS MediPlus

Disease Analyzer database includes date of service, diagnoses (in ICD-10 format), actions taken (e.g., referrals to other providers [i.e., specialists], dispensing of sick notes [physician-excused absences from work]), and medications dispensed, including the dispensing date, the quantity dispensed, the number of days of therapy supplied, and the associated diagnosis (available for about 60% of all prescription records). Selected demographic information is also available, including patient age and gender. All patient-level data can be arrayed chronologically to provide a detailed, longitudinal profile of all medical and pharmacy services rendered by participating GPs. Because this study was retrospective in nature, used completely anonymized data, and did not involve patient contact, institutional review board (IRB) approval was not sought. The database for this study encompassed the period, October 1, 2003 through September 30, 2004 ("study period").

The study sample consisted of all patients with any GP office visits or dispensed prescriptions with a diagnosis of GAD (ICD-10 diagnosis code F41.1) during the study period. Persons aged <65 years as of their first-noted encounter for GAD were excluded from the study sample. All GP encounters were then compiled for all subjects over the one-year period of study.

The prevalence of a number of medically attended comorbidities (i.e., noted by GPs during office or clinic visits) was examined for patients in the study sample, including: (1) neoplasms; (2) anemia and other blood antibody disorders; (3) diabetes; (4) circulatory system disorders; (5) respiratory system disorders; (6) eye, nose, and throat disorders; (7) digestive system disorders; (8) painful neuropathic disorders (e.g., diabetic peripheral neuropathy, post-herpetic neuralgia, causalgia, neuropathic back pain); (9) musculoskeletal system disorders; (10) symptom, signs, ill-defined conditions; (11) somatoform disorders; (12) neurasthenia; (13) substance use disorders; and (14) sleep disorders. Patients were deemed to have these conditions if they had any encounters during the study period with the corresponding diagnosis code(s).

The numbers of patients receiving various medications often used to treat GAD ("GAD-related medications") were tabulated. Medications (and corresponding dosages, if relevant [see below]) were deemed "potentially inappropriate" based on their inclusion in Beers' 1997 criteria and/or in subsequent updates to these criteria [11,13,14]. Medication regimens designated as potentially inappropriate were as follows: (1) alprazolam (>2 mg daily); (2) amitriptyline; (3) chlorazepate; (4) chlordiazepoxide; (5) diazepam; (6) doxepin; (7) flurazepam; (8) halazepam; (9) lorazepam (>3 mg daily); (10) oxazepam (>60 mg daily); (11) temazepam (>15 mg daily); (12) triazolam (>0.25 mg daily); and (13) zolpidem (>5 mg daily) [Table 1]. Daily dose was calculated using information in the database; in instances where such information was missing, daily dose was assumed equivalent to the modal value from all other prescriptions for the same product with non-missing values.

Table 1. Medications for treatment of generalized anxiety disorder

Medication Type	Comment
Benzodiazepines	
Short-acting	
Alprazolam	>2 mg/d deemed potentially inappropriate
Bromazepam	
Lorazepam	>3 mg/d deemed potentially inappropriate
Lormetazepam	
Nitrazepam	
Oxazepam	>60 mg/d deemed potentially inappropriate
Temazepam	>15 mg/d deemed potentially inappropriate
Tetrazepam	
Triazolam	>0.25 mg/d deemed potentially inappropriate
Zolpidem	>5 mg/d deemed potentially inappropriate
Long-acting	
Chlordiazepoxide	All long-acting benzodiazepines deemed potentially inappropriate
Chlorazepate	All long-acting benzodiazepines deemed potentially inappropriate
Clobazam	All long-acting benzodiazepines deemed potentially inappropriate
Clonazepam	All long-acting benzodiazepines deemed potentially inappropriate
Diazepam	All long-acting benzodiazepines deemed potentially inappropriate
Flunitrazepam	All long-acting benzodiazepines deemed potentially inappropriate
Flurazepam	All long-acting benzodiazepines deemed potentially inappropriate
Halazepam	All long-acting benzodiazepines deemed potentially inappropriate
Medazepam	All long-acting benzodiazepines deemed potentially inappropriate
Nordazepam	All long-acting benzodiazepines deemed potentially inappropriate
Prezepam	All long-acting benzodiazepines deemed potentially inappropriate
Tricyclic antidepressants	
Amitriptyline	Deemed potentially inappropriate
Amitriptylinoxide	
Clomipramine	
Dibenzepin	
Desipramine	
Doxepin	Deemed potentially inappropriate
Imipramine	
Maprotiline	
Nortiptyline	
Opipramol	
Trimipramine	
Selective serotonin reuptake inhibitor	
Escitalopram	
Citalopram	
Fluoxetine	
Fluvoxamine	
Paroxetine	
Sertraline	
Venlafaxine	
Hydroxyzine	Deemed potentially inappropriate
Buspirone	
Pregabalin	
Trifluoperazine	

Medication receipt was ascertained for the overall study sample, as well as within strata defined on the basis of age, sex, and selected comorbidities (e.g., depression, neoplasms, respiratory disorders). The statistical significance of differences within strata was ascertained using chi-square tests and Fisher's exact tests, as appropriate. All analyses were conducted using PC-SAS® v.9.1 [23].

Results

The study sample consisted of 975 patients, aged ≥65 years, with diagnoses of GAD; mean (± SD) age was 75.0 (± 7.3) years, and 71.6% were women [Table 2]. Twenty-four percent of study subjects had at least one encounter during the study period at which a diagnosis of another anxiety disorder was noted; 29.2% had at least one encounter resulting in a diagnosis of depression. The prevalence of other medically attended comorbidities was high, including circulatory disorders (87.5%), digestive disorders (56.0%), respiratory disorders (42.5%), musculoskeletal disorders (69.5%), painful neuropathies (28.8%), and sleep disorders (24.7%). Ninety-eight percent of patients had at least one comorbidity, and 87.0% had three or more.

Table 2. Demographic and clinical characteristics of study subjects (N = 975)*

Characteristic	
Age, years	
65–74	507 (52.0)
75–84	365 (37.4)
≥85	103 (10.6)
Mean (SD)	75.0 (7.3)
Females	698 (71.6)
Males	277 (28.4)
Comorbid conditions	
Anxiety disorders	
Panic disorder	17 (1.7)
OCD	2 (0.2)
PTSD	0 (0.0)
Phobias	
Social phobia	0 (0.0)
Agoraphobia	0 (0.0)
All other phobias	10 (1.0)
Any phobia	10 (1.0)
Other anxiety disorders	215 (22.1)
Any anxiety disorder (other than GAD)	237 (24.3)
Depression disorders	
Dysthymic disorder	10 (1.0)
Adjustment disorder with depression	4 (0.4)
Bipolar depression	1 (0.1)
MDD	36 (3.7)
Unspecified depression	256 (26.3)
Any depression	285 (29.2)
Neoplasms	158 (16.2)
Anemia and other blood/antibody disorders	136 (13.9)
Diabetes	269 (27.6)
Circulatory system disorders	853 (87.5)
Respiratory system disorders	414 (42.5)
Eyes, nose, and throat	262 (26.9)
Digestive system disorders	546 (56.0)
Painful neuropathic disorders	281 (28.8)
Musculoskeletal system disorders	679 (69.6)
Symptoms, signs, ill-defined conditions	610 (62.6)
Somatoform disorders	130 (13.3)
Neurasthenia	39 (4.0)
Substance use disorders	30 (3.1)
Sleep disorders	241 (24.7)
Number of comorbidities	
0	17 (1.7)
1	44 (4.5)
2	66 (6.8)
≥3	848 (87.0)

*Unless otherwise indicated, all values represent number (%)
GAD: Generalized anxiety disorder; OCD: Obsessive-compulsive disorder; PTSD: Post-traumatic stress disorder; MDD: Major depressive disorder

A total of 607 study subjects (62.3%) had received one or more GAD-related medications (both those deemed potentially inappropriate and all others) from their GPs—most commonly, benzodiazepines (43.7%), including both short-acting (24.8%) and long-acting (23.1%) formulations, and TCAs (25.6%) [Table 3]. Nine percent of patients received SSRIs, and 0.9% received venlafaxine.

Table 3. Numbers of study subjects receiving potentially inappropriate medications for treatment of GAD*

Medication	Potentially Inappropriate	Possibly Appropriate	Total
Benzodiazepines			
Short-acting			
Alprazolam[1]	0 (0.0)	29 (3.0)	29 (3.0)
Lorazepam[2]	50 (5.1)	21 (2.2)	71 (7.3)
Lormetazepam	14 (1.4)	0 (0.0)	14 (1.4)
Oxazepam[3]	0 (0.0)	88 (9.0)	88 (9.0)
Temazepam[4]	8 (0.8)	0 (0.0)	8 (0.8)
Tetrazepam	0 (0.0)	29 (3.0)	29 (3.0)
Triazolam[5]	0 (0.0)	1 (0.1)	1 (0.1)
Zolpidem[6]	32 (3.3)	3 (0.3)	34 (3.5)
Any of above	97 (9.9)	162 (16.6)	242 (24.8)
Long-acting**			
Chlordiazepoxide	6 (0.6)	---	6 (0.6)
Chlorazepate	11 (1.1)	---	11 (1.1)
Clobazam	2 (0.2)	---	2 (0.2)
Clonazepam	3 (0.3)	---	3 (0.3)
Diazepam	91 (9.3)	---	91 (9.3)
Flunitrazepam	7 (0.7)	---	7 (0.7)
Flurazepam	6 (0.6)	---	6 (0.6)
Halazepam	0 (0.0)	---	0 (0.0)
Medazepam	5 (0.5)	---	5 (0.5)
Nordazepam	2 (0.2)	---	2 (0.2)
Prezepam	0 (0.0)	---	0 (0.0)
Bromazepam	93 (9.5)	---	93 (9.5)
Nitrazepam	14 (1.4)	---	14 (1.4)
Midazolam	0 (0.0)	---	0 (0.0)
Any of above	225 (23.1)	---	225 (23.1)
Any of above	298 (30.6)	162 (16.6)	426 (43.7)
Tricyclic antidepressants			
Amitriptyline	52 (5.3)	---	52 (5.3)
Amitriptylinoxide	0 (0.0)	3 (0.3)	3 (0.3)
Clomipramine	0 (0.0)	4 (0.4)	4 (0.4)
Dibenzepin	0 (0.0)	3 (0.3)	3 (0.3)
Desipramine	0 (0.0)	0 (0.0)	0 (0.0)
Doxepin	70 (7.2)	---	70 (7.2)
Imipramine	0 (0.0)	0 (0.0)	0 (0.0)
Maprotiline	0 (0.0)	4 (0.4)	4 (0.4)
Nortriptyline	0 (0.0)	0 (0.0)	0 (0.0)
Opipramol	0 (0.0)	105 (10.8)	105 (10.8)
Trimipramine	0 (0.0)	29 (3.0)	29 (3.0)
Any of above	119 (12.2)	142 (14.6)	250 (25.6)
Selective serotonin reuptake inhibitors			
Escitalopram	0 (0.0)	7 (0.7)	7 (0.7)
Citalopram	0 (0.0)	40 (4.1)	40 (4.1)
Fluoxetine	11 (1.1)	0 (0.0)	11 (1.1)
Fluvoxamine	0 (0.0)	3 (0.3)	3 (0.3)
Paroxetine	0 (0.0)	16 (1.6)	16 (1.6)
Sertraline	0 (0.0)	13 (1.3)	13 (1.3)
Any of above	11 (1.1)	77 (7.9)	86 (8.8)
Venlafaxine	0 (0.0)	9 (0.9)	9 (0.9)
Hydroxyzine	4 (0.4)	---	4 (0.4)
Buspirone	0 (0.0)	5 (0.5)	5 (0.5)
Pregabalin	0 (0.0)	0 (0.0)	0 (0.0)
Trifluoperazine	0 (0.0)	0 (0.0)	0 (0.0)
Any of above	392 (40.2)	345 (35.4)	607 (62.3)

*All values are number of patients (%); dose calculated using all values found in database
**Any use of long-acting benzodiazepines deemed potentially inappropriate
[1] Dose >2 mg/d deemed potentially inappropriate
[2] Dose >3 mg/d deemed potentially inappropriate
[3] Dose >60 mg/d deemed potentially inappropriate
[4] Dose >15 mg/d deemed potentially inappropriate
[5] Dose > 0.25 mg/d deemed potentially inappropriate
[6] Dose >5 mg/d deemed potentially inappropriate
GAD: Generalized anxiety disorder

About forty percent of study subjects—or two-thirds of patients (64.6%) with evidence of receipt of GAD-related medications—were dispensed at least one therapy considered potentially inappropriate for use in the elderly. Ten percent of patients received short-acting benzodiazepines at dosages that rendered their use potentially inappropriate, 23.1% received long-acting benzodiazepines, and 12.2% received amitriptyline or doxepin (primarily doxepin).

Receipt of potentially inappropriate therapy did not differ by age (39.1% for patients aged 65–74 years vs 41.5% for patients aged ≥75 years; p = 0.45) or gender (41.1% of men vs 37.9% of women; p = 0.36). Patients with comorbid depression were more likely to receive potentially inappropriate therapy (51.6% vs 35.5% for those without comorbid depression; p < 0.01), as were those with digestive system disorders (43.0% vs 36.6% for those without these disorders; p = 0.04), and those with sleep disorders (62.2% vs 33.0% for those without these disorders; p < 0.01); receipt of these therapies did not differ significantly for patients with versus without any of the other selected comorbidities.

Discussion

GAD can be difficult to treat, and several different medications—including benzodiazepines, buspirone, TCAs, and SSRIs—are recommended for use in these patients [5-7]. Although these therapies are often of benefit, they also can confer significant risks in older adults. Because of these risks, some of these drugs have been designated potentially inappropriate for use in persons aged 65 years or older.

Using the 1997 Beers criteria [11] and subsequent updates [13,14], we found that four out of every 10 patients—and two out of every three of those who received any GAD-related therapy—received medications that have been designated as potentially inappropriate for use among persons aged 65 years and older. Notably, use of these medications did not differ between the "young" old versus the "old" old (i.e., aged 65–74 years vs ≥ 75 years).

Benzodiazepines have long been the mainstay of pharmacologic treatment for GAD, and they were the medication most commonly dispensed among subjects in our study: 44% of patients in our study received these agents. Paradoxically, they also comprise the majority of agents deemed potentially inappropriate for use in the elderly, due in part to an increased risk of falls, hip fractures, drug-induced disorders of cognition, and motor vehicle accidents [9,24-27]. We note that GPs dispensed short-acting benzodiazepines at daily doses deemed potentially inappropriate to 10% of study subjects; they also dispensed long-acting benzodiazepines—which are deemed potentially inappropriate in older patients regardless of daily dose—to 31% of patients. About one in 10 patients in our study received TCAs deemed potentially inappropriate—either amitriptyline or doxepin—despite the fact that several other TCAs (e.g., nortriptyline) are available with similar efficacy that are better tolerated by older patients. We are unaware of any other study of potentially inappropriate prescribing in patients (of any age) with GAD. Compared with rates of potentially inappropriate prescribing reported among the elderly in general [13,16-21,28], the rates we report are considerably higher. This finding is consistent, however, with the fact that many of the medications that have been designated by some experts as potentially inappropriate in the elderly are also commonly recommended by other experts to treat GAD.

As people age, their ability to metabolize drugs decreases, and receptor sensitivity to the effects of pharmacotherapy changes [29,30]. Unfortunately, older adults are often underrepresented in randomized controlled trials [15], an important source of information for prescribers. Consequently, physicians may sometimes prescribe therapies on presumptions of efficacy and safety based on clinical trial results that are not necessarily generalizable to persons of advanced age. This may explain why we found that so many patients with GAD received medications that were potentially inappropriate. Inappropriate prescribing also may result from the use of inappropriate doses of otherwise age-appropriate medications and/or not prescribing medications that may be of benefit (e.g., not prescribing GAD-related therapies for patients with this condition) [31]. Accordingly, our findings probably underestimate the magnitude of potentially inappropriate prescribing in Germany among older patients with GAD. Further study is needed to ascertain the extent to which these other aspects of potentially inappropriate prescribing are evident in this patient population.

The clinical implication of our findings appears to be that benzodiazepines, amitriptyline, and doxepin are being overprescribed in Germany among elderly patients with GAD, especially among those with comorbid depression and/or sleep disorders. A major limitation of our study, however, is that we cannot assess the actual extent to which prescribing was truly clinically inappropriate. The treatment of GAD in the elderly is clinically complex and presents many challenges. Since we did not employ a "therapeutic timeline" as part of our research design, patients could have received potentially inappropriate agents either as initial treatment or following failure of safer, more appropriate agents. To the extent that the latter occurred, providers may have known about the risks associated with these medications and nonetheless used them only after a careful balancing of these risks against their potential benefits in patients with severe symptoms [29]. Further study is needed to clarify the extent of truly clinically inappropriate prescribing.

A second limitation of our study is that the list of potentially inappropriate agents may be too rigid. For example, while Beers classified amitriptyline as potentially inappropriate, Zhan et al. [13] deemed the use of low doses of this agent as appropriate in older adults in some instances, such as for the treatment of neuropathic pain. We note, however, that the Beers criteria are not disease-specific; these agents are deemed potentially inappropriate based exclusively on age ≥65 years. Furthermore, the panel convened by Zhan et al. concluded that drugs such as amitriptyline "are often misused in clinical practice" [13].

Finally, we note that the database was limited to information from encounters with GPs. Patients who were seen for the treatment of GAD by psychiatrists may not have been included in our study (unless they also received a diagnosis of GAD from their GPs). Moreover, all information on prescription drugs is limited to those that are dispensed by GPs. The generalizability of our findings to all older patients with GAD is therefore unknown.

Conclusion

GPs in Germany often prescribe medications that have been designated as potentially inappropriate to their elderly patients with GAD—especially those with comorbid depressive disorders. Further research is needed to ascertain whether there are specific subgoups of elderly patients with GAD for whom the benefits of these medications outweigh their risks.

Competing Interests

Funding for this research was provided by Pfizer Inc., New York, NY. Marko Mychaskiw, R.Ph., Ph.D., and Ellen Dukes, Ph.D., are employees of Pfizer Inc. as well as co-authors of this manuscript; they were involved with the design of the study, data analysis and interpretation, manuscript preparation, and publication decisions. Ariel Berger, M.P.H., John Edelsberg, M.D., M.P.H., and Gerry Oster, Ph.D., are employees of Policy Analysis, Inc., who were paid consultants to Pfizer in consultation with the development of this manuscript.

Authors' Contributions

AB, MM, ED, JE, and GO all contributed to the conceptualization and design of study, data interpretation, and manuscript preparation. In addition, AB, JE, and GO undertook all statistical analyses of the data. All authors read and approved the final manuscript.

Acknowledgements

ED and MM, were both employees of Pfizer at the time the work was conducted. The analyses were conducted by AB, JE, and GO, employees of Policy Analysis, Inc. Policy Analysis, Inc received financial support from Pfizer, Inc. for the conduct of this analysis and development of this manuscript.

References

1. American Psychiatric Association: Diagnostic and Statistical Manual of Mental Disorders, Fourth Edition, Text Revision. Washington DC, American Psychiatric Association; 2000.

2. Kessler R, Berglund P, Demler O, et al.: Lifetime prevalence and age of onset distributions of DSM-IV disorders in the National Comorbidity Survey Replication. Arch Gen Psychiatry 2005, 62:593–602.

3. Ormel J, VonKorff M, Ustun T, et al.: Common mental disorders and disability across cultures: Results from the WHO collaborative study on psychological problems in general health care. JAMA 1994, 272:1741–1748.

4. Wittchen HU, Kessler RC, Beesdo K, et al.: Generalized anxiety and depression in primary care: Prevalence, recognition, and management. J Clin Psychiatry 2002, 63(Suppl 8):24–34.

5. Baldwin DS, Anderson IM, Nutt DJ, et al.: Evidence-based guidelines for the pharmacological treatment of anxiety disorders: Recommendations from the British Association for Psychopharmacology. J Psychopharmacol 2005, 19:567–596.

6. Gorman JM: Treating generalized anxiety disorder. J Clin Psychiatry 2003, 64(Suppl 2):24–29.

7. Rickels K, Schweizer E: The spectrum of generalised anxiety in clinical practice: The role of short-term, intermittent treatment. Br J Psych 1998, 34:49–54.

8. Shader RI, Greenblatt DJ: Use of benzodiazepines in anxiety disorders. N Engl J Med 1993, 328:1398–1405.

9. Oster G, Huse DM, Adams SF, et al.: Benzodiazepine tranquilizers and the risk of accidental injury. Am J Pub Health 1990, 90:1467–1470.

10. Beers MH, Ouslander JG, Rollinger I, et al.: Explicit criteria for determining inappropriate medication use in nursing home residents. Arch Intern Med 1991, 151:1825–1832.

11. Beers MH: Explicit criteria for determining potentially inappropriate medication use by the elderly: An update. Arch Intern Med 1997, 157:1531–1536.

12. Anderson GM, Beers MH, Kerluke K: Auditing prescription practice using explicit criteria and computerized drug benefit claims data. J Eval Clin Pract 1997, 3:283–294.

13. Field TS, Gurwitz JH, Glynn RJ, et al.: The renal effects of nonsteroidal anti-inflammatory drugs in older people: Findings from the Established Populations for Epidemiologic Studies of the Elderly. J Am Geriatr Soc 1999, 47:507–511.

14. Fick DM, Cooper JW, Wade WE, et al.: Updating the Beers Criteria for potentially inappropriate medication use in older adults: Results of a US consensus panel of experts. Arch Intern Med 2003, 163:2716–2724.

15. Gurwitz JH, Avorn J: The ambiguous relation between aging and adverse drug reactions. Ann Intern Med 1991, 114:956–966.

16. Spore DL, Mor V, Larrat P, et al.: Inappropriate drug prescriptions for elderly residents of board and care facilities. Am J Public Health 1997, 87:404–409.

17. Aparasu RR, Mort JR: Inappropriate prescribing for the elderly: Beers criteria-based review. Ann Pharmacother 2000, 34:338–346.

18. Liu GG, Christensen DB: The continuing challenge of inappropriate prescribing in the elderly: An update of the evidence. J Am Pharm Assoc 2002, 42:847–857.

19. Caterino JM: Administration of inappropriate medications to elderly emergency department patients: Results of a national survey. Acad Emerg Med 2003, 10:493–494.

20. Willcox SM, Himmelstein DU, Woolhandler S: Inappropriate drug prescribing for the community-dwelling elderly. JAMA 1994, 272:292–296.

21. Fialova D, Topinkova E, Gambassi G, et al.: Potentially inappropriate medication use among elderly home care patients in Europe. JAMA 2005, 293:1348–1358.

22. Dietlein G, Schroder-Bernhardi D: Use of the Mediplus® patient database in healthcare research. Int J Clin Pharm Ther 2002, 40:130–133.

23. SAS® Proprietary Software, Release 8.4. SAS Institute Inc., Cary, NC;

24. Cumming RG, LeCouteur DG: Benzodiazepines and risk of hip fractures in older people: A review of the evidence. CNS Drugs 2003, 17:825–837.

25. Petrovic M, Mariman A, Warie H, et al.: Is there a rationale for prescription of benzodiazepines in the elderly? Review of the literature. Acta Clin Belg 2003. 58-27-36.

26. Gray SL, Lai KV, Larson EB: Drug-induced cognition disorders in the elderly: Incidence, prevention and management. Drug Saf 1999, 21:101–122.

27. Thomas RE: Benzodiazepine use and motor vehicle accidents. Systematic review of reported association. Can Fam Physician 1998, 44:799–808.

28. Gurwitz JH: Suboptimal medication use in the elderly. The tip of the iceberg. JAMA 1994, 272:316–317.

29. AGS Panel on Persistent Pain in Older Persons. The management of persistent pain in older persons J Am Geriatr Soc 2002, 50(Suppl 6):S205–S224.

30. Knight EL, Avorn J: Quality indicators for appropriate medication use in vulnerable elders. Ann Intern Med 2001, 135:703–710.

31. Hamilton HJ, Gallagher PF, O'Mahony D: Inappropriate prescribing and adverse drug events in older people. BMC Geriatrics 2009, 9:5.

CITATION

Originally published under the Creative Commons Attribution License. Berger A, Mychaskiw M, Dukes E, Eldersberg J, Oster, G. "Magnitude of potentially inappropriate prescribing in Germany among older patients with generalized anxiety disorder," in BMC Geriatrics, 2009 Jul 27;9:31. © 2009 Berger et al; licensee BioMed Central Ltd. doi:10.1186/1471-2318-9-31.

Social Vulnerability, Frailty and Mortality in Elderly People

Melissa K. Andrew, Arnold B. Mitnitski and Kenneth Rockwood

ABSTRACT

Background

Social vulnerability is related to the health of elderly people, but its measurement and relationship to frailty are controversial. The aims of the present study were to operationalize social vulnerability according to a deficit accumulation approach, to compare social vulnerability and frailty, and to study social vulnerability in relation to mortality.

Methods and Findings

This is a secondary analysis of community-dwelling elderly people in two cohort studies, the Canadian Study of Health and Aging (CSHA, 1996/7–2001/2; N = 3707) and the National Population Health Survey (NPHS, 1994–2002; N = 2648). Social vulnerability index measures that used self-reported items (23 in NPHS, 40 in CSHA) were constructed. Each measure ranges from 0 (no vulnerability) to 1 (maximum vulnerability). The primary outcome measure was mortality over five (CHSA) or eight (NPHS) years. Associations with age, sex, and frailty (as measured by an analogously constructed frailty index) were also studied. All individuals had some degree of social vulnerability. Women had higher social vulnerability than men, and vulnerability increased with age. Frailty and social vulnerability were moderately correlated. Adjusting for age, sex, and frailty, each additional social 'deficit' was associated with an increased odds of mortality (5 years in CSHA, odds ratio = 1.05, 95% confidence interval: 1.02–1.07; 8 years in the NPHS, odds ratio = 1.08, 95% confidence interval: 1.03–1.14). We identified a meaningful survival gradient across quartiles of social vulnerability, and although women had better survival than men, survival for women with high social vulnerability was equivalent to that of men with low vulnerability.

Conclusions

Social vulnerability is reproducibly related to individual frailty/fitness, but distinct from it. Greater social vulnerability is associated with mortality in older adults. Further study on the measurement and operationalization of social vulnerability, and of its relationships to other important health outcomes, is warranted.

Introduction

As people age and become more vulnerable, their social circumstances particularly impact their health.[1]–[8] Even so, the many descriptions of how social factors, aging and health relate to each other employ various terms. Social inequalities, social environments, sense of life control and coherence, social support, social networks, social engagement, social capital, social cohesion, and socioeconomic status each have been associated with health status.[9]–[11] While the varying terminology reflects different traditions and fields of study, a useful discipline is imposed by aiming for an approach that is feasible, valid, rooted in clinical practice and summarizable for policy-making.

Recent work on the quantification of fitness and frailty might provide a guide to quantifying social vulnerability.[12], [13] A series of studies has shown that health status can be summarized by a deficit accumulation approach, i.e. counting the deficits present in an individual.[14]–[16] The underlying idea is that the more deficits (or problems) an individual has (or accumulates), the more vulnerable she or he will be to insults that an individual with fewer deficits might be able to keep at bay. This has proved to be a robust enough approach to yield comparable estimates of the rate of deficit accumulation of health-related deficits—adding about 3 percent of a list of deficits with each increasing year of age—across several surveys,[17] and to demonstrate replicable limits to frailty.[18]

If a series of individual deficits could be combined to estimate not just relative fitness/frailty, but also social vulnerability, the resulting social vulnerability index variable would offer insights into understanding the complex health and social care needs of older adults. Especially as people become very old, "social" and "medical" factors have a complex inter-play that affects important health outcomes and is important for both clinical care and policy-making, but how to consider so many factors has been a challenge. The aims of the present study were to operationalize social vulnerability according to a deficit accumulation approach, to compare social vulnerability and frailty, and to study social vulnerability in relation to mortality.

Materials and Methods

Study Samples

The Canadian Study of Health and Aging (CSHA) is a representative study of dementia and related conditions in older Canadians (age ≥65 years). Details of the methods are described elsewhere.[19] Briefly, sampling was population-based and representative of English- and French-speaking older Canadians (age ≥65). The sample of 10,263 individuals was clustered within five Canadian regions and stratified by age, with over-sampling of those aged 75 and older. In CSHA-1 (1991–92), a screening interview was conducted with 9,008 community-dwelling participants. Follow-up at 5 years (CSHA-2) and 10 years (CSHA-3) included repeat screening assessments. Of the 10,263 initial CSHA study participants, 9,998 individuals were accounted for at CSHA-2 (of whom 2,982 had died) and 9,578 were accounted for at CSHA-3 (of whom 5,150 had died). Here, baseline data were drawn from the CSHA-2 screening interview, which had included additional information about social factors. As such the study sample comprised community dwelling adults aged 70+ at baseline (Figure 1).

The National Population Health Survey (NPHS) is a panel survey sampling Canadian residents of all ages and administered by Statistics Canada. Survey waves were completed every two years starting in 1994; the most recent available follow-up wave was done in 2002, yielding eight year follow-up. The sampling design included multistage stratification by geographic and socio-economic characteristics and clustering by Census Enumeration Areas.[20] Eight-year follow-up was available for 2468 individuals aged 65+ at baseline who had completed all items in the social vulnerability measure (Figure 1).

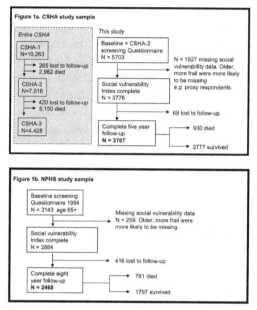

Figure 1. Composition of the Canadian Study of Health and Aging sample (Panel a) and the National Population Health Survey (Panel b).

Measures

Social Vulnerability

Self-report variables relating to social factors that could be considered as deficits were identified separately in the CSHA and the NPHS (Table 1). Deficit selection was guided by two imperatives. First, we aimed to include a broad representation of factors that influence and describe an individual's social circumstance. These factors were based on previous studies which have suggested that they are relevant (e.g. social support, social engagement, sense of mastery/control over one's life circumstances).[1]–[4], [8], [21] As part of a holistic description of social vulnerability, socioeconomic status (e.g. income adequacy, home ownership—addressing both wealth and housing security—and educational attainment) was also included,[22], [23] as these factors also influence vulnerability to insults with the potential to impact their health status. The two instrumental activities of daily living items included (ability to use the telephone and to get to places outside of walking distance) explicitly relate to an individual's ability to maintain social ties and participate in their community, and were therefore included in the social index. Second, working within the constraints of secondary data analysis, we aimed to make the measure of social vulnerability as sensible and as broadly applicable and comparable between datasets as possible.

Table 1. Items aggregated in the social vulnerability index used in each survey.

Communication to engage in wider community	
1	Read English or French
2	Write English or French
Living situation	
3	Marital status
4	Lives alone
Social support	
5	Someone to count on for help or support
6	Feel need more help or support
7	Someone to count on for transportation
8	Feel need more help with transportation
9	Someone to count on for help around the house
10	Feel need more help around the house
11	Someone to count on to listen
12	Feel need more people to talk with
13	Number of people spend time with regularly
14	Feel need to spend more time with friends/family
15	Someone to turn to for advice
16	Feel need more advice about important matters
Socially oriented Activities of Daily Living	
17	Telephone use
18	Get to places out of walking distance
Leisure activities	
19	How often visit friend or relatives
20	How often work in garden
21	How often golf or play other sports
22	How often go for a walk
23	How often go to clubs, church, community centre
24	How often play cards or other games
Ryff scales	
25	Feel empowered, in control of life situation
26	Maintaining close relationships is difficult and frustrating
27	Experience of warm and trusting relationships
28	People would describe me as a giving person
How do you feel about your life in terms of ...	
29	Family relationships
30	Friendships
31	Housing
32	Finances
33	Neighbourhood
34	Activities
35	Religion
36	Transportation
37	Life generally
Socio-economic status	
38	Does income currently satisfy needs
39	Home ownership
40	Education
A) Canadian Study of Health and Aging	
Communication to engage in wider community	
1	Can speak English or French
Living situation	
2	Marital status
3	Lives alone
Social support	
4	Someone to count on for help in crisis
5	Someone to confide in
6	Someone to count on for advice in personal decisions
7	Someone to make you feel loved and cared for
8	Frequency of contact with friends
9	Frequency of contact with relatives
10	Frequency of contact with neighbours
Social engagement and leisure	
11	How often participate in groups
12	How often attend religious services
13	Member of voluntary organisations
14	Participation in physical leisure activities (list of 20)
Empowerment, life control	
15	Too much is expected of you by others
16	You would like to move but cannot (control/empowerment)
17	Neighbourhood or community is too noisy or polluted
18	You have little control over the things that happen to you
19	Feel that you are a person of worth at least equal to others
20	You take a positive attitude towards yourself
21	How often have people you counted on let you down?
Socio-economic status	
22	Not enough money to buy the things you need (income)
23	Educational attainment
B) National Population Health Survey	

Each respondent was assigned a score of 0 if a binary social deficit was absent and 1 if it was present; intermediate values were applied in cases of ordered response categories. For example, an individual scored 1 on the "lives alone" deficit if he/she reported living alone, and 0 if he/she did not. On the "do you ever feel you need more help" deficit, which had three response categories, possible scores were 0 if the answer was "never," 0.5 for "sometimes" and 1 for "often." As such, vulnerability on each deficit was mapped to the 0–1 interval. For each individual, a social vulnerability index was constructed using the sum of the deficit scores, yielding a theoretical range of 0–40 in the CSHA and 0–23 in the NPHS. To allow better comparison between the datasets, each with a different number of social deficits, the social vulnerability index was also calculated as a proportion of the total number of deficit items by dividing the sum of deficit scores by the number of deficits considered (40 in the CSHA and 23 in the NPHS), yielding an index with a theoretical range of 0–1.

Frailty

Frailty was operationalized analogously to the social vulnerability index, in both the CSHA and NPHS, as described elsewhere.[17], [24] In brief, deficits representing self-reported symptoms, health attitudes, illnesses, and impaired functions (Table S1) were identified and given scores mapping to the 0–1 interval as described above, with a greater score corresponding to worse health status. The social vulnerability and frailty indices were mutually exclusive; no deficits overlapped.

Statistical Analysis

Distributions and properties of both the social vulnerability index and the frailty index were explored using descriptive techniques, including graphs (histograms and scatter/ correlation plots), and descriptive statistics (mean and variance values). ANOVA was used for differences in means, and Chi-square testing for proportions. The characteristics (distributions, means, and ranges) of the frailty and social vulnerability indices were compared and correlations calculated.

Logistic regression modeling was used to determine the association between social vulnerability (explanatory variable) and the primary study outcome of survival at follow-up (five years in CSHA, eight years in NPHS). Survival time was determined by vital status at follow-up and date of death, if the respondent died during the follow-up period, Survival analyses were done using Kaplan Meier curves and Cox proportional hazards regression. All models exploring associations between social vulnerability and survival were adjusted for age, sex, and frailty. Statistical significance of survival differences was assessed using log-rank testing. Proportional sampling weights used where possible to account for sample design.

To investigate the robustness of the composition of the social vulnerability index in respect to individual items, and whether mortality was driven by one or a few of the index's constituent variables, we employed a multi-stage approach. At the design level, we investigated the social vulnerability index in two separate samples, as described. At the instrumental level, we employed two different social vulnerability measures, as also detailed above. At the analytical level, we employed two techniques, each based on repeated re-sampling within the index. Established repeated re-sampling techniques such as "jackknifing" and "bootstrapping" are used to estimate variance and confidence intervals.[25] In most applications, the re-sampling is based on observations, or individuals within the sample. Here, as we have done elsewhere with respect to the frailty index,[26]–[28] we have employed these techniques by applying the re-sampling procedure to a group of variables rather than to a group of observations. The earlier analyses with the frailty index have suggested that a greater number of variables is required to ensure stability in the modeling,[28] so these techniques were applied to the CSHA data, which had a high enough number of variables to yield stable estimates. In the first, a "jackknife by variables" procedure, the social vulnerability index was reconstructed n times (where n is the number of variables in the index), each time leaving out a different variable, such that the total number of included variables in each reconstruction was n-1. In the second, a "bootstrap by variables" procedure, the index was reconstructed 100 times, each time randomly sampling 80 percent of the variables such that on each iteration 20 percent of the constituent variables were randomly left out of the index.[28] For both the "jackknife" and "bootstrap by variables" techniques, associations with survival were tested with each resampled and reconstructed version of the social vulnerability index to assess the impact of leaving out single variables or randomly selected groups of variables from the index.

Statistical analyses were done using STATA 8 and Matlab 7.1 software packages.

Results

Descriptive Analyses

Mean age was 77.9 (95% CI: 77.8–78.1) in the CSHA and 73.4 years (95% CI: 73.0–73.7) in the NPHS. The samples comprised 60% women in the CSHA and 58% women in the NPHS. 41% of CSHA participants lived alone, compared with 35% in the NPHS. 66% of CSHA participants had less than secondary school education (<12 years of formal schooling); this was true of 52% in the NPHS. While a few items were strongly correlated (e.g. in the CSHA, reading correlated strongly with writing (r = 0.60), and marital status correlated with living alone (r = 0.77)), correlation among the items in the social vulnerability indices was generally weak: CSHA median correlation 0.085, IQR = 0.04–0.14. (Statistics Canada confidentiality agreement for data release does not allow the NPHS correlations to be released or published). The distributions of the social vulnerability and frailty indices were similar in the CSHA and NPHS (Figure 2a–d). Median social vulnerability was 0.25 (0.20, 0.31) in the CSHA and 0.28 (IQR 0.21,0.35) in the NPHS. While some people showed no degree of frailty, no individual was completely free of social vulnerability in either dataset. In both samples, social vulnerability increased weakly but significantly with age; women had higher index scores than men at all ages in the CSHA and this trend was present in the NPHS (Figure 3). The social vulnerability and frailty indices were weakly to moderately correlated with each other. The correlations were higher for women than for men (CSHA r = 0.37 for men and 0.47 for women; NPHS r = 0.13 for men and r = 0.24 for women).

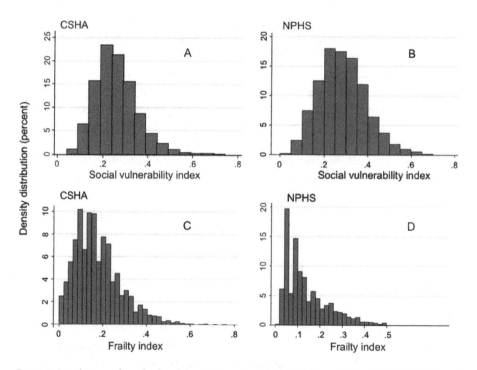

Figure 2. Distributions of social vulnerability: A) Canadian Study of Health and Aging (CSHA), B) National Population Health Survey (NPHS) and frailty: C) CSHA, D) NPHS. While some individuals scored "zero" on the frailty index, no individual was completely free of social vulnerability.

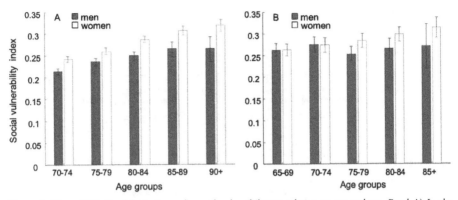

Figure 3. Mean (95% Confidence Interval) social vulnerability in relation to age and sex. Panel A) In the Canadian Study of Health and Aging, social vulnerability increased with age and women had higher index scores than men at all ages. Panel B) In the National Population Health Survey, women showed a trend towards higher scores at older ages.

Mortality

Adjusting for age, sex, and frailty, each additional social deficit in the index was associated with an increased odds of death over five years in the CSHA (OR = 1.05, 95% CI: 1.02, 1.07) and eight years in the NPHS (OR = 1.08, 95% CI: 1.03, 1.14). Cox regression modeling yielded similar results: adjusting for age, sex, and frailty, each additional social deficit increased the risk of death by 3% in the CSHA (HR 1.03, 95% CI: 1.01–1.05) and 4% in the NPHS (HR 1.04, 95% CI: 1.01–1.07). Using the index operationalization that scales each index to values between 0 and 1, thereby adjusting for the different number of deficits included in the two indices and allowing direct comparison between the two datasets, the strength of association was similar in the CSHA and NPHS. For this hypothetical comparison of no social vulnerability (index = 0) vs. maximal vulnerability (index = 1), adjusting for age, sex, and frailty, maximal vulnerability would confer six times the odds of mortality: OR = 6.22 (95% CI: 2.30, 16.83) in the CSHA and OR = 6.22 (95% CI: 1.82, 21.21) in the NPHS.

Survival decreased progressively in each quartile of increasing social vulnerability (Figure 4a&b). This survival gradient remained statistically significant when adjusted for age and sex (stratified log-rank test p<0.001 in both the CSHA the NPHS). Further adjusting for frailty, the survival gradient remained statistically significant in the NPHS (stratified log-rank test p = 0.04) but not in the CSHA (p = 0.15). Although women had better survival than men, survival for women with high social vulnerability was equivalent to that of men with low vulnerability (Figure 4c&d).

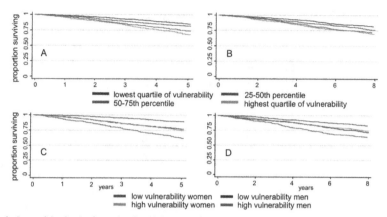

Figure 4. Survival by level of social vulnerability. Panels A (Canadian Study of Health and Aging) and B (National Population Health Survey) show decreasing survival by increasing quartile of social vulnerability. Panels C (CHSA) and D (NPHS) show that although women had better survival than men, survival for women with high social vulnerability was equivalent to that of men with low vulnerability.

Re-Sampling Techniques

Associations with mortality, adjusted for age, sex, and frailty, remained unchanged as each individual social deficit was left out of the CSHA social vulnerability index in the "jackknife by variables" procedure. Survival analysis results using the "bootstrap by variables" technique are shown in Figures 5a&b. The separation between quartiles of social vulnerability remains clear for men despite random omission of 20% of the index variables in each iteration. For women, the separation was clear for the two quartiles indicating the highest social vulnerability, but less so for those with lower vulnerability according to the index.

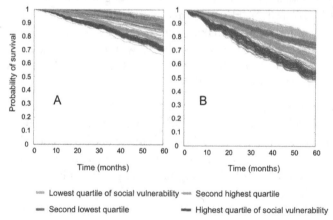

Figure 5. 'Bootstrap by variables' analyses. Survival curves show 100 replications of 80% re-sampling within the Canadian Study of Health and Aging social vulnerability index. Panel A—women, Panel B—men.

Discussion

We used a social vulnerability index to evaluate social factors as they relate to older adults' health. The distribution of social vulnerability was such that no individual was free of social deficits. Social vulnerability increased with age, and women had higher index values than men. Social vulnerability was weakly to moderately correlated with frailty; while the two may be related, they are clearly distinct, particularly since each contributes independently to mortality. Increasing social vulnerability was associated with reduced medium-term survival (5–8 years).

Our findings must be interpreted with caution. Our operationalization of social vulnerability was based on self-report data rather than on objectively defined social factors. Thus it is possible that some individuals over-report and others under-report vulnerability. Further study of distinctions between subjective and objective aspects of social vulnerability is warranted. It is, however, conceivable that older adults' self-perceived vulnerability may be more relevant to their health and quality of life than more objective measures. While we found that social vulnerability increases with age, it was not possible to distinguish between accumulation of deficits with age and the possibility of cohort differences in vulnerability. This would require a different study design (follow-up of different age-based cohorts over time) but warrants further study. In addition, each study had important non-response, and we found that people who did not respond, or who did not have information on social factors were frailer and older. They had higher mortality rates and were more likely to be institutionalized. As both increasing age and increasing frailty are associated with increased social vulnerability, exclusion of these individuals may have led to underestimates of the levels of social vulnerability in the populations of older Canadians represented by the samples, and may have resulted in conservative estimates of associations between social vulnerability and mortality. Little is known about social vulnerability in institutional settings, but given that institutional living would affect social vulnerability in important ways (e.g. not living alone, access to social support, networks, and activities), further research is warranted.

We devised tests to address another potential critique of our approach: whether some individual items included in the index drive the identified associations with mortality. If this were the case for one or more variables, an argument might be made it/that they should not be combined in the index. For example, items such as income and education could be treated as separate confounders rather than being included in the index. For this reason, we investigated whether inclusion or exclusion of individual variables (using the "jackknife by variables" technique), and groups of variables (using the "bootstrap by variables technique") materially affect the analysis results. We have demonstrated that inclusion or exclusion of single variables in the index does not affect the results, and that the same may be true for randomly selected groups of variables (particularly for men), when a sufficient number of variables are included in the index. Of course, the unit of observation is important: for individuals, knowing exactly which deficits are present is likely to be important, but at a population level we find that the number of deficits (rather than the content of these deficits) is more predictive of mortality.

Our findings are consistent with previous research associating various social factors, generally studied one at a time, with health and survival. For example, increased social ties, participation in groups, contact with friends and family, and perceived social support have been associated with survival.[1]–[4] Social disengagement, low participation in leisure activities, and limited social networks have been associated with cognitive decline and dementia[5], [7], [29], [30] and disability.[8] Trust and voluntary sector participation are associated with survival at state and neighbourhood levels.[11], [31] Weak social cohesion has been proposed as an explanation for observed links between poor health and income inequalities[32] and social status inequalities.[21]

The social vulnerability index is a new measure which allows pragmatic quantification of important health information. It appears to be a valid measure, as it predicts mortality and has preserved properties in two independent samples, though further study is warranted to strengthen understanding of its validity and properties. Validation in further independent samples is warranted. Content and construct validity are addressed by embedding the index in a theoretical framework[33], [34] and including social factors that have been found to be relevant in characterizing individuals' social situations and to health outcomes. The weak to moderate correlation seen with frailty is evidence for criterion and convergent validity, as some relationship between social vulnerability and frailty is reasonable, though the two are distinct measures. The remarkable conservation of the properties of the social vulnerability index approach and associations (albeit of two indices differing in the details of their construction but sharing a common approach and theoretical basis) with health in two different cohorts of older Canadians suggests generalizability and reproducibility, though replication in other populations and settings is needed. As the social vulnerability index is a new measure, its reliability (within and between raters, and over time) has yet to be quantified, but is of interest, particularly in considering potential applicability to clinical settings.

The social vulnerability index is an aggregate of items that each have been put forward as reflecting particular aspects of how social factors interact with health. They were not proposed to be combined in the way that we have done, so it is reasonable to ask whether it is fair to combine these many factors into a single index. Two considerations motivated our combining individual factors, even though we recognize that the factors come from different theoretical backgrounds and not all were intended to be combined as we have done. The first is entirely pragmatic. Large numbers of factors are difficult to handle in multivariable models, and require impracticably large sample sizes, especially if interactions are to be modeled. The second motivation for combining factors was so that we could study the properties of the social vulnerability index. In working with the frailty index, we have been struck by the insights that it allows regarding the complexity of frailty. Analyzing the properties of the frailty index has allowed us to employ tools from mathematics which allow us to consider complexity more formally, and not just as a synonym for 'complicated.' For example, the frailty index appears to accumulate at a characteristic rate across studies (at about 0.03/year on a log scale). [17] Here, accumulation of social vulnerability with age was seen chiefly with women. The frailty index has a characteristic sub maximal limit (about 0.67), i.e. people generally do not have more than two thirds of the deficits included in a frailty index—in other words, when the limit has been achieved, no further deficit accumulation is possible, as further deficits would result in death.[18] This is an intriguing observation and we aim to investigate whether there is a maximal limit to how socially vulnerable an individual can become and still survive. Additionally, the frailty index shows reproducible transitions between health states,[35] pointing to additional studies of how individuals transition between levels of social vulnerability—i.e. how they accumulate deficits as they move from lower to higher vulnerability.[35]

It is possible that, for example, a principal components analysis might suggest separable domains of social vulnerability. Though such analyses are traditional, they are not without arbitrariness (for example, allowing the operator to specify the number of dimensions to be 'discovered'), and there are reasons to be skeptical about the approach. It is more instrument-dependent, and thus less generalizable. Many single items that are readily measured in younger people—socioeconomic status in relation to occupation, income and address, for example, are less well measured in people post-retirement, or in neighborhoods in transition.[22] In general, psychometric reductionist techniques consider fewer variables, but lose information. Here we achieved analytical parsimony with just one variable, without losing items that were individually informative. What is more, the index also allows an essentially continuous distribution of risk rather than the artificially small number of risk groups possible with ordinal variables.

Our approach has certain strengths. Several estimates were closely replicable, despite the social vulnerability indexes being constructed differently in the two samples. This suggests that the social vulnerability index has potentially wide applicability: the constituent variables can differ in different settings as long as the basic tenant of including multiple social factors relating to important broad domains is met. The holistic quantification and measurement of social vulnerability has great potential relevance for health and social policy. Being able to identify individuals and groups who are social vulnerable could be useful for prediction of health outcomes as well as for targeting of interventions and design of specialized programs. While it is certainly possible to study the health influence of individual social factors, this "one thing at once" approach is limited, especially for older adults in whom complex sets of social circumstances may exist and interact in different (possibly unpredictable) ways to contribute to vulnerability in an aggregate sense. Even older adults who have a particular deficit (e.g. who live alone) would still be differentially susceptible to insults of circumstance (i.e. those insults that perturb the delicate balance of assets and deficits, strengths and weaknesses, which has thus allowed them to maintain their heath), depending on their profile of other deficits and strengths.

Several of our findings point to interesting sex differences in social vulnerability. In both the CSHA and the NPHS, women had higher social vulnerability index values than men. One might wonder whether this is due to older age among women, but the finding was independent of age. Correlations with frailty also differed, and were higher for women than for men. Additionally, although women had better survival than men (consistent with many other epidemiological studies), high social vulnerability in women seems to negate this sex benefit, reducing their survival to equal that of less vulnerable men (Figure 4). The index's composition also seems to matter differently between the sexes. For men, survival analyses using re-sampling techniques maintained clear separation into quartiles of social vulnerability, suggesting that for men the specific variables included in the index are not as important as the overall impression of vulnerability. For women, separation was quite clear for the two highest quartiles of social vulnerability, although there was overlap in the survival curves of those less vulnerable. This suggests that for the most socially vulnerable women, the specific individual variables included in the index are less important than they are for those less vulnerable. The reasons for sex differences in the characterization of social vulnerability and in associations with survival are unclear, suggesting a need for further research. Possible contributing factors include sex differences in self-reporting behavior or coping strategies.

Conclusions

In two separate samples, we have found that social vulnerability is higher amongst people who are frailer, and that social vulnerability is associated with higher mortality, independent of frailty. Although much work needs to be done in characterizing social vulnerability, clinical and public health services for older people need to recognize that attention to social factors is integral to the provision of care.

Acknowledgements

The authors thank Dr. Xiaowei Song for her assistance with MATLAB figure production.

Authors' Contributions

Conceived and designed the experiments: KR MA. Analyzed the data: MA. Contributed reagents/materials/analysis tools: AM. Wrote the paper: AM KR MA.

References

1. Seeman TE, Berkman LF, Kohout F, Lacroix A, Glynn R, et al. (1993) Inter-community variations in the association between social ties and mortality in the elderly. A comparative analysis of three communities. Ann Epidemiol 3(4): 325–335.

2. Seeman TE, Kaplan GA, Knudsen L, Cohen R, Guralnik J (1987) Social network ties and mortality among the elderly in the Alameda County Study. Am J Epidemiol 126(4): 714–723.

3. Blazer DG (1982) Social support and mortality in an elderly community population. Am J Epidemiol 115(5): 684–694.

4. Schoenbach VJ, Kaplan BH, Fredman L, Kleinbaum DG (1986) Social ties and mortality in Evans County, Georgia. Am J Epidemiol 123(4): 577–591.

5. Bassuk SS, Glass TA, Berkman LF (1999) Social disengagement and incident cognitive decline in community-dwelling elderly persons. Ann Intern Med 131(3): 165–173.

6. Andrew MK (2005) Social capital, health, and care home residence among older adults: A secondary analysis of the Health Survey for England 2000. Eur J Ageing 2(2): 137–148.

7. Fratiglioni L, Wang HX, Ericsson K, Maytan M, Winblad B (2000) Influence of social network on occurrence of dementia: a community-based longitudinal study. Lancet 355(9212): 1315–1319.

8. Mendes de Leon CF, Glass TA, Berkman LF (2003) Social engagement and disability in a community population of older adults: the New Haven EPESE. Am J Epidemiol 157(7): 633–642.

9. McCulloch A (2001) Social environments and health: cross sectional national survey. Bmj 323(7306): 208–209.

10. Kawachi I, Berkman LF (2000) Social cohesion, social capital, and health. In: Berkman LF, Kawachi I, editors. Social Epidemiology. Oxford: Oxford University Press. pp. 174–190.

11. Kawachi I, Kennedy BP, Lochner K, Prothrow-Stith D (1997) Social capital, income inequality, and mortality. Am J Public Health 87(9): 1491–1498.

12. Ahmed N, Mandel R, Fain MJ (2007) Frailty: an emerging geriatric syndrome. Am J Med 120(9): 748–753.

13. Andrew MK, Mitnitski A (2008) Different ways to think about frailty? Am J Med 121(2): e21.

14. Mitnitski AB, Mogilner AJ, Rockwood K (2001) Accumulation of deficits as a proxy measure of aging. Scientific World Journal 1: 323–336.

15. Kulminski A, Yashin A, Ukraintseva S, Akushevich I, Arbeev K, et al. (2006) Accumulation of heath disorders as a systemic measure of aging: Findings from the NLTCS data. Mech Ageing Dev.

16. Woo J, Goggins W, Sham A, Ho SC (2006) Public health significance of the frailty index. Disabil Rehabil 28(8): 515–521.

17. Mitnitski A, Song X, Skoog I, Broe GA, Cox JL, et al. (2005) Relative fitness and frailty of elderly men and women in developed countries, in relation to mortality. J Am Geriatr Soc 53: 2184–2189.

18. Rockwood K, Mitnitski A (2006) Limits to deficit accumulation in elderly people. Mech Ageing Dev 127(5): 494–496.

19. Rockwood K, McDowell I, Wolfson C (2001) Canadian Study of Health and Aging. International Psychogeriatrics 13: (Suppl. 1)1–237.

20. Singh MP, Tambay JL, Krawchuk S (1994) The National Population Health Survey: Design and issues. Ottawa: Statistics Canada.

21. Marmot M (2004) Status syndrome: How your social standing directly affects your health and life expectancy. London: Bloomsbury Publishing.

22. Grundy E, Holt G (2001) The socioeconomic status of older adults: how should we measure it in studies of health inequalities? J Epidemiol Community Health 55(12): 895–904.

23. Grundy E, Sloggett A (2003) Health inequalities in the older population: the role of personal capital, social resources and socio-economic circumstances. Soc Sci Med 56(5): 935–947.

24. Mitnitski AB, Song X, Rockwood K (2004) The estimation of relative fitness and frailty in community-dwelling older adults using self-report data. J Gerontol A Biol Sci Med Sci 59(6): M627–632.

25. Armitage P, Berry G, Matthews JNS (2002) Statistical methods in medical research. Fourth edition 298–306.

26. Rockwood K, Abeysundera MJ, Mitnitski A (2007) How should we grade frailty in nursing home patients? J Am Med Dir Assoc 8(9): 595–603.

27. Rockwood K, Andrew M, Mitnitski A (2007) A comparison of two approaches to measuring frailty in elderly people. J Gerontol A Biol Sci Med Sci 62(7): 738–743.

28. Rockwood K, Mitnitski A, Song X, Steen B, Skoog I (2006) Long-term risks of death and institutionalization of elderly people in relation to deficit accumulation at age 70. J Am Geriatr Soc 54(6): 975–979.

29. Fratiglioni L, Paillard-Borg S, Winblad B (2004) An active and socially integrated lifestyle in late life might protect against dementia. Lancet Neurol 3(6): 343–353.

30. Wang HX, Karp A, Winblad B, Fratiglioni L (2002) Late-life engagement in social and leisure activities is associated with a decreased risk of dementia: a longitudinal study from the Kungsholmen project. Am J Epidemiol 155(12): 1081–1087.

31. Lochner KA, Kawachi I, Brennan RT, Buka SL (2003) Social capital and neighborhood mortality rates in Chicago. Soc Sci Med 56(8): 1797–1805.

32. Wilkinson RG (1996) Unhealthy societies: The afflictions of inequality. London: Routledge.

33. Andrew MK (2005) Le capital social et la santé des personnes âgées. Retraite et Société 46: 129–143.

34. Hepburn KW (2003) Social Gerontology. In: Tallis R, Fillit H, editors. Brocklehurst's textbook of geriatric medicine and gerontology. 6th ed. London: Churchill Livingston. pp. 183–191.

35. Mitnitski A, Bao L, Rockwood K (2006) Going from bad to worse: a stochastic model of transitions in deficit accumulation, in relation to mortality. Mech Ageing Dev 127(5): 490–493.

CITATION

Impact of Exercise in Community-Dwelling Older Adults

Ruth E. Hubbard, Nader Fallah, Samuel D. Searle, Arnold Mitnitski and Kenneth Rockwood

ABSTRACT

Background

Concern has been expressed that preventive measures in older people might increase frailty by increasing survival without improving health. We investigated the impact of exercise on the probabilities of health improvement, deterioration and death in community-dwelling older people.

Methods and Principal Findings

In the Canadian Study of Health and Aging, health status was measured by a frailty index based on the number of health deficits. Exercise was classified as either high or low/no exercise, using a validated, self-administered questionnaire. Health status and survival were re-assessed at 5 years. Of 6297 eligible participants, 5555 had complete data. Across all grades of frailty, death rates for both men and women aged over 75 who exercised were similar to their peers aged 65 to 75 who did not exercise. In addition, while all those who exercised had a greater chance of improving their health status, the greatest benefits were in those who were more frail (e.g. improvement or stability was observed in 34% of high exercisers versus 26% of low/no exercisers for those with 2 deficits compared with 40% of high exercisers versus 22% of low/no exercisers for those with 9 deficits at baseline).

Conclusions

In community-dwelling older people, exercise attenuated the impact of age on mortality across all grades of frailty. Exercise conferred its greatest benefits to improvements in health status in those who were more frail at baseline. The net effect of exercise should therefore be to improve health status at the population level.

Introduction

The benefits of exercise have long been recognized. Joseph Addison wrote in 1711 that without exercise "the body cannot subsist in its vigor, nor the soul act with cheerfulness" [1]. Exercise programs of varying design have diverse positive effects in community-dwelling older people including improved muscle strength and gait speed [2], reduction in falls [3] improved balance [4] and increased bone mineral density [5]. In longitudinal cohort studies, physical activity is protective of impaired physical function [6] and modifies the effect of disability on depression [7].

Exercise programs for frail older people, however, have yielded conflicting results. A systematic review of physical training in institutionalised elderly indicated positive effects on muscle strength but effects on gait, disability, balance and endurance were inconclusive [8]. In some studies, exercise programs in very frail older people result in no improvements in physical health or function [9] and increase musculoskeletal injury [10] and falls [11]. In contrast, other studies conclude that exercise improves physical performance scores [12] and reduces falls [13]. In an international observational study, physical activity in frail older people seemed to slow further functional decline [14].

The concern that preventive care in older people merely creates a different set of health problems has been expressed with some vigour both in the lay [15] and in the medical press [16] (e.g. "...preventive interventions are encouraged regardless of age, and thus can be harmful to the patient and expensive to the health service" [16]. Since exercise is associated with increased longevity [17] and frailty is inextricably linked with aging [18] exercise could, in theory, increase the overall burden of frailty by allowing more people to live to advanced old age where frailty is most common.

How exercise affects the health of older people over the long term is unlikely to be the subject of a randomized, controlled trial, given the many benefits known to be associated with physical activity in community-dwelling older people. In consequence, longitudinal, observational studies are essential if we are to understand whether the benefits of exercise extend to all older people, regardless of their frailty status or whether there is a certain age or physiological threshold beyond which exercise may not have a positive effect [19].

The aims of this study were to examine how exercise impacts the health of older people and how these impacts might differ by individual health status—i.e. by level of fitness or frailty. We also aimed to discriminate whether effects are due to slower decline, more frequent improvement, or differing mortality rates.

Methods

Ethics Statement

The Canadian Study of Health and Aging was approved by each of the Research Ethics Committees at the 36 participating centers. Approval for these analyses came from the Research Ethics Committee of the Capital District Health Authority, Halifax, Canada.

This is a secondary analysis of the Canadian Study of Health and Aging (CSHA), a nationally representative cohort study of people age 65 years and over at baseline [20]. Briefly, 9008 community-dwelling elderly people were randomly sampled from 36 communities in all 10 Canadian provinces. In this study, we examined the 6297 participants able to fully complete a self-administered risk factor questionnaire, investigating their frailty status at baseline (CSHA-1, conducted in 1990–1991) and at 5-year follow up (CSHA-2). Decedent information was collected at follow-up to assess date of death.

The risk factor questionnaire addressed demographics, health attitudes, medical and family histories, activities of daily living and current health problems. Two questions based on the frequency and intensity of exercise assessed the level of physical activity, as validated elsewhere [21]. Subjects were classified as participating in 'high exercise' (three or more times per week, at least as intense as walking) or low/no exercise (all other exercisers and non exercisers). Of all 6297 eligible participants, 742 were lost to follow up. People with known frailty status at CSHA-2 (n = 4491) or those who died between CSHA-1 and CSHA-2 (n = 1064) were included in our sample (Figure 1).

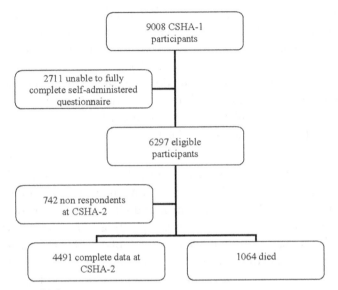

Figure 1. Derivation of Cohort.

Frailty Index

In general, frailty is understood as an increased vulnerability to a range of adverse outcomes, including death, institutionalisation and worse health [22]. It can be operationalized in many ways. A variety of tools identify phenotypic frailty as a clinical syndrome (a set of signs and symptoms that tend to occur together, thus characterizing a specific medical condition) [23]. The most well known of these is Fried et al.'s frailty phenotype identifying someone as frail when they meet ≥3 of 5 criteria (unintentional weight loss of 10 lbs or more in past year, self reported exhaustion, weak grip strength, slow walking speed and low physical activity) [24]. The other widely-used approach conceptualizes frailty as the result of multiple interacting factors, to create an index as a proportion of deficits [25], [26]. Deficit accumulation or frailty indices can be constructed from different numbers and types of variables, allowing comparisons between datasets [27]. For example, analysis of data for 36, 424 older people in four developed countries found frailty index values to be closely comparable across countries, increasing with age at approximately 3% per year in community-dwellers and correlating highly with mortality [28]. The Frailty Index approach has also recently been adopted by developing countries, exploring the affect of health status on type of death [29]. Further studies confirm that the risk of adverse outcomes is defined more precisely by deficit indices than by phenotypic definitions of frailty [30].

As in earlier reports, [31], [32] this frailty index was determined from 40 variables, selected as representing a range of health conditions and disabilities. No variable had more than 5% missing values; where missing values existed, they were inputed using the relevant mean [33]. Each variable represents a potential health deficit (e.g. symptoms, signs, functional impairments, co-morbidities, poor health attitudes). For any individual, the Frailty Index is the number of deficits present, divided by the number of deficits counted, here 40: hence someone with 6 deficits would have a Frailty Index of 0.15.

With respect to clinical translation, the Frailty Index can capture gradations in health status and the risk of adverse outcomes. It has been contextualised against a Clinical Frailty Scale in 2305 participants of the Canadian Study of Health and Ageing; this describes the functional and clinical characteristics related to different Frailty Index scores [34]. For example, those who are well with treated co-morbid disease have a mean of 6.4 deficits out of 40 (FI 0.16) whereas an FI of 0.36 (14 deficits) tends to describe those who need help with both instrumental and non-instrumental activities of daily living [31]. Note too that most people with more than 9 deficits out of 40 (Frailty Index score >0.22) are frail by any definition [34], [35].

Analysis

To distinguish the impact of exercise in relation to graded exposures (i.e. the different levels of health graded in the frailty index) with four different outcomes (improved health status, same status, worse status or death) we employed a multi-state model [36], [37], [38] (Appendix S1). The model allows all possible outcomes at all relevant health states to be summarized with just four parameters, and for the influence of co-variates (age, sex) on these parameters to be estimated. To minimize the inaccuracies of predicting outcomes for very small numbers of participants, 23 people with 18 or more deficits (i.e. Frailty Index scores >0.45) were combined in a single group.

Results

We have complete data, including data for frailty status and exercise participation at baseline as well as frailty status or death at 5 years, on 5555 participants (Figure 1). Compared to the low/no exercisers, the 2708 participating in regular exercise tended to be younger and comprised a higher proportion of men. The high exercise group was significantly fitter than low/no exercisers, with a mean Frailty Index (FI) values of 0.08 (SD 0.06) compared to 0.11 (0.09) (Table 1).

Table 1. Demographics, mean frailty index and mean 3MS cognitive scores.

	High Exercise n = 2708	Low/No Exercise n = 2847	Non Respondents at CSHA-2 n = 742
Age, mean (SD)	73.5 (6.2)	75.3 (6.8)	75.4 (6.8)
Exercise (%)	100	0	45.9
Sex (% female)	54.2	63.0	63.8
3MS total score, mean (SD)	91.1 (5.9)	89.7 (6.0)	88.0 (6.0)
Frailty Index at CSHA-1, mean (SD)	0.08(0.06)	0.11(0.09)	0.10 (0.08)

As one might expect, baseline frailty and participation in little or no exercise were each associated with an increased risk of death. Using logistic regression techniques, the risk ratio for frailty was 1.21 (95% CI 1.19 to 1.24) and for low exercise 1.95 (1.73 to 2.28).

Exercise had an impact on both mortality and on health status that was highly fit with a Markov model. For both men and women, whether older or younger, mortality increased as number of health deficits at baseline increased (Figure 2). The effect of exercise was to attenuate the impact of age on mortality i.e. for both

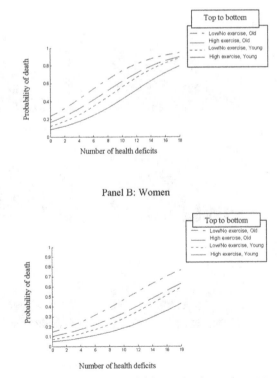

Panel B: Women

Figure 2. Probability of death within 5 years by number of deficits at baseline with participants grouped by age (<75 years, ≥75 years) and exercise status (high exercise: three or more times per week, at least as intense as walking or low/no exercise: all other exercisers and non exercisers). Panel A: Men, Panel B: Women.

men and women; those aged ≥75 years who exercised had a similar probability of death to those aged <75 who did no exercise. While exercise reduced the risk of death in all participants, it conferred its greatest mortality benefit in those with lower baseline frailty. For example, using unadjusted data the relative risk of death for low exercisers was 2.39 (95% CI 1.18–4.81) for the fittest older people (with 0 deficits at baseline) compared to a relative risk of 2.11 (0.92–4.77) for older people who would be considered frail (those with 9 deficits at baseline, a Frailty Index of 0.225).

With respect to changes in health status, and noting that people with 0 deficits at baseline have no opportunity to improve, there was no reduction of benefit across the frailty states studied. Rather, all those who exercised had a greater chance of improving their health status, which was enhanced as baseline frailty worsened (Figure 3). For example, improvement or stability was observed in 34% of high exercisers versus 26% of low/no exercisers for those with 2 deficits (FI 0.05) compared with 40% of high exercisers versus 22% of low/no exercisers for those with 9 deficits at baseline (FI 0.225).

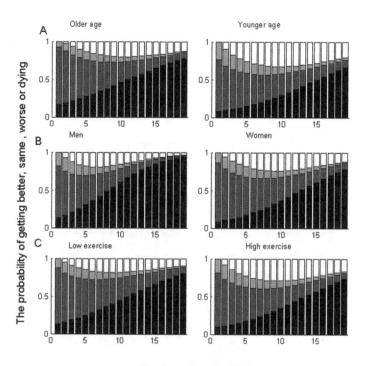

Figure 3. The probability of getting better (beige), remaining the same (tan), getting worse (dark brown) and mortality (black) as a function of number of health deficits at baseline. Panel A: <75 years vs. ≥75 years Panel B: men vs. women Panel C: low/no exercise vs. high exercise.

Discussion

In this secondary analysis of the Canadian Study of Health and Aging, we evaluated the impact of exercise on health status and near term (up to five year) survival of older people. We found that people who participated in high levels of physical activity had a lower risk of death than those who did little or no exercise. Death rates for both men and women aged over 75 who exercised were similar to their peers aged 65 to 75 who did not exercise. By mapping the transitions in numbers of health deficits in relation to exercise and adjusting for age and sex, we found that people who exercised had a greater chance of improving their health status than those who did not exercise. Interestingly, the absolute benefits in health status were greatest for those with the highest number of health deficits, i.e. the most frail.

Our data must be interpreted with caution. First, the measurement of frailty is an area of ongoing debate and the definition of frailty used here is not the only one available [22]. Phenotypic definitions of frailty tend to dichotomise or trichotomise participants (e.g. "frail," "pre-frail" or "non frail" [24]) and may exclude many of the "frailest" participants who are unable to complete performance based tests [39]. Here we were interested in quantifying risks of death and in capturing subtle changes in health status across the whole health continuum. The use of the deficit count therefore seems most appropriate for this study. Second, the follow up period was only 5 years and the effects of exercise on health transitions needs to be examined for longer follow up periods. Note too that we used self reported data. Self-report of physical activity has limitations and correlation with objective assessment is variable [40]. On the other hand, more objective physical performance measures tend to under-estimate the impact of poor function, as they commonly exclude people with the worst performance [41] and by using broad groups of activity levels, we minimize the impact of self-reporting bias. Finally, although the CSHA is a large, representative sample, 11.8% of eligible participants were lost to follow up. Since these non respondents did not appear to be systematically different to those remaining in the study at CSHA-2, this is unlikely to have materially affected our conclusions.

These results cannot be extrapolated to those in residential or nursing homes or to those with significant cognitive impairment. In this study, we investigated only community-dwelling older people who were able to complete a self-administered questionnaire. CSHA participants unable to do so were older (mean age 78.5 y [SD 7.4]) and more likely to be cognitively impaired (mean 3MS score 75.2 [15.1]). However, in longitudinal cohort studies low exercise is a risk factor for dementia [42] and analysis of this CSHA cohort showed exercise to be strongly associated with improving cognition [38]. In a recent randomized controlled trial of adults with subjective memory impairment, a 6-month program of physical activity provided a modest improvement in cognition over an 18-month follow-up period [43] and cognitively impaired older adults who participate in exercise rehabilitation programs have similar strength and endurance training outcomes as age and gender matched cognitively intact older participants [44]. There is therefore reason to be optimistic that the benefits of exercise do extend to those with cognitive impairment at baseline.

Our results suggest that older people benefit from exercise with lower mortality rates and increased likelihood of improvement in health status. These benefits extend to those with higher numbers of health deficits at baseline. Our study provides some evidence to relieve the concern that health prevention in older people extends longevity by prolonging time in impaired states. Exercise conferred its greatest benefits to improvements in health status to those who were more frail. At a population level, age and frailty should be reasons to promote rather than to limit physical activity.

Authors' Contributions

Conceived and designed the experiments: KR. Analyzed the data: NF SDS. Contributed reagents/materials/analysis tools: ABM. Wrote the paper: REH KR. Verified the analyses and developed procedures for model adjustment: NF ABM. Read and approved the final version of he manuscript: NF SDS ABM KR. Secured funding for the study and these analyses: KR. Senior investigator of CSHA, collected data and supervised local data collection at each wave: KR.

References

1. Addison JThe Spectator, July 12, 1711. Available: http://www.fullbooks.com/The-Spectator-Volume-112.html Accessed March 3rd 2009.

2. Latham NK, Bennett DA, Stretton CM, Anderson CS (2004) Systematic review of progressive resistance strength training in older adults. J Gerontol A Biol Sci Med Sci 59: 48–61.

3. Weerdesteyn V, Rijken H, Geurts AC, Smits-Engelsman BC, Mulder T, et al. (2006) A five-week exercise program can reduce falls and improve obstacle avoidance in the elderly. Gerontology 52: 131–141.

4. Bulat T, Hart-Hughes S, Ahmed S, Quigley P, Palacios P, et al. (2007) Effect of a group-based exercise program on balance in elderly. Clin Interv Aging 2: 655–660.

5. Nelson ME, Fiatarone MA, Morganti CM, Trice I, Greenberg RA, et al. (1994) Effects of high-intensity strength training on multiple risk factors for osteoporotic fractures. A randomized controlled trial. JAMA 272: 1909–1914.

6. Lang IA, Guralnik JM, Melzer D (2007) Physical activity in middle-aged adults reduces risks of functional impairment independent of its effect on weight. J Am Geriatr Soc 55: 1836–41.

7. Lee Y, Park K (2007) Does physical activity moderate the association between depressive symptoms and disability in older adults? Int J Geriatr Psychiatry 23: 249–256.

8. Rydwik E, Frandin K, Akner G (2004) Effects of physical training on physical performance in institutionalised elderly patients (70+) with multiple diagnoses. Age Ageing 33: 13–23.

9. Gill TM, Baker DI, Gottschalk M, Peduzzi PN, Allore H, et al. (2002) A program to prevent functional decline in physically frail, elderly persons who live at home. N Engl J Med 347: 1068–1074.

10. Latham NK, Anderson CS, Lee A, Bennett DA, Moseley A, et al. (2003) A randomized, controlled trial of quadriceps resistance exercise and vitamin D in frail older people: the Frailty Interventions Trial in Elderly Subjects (FITNESS). J Am Geriatr Soc 51: 291–299.

11. Faber MJ, Bosscher RJ, Chin A, Paw MJ, van Wieringen PC (2006) Effects of exercise programs on falls and mobility in frail and pre-frail older adults: A multicenter randomized controlled trial. Arch Phys Med Rehabil 87: 885–896.

12. Binder EF, Schechtman KB, Ehsani AA, Steger-May K, Brown M, et al. (2002) Effects of exercise training on frailty in community-dwelling older adults: results of a randomized controlled trial. J Am Geriatr Soc 50: 2089–2091.

13. Wolf SL, O'Grady M, Easley KA, Guo Y, Kressig RW, et al. (2006) The influence of intense Tai Chi training on physical performance and hemodynamic outcomes in transitionally frail, older adults. J Gerontol A Biol Sci Med Sci 61: 184–189.

14. Landi F, Onder G, Carpenter I, Cesari M, Soldato M, et al. (2007) Physical activity prevented functional decline among frail community-living elderly subjects in an international observational study. J Clin Epidemiol 60: 518–524.

15. Footman T (2008) Postponing the inevitable. February 11th 2008. Available at www.guardian.co.uk Accessed January 7th 2009.

16. Mangin D, Sweeney K, Heath I (2007) Preventive health care in elderly people needs rethinking. BMJ 335: 285–287.

17. Manini TM, Everhart JE, Patel KV, Schoeller DA, Colbert LH, et al. (2006) Daily activity energy expenditure and mortality among older adults. JAMA 296: 171–179.

18. Izaks GJ, Westendorp RG (2003) Ill or just old? Towards a conceptual framework of the relation between aging and disease. BMC Geriatrics 3: 7.

19. Reznick AZ, Witt EH, Silbermann M, Packer L (1992) The threshold of age in exercise and antioxidants action. EXS 62: 423–427.

20. CSHA Working Group (1994) Canadian Study of Health and Aging: study methods and prevalence of dementia. CMAJ 150: 899–913.

21. Davis HS, MacPherson K, Merry HR, Wentzel C, Rockwood K (2001) Reliability and validity of questions about exercise in the Canadian Study of Health and Aging. Int Psychogeriatr 13: Supp 1177–182.

22. Pel Littel RE, Schuurmans MJ, Emmelot Vonk MH, Verhaar HJ (2009) Frailty: defining and measuring of a concept. J Nutr Health Aging 13(4): 390–4.

23. Abate M, Di Iorio A, Di Renzo D, Paganelli R, Saggini R, Abate G (2007) Frailty in the elderly: The physical dimension. Eura Medicophys 43(3): 407–15.

24. Fried LP, Tangen CM, Walston J, Newman AB, Hirsch C, et al. (2001) Frailty in older adults: evidence for a phenotype. J Gerontol A Biol Sci Med Sci 56: 146–156.

25. Mitnitski AB, Mogilner AJ, Rockwood K (2001) Accumulation of deficits as a proxy measure of aging. ScientificWorldJournal 1: 323–36.

26. Goggins WB, Woo J, Sham A, Ho SC (2005) Frailty index as a measure of biological age in a Chinese population. J Gerontol A Biol Sci Med Sci 60: 1046–1051.

27. Rockwood K, Mitnitski A (2007) Frailty in relation to the accumulation of deficits. J Gerontol A Biol Sci Med Sci 62: 722–7.

28. Mitnitski A, Song X, Skoog I, Broe GA, Cox JL, et al. (2005) Relative fitness and frailty of elderly men and women in developed countries and their relationship with mortality. J Am Geriatr Soc 53: 2184–9.

29. Dupre ME, Gu D, Warner DF, Yi Z (2009) Frailty and type of death among older adults in China: prospective cohort study. BMJ 338: b1175.

30. Kulminski AM, Ukraintseva SV, Kulminskaya IV, Arbeev KG, Land K, et al. (2008) Cumulative deficits better characterize susceptibility to death in elderly people than phenotypic frailty: lessons from the Cardiovascular Health Study. J Am Geriatr Soc 56: 898–903.

31. Song X, MacKnight C, Latta R, Mitnitski AB, Rockwood K (2007) Frailty and survival of rural and urban seniors: results from the Canadian Study of Health and Aging. Aging Clin Exp Res 19: 145–153.

32. Mitnitski AB, Song X, Rockwood K (2004) The estimation of relative fitness and frailty in community-dwelling older adults using self-report data. J Gerontol A Biol Sci Med Sci 59: M627–632.

33. Searle SD, Mitnitski A, Gahbauer EA, Gill TM, Rockwood K (2008) A standard procedure for creating a frailty index. BMC Geriatr 8: 24.

34. Rockwood K, Song X, MacKnight C, Bergman H, Hogan DB, et al. (2005) A global clinical measure of fitness and frailty in elderly people. CMAJ 173: 489–495.

35. Rockwood K, Andrew M, Mitnitski A (2007) A comparison of two approaches to measuring frailty in elderly people. J Gerontol A Biol Sci Med Sci 62: 738–743.

36. Mitnitski A, Bao L, Rockwood K (2006) Going from bad to worse: a stochastic model of transitions in deficit accumulation, in relation to mortality. Mech Ageing Dev 127: 490–493.

37. Mitnitski A, Song X, Rockwood K (2007) Improvement and decline in health status from late middle age: modeling age-related changes in deficit accumulation. Exp Gerontol 42: 1109–1115.

38. Middleton LE, Mitnitski A, Fallah N, Kirkland SA, Rockwood K (2008) Changes in cognition and mortality in relation to exercise in late life: a population based study. PLoS ONE 3: e3124.

39. Hubbard RE, O'Mahony MS, Woodhouse KW (2009) Characterising frailty in the clinical setting—a comparison of different approaches. Age Ageing 38(1): 115–9.

40. Tudor-Locke C, Williams JE, Reis JP, Pluto D (2002) Utility of pedometers for assessing physical activity: convergent validity. Sports Med 32: 795–808.

41. Rockwood K, Jones D, Wang Y, Carver D, Mitnitski A (2007) Failure to complete performance-based measures is associated with poor health status and an increased risk of death. Age Ageing 36: 225–228.

42. Rockwood K, Middleton L (2007) Physical activity and the maintenance of cognitive function. Alzheimers Dement 3: S38–S44.

43. Lautenschlager NT, Cox KL, Flicker L, Foster JK, van Bockxmeer FM, et al. (2008) Effect of physical activity on cognitive function in older adults at risk for Alzheimer disease: a randomized trial. JAMA 300: 1027–1037.

44. Heyn PC, Johnson KE, Kramer AF (2008) Endurance and strength training outcomes on cognitively impaired and cognitively intact older adults: a meta-analysis. J Nutr Health Aging 12: 401–409.

CITATION

Originally published under the Creative Commons Attribution License. Hubbard RE, Fallah N, Searle SD, Mitnitski A, Rockwood K. "Impact of Exercise in Community-Dwelling Older Adults," in PLoS ONE 4(7): e6174.© 2009 Hubbard et al. doi:10.1371/journal.pone.0006174.

Social Participation and Independence in Activities of Daily Living: A Cross Sectional Study

Encarnación Rubio, Angelina Lázaro and Antonio Sánchez-Sánchez

ABSTRACT

Background

It is today widely accepted that participation in social activities contributes towards successful ageing whilst, at the same time, maintaining independence in the activities of daily living (ADLs) is the sine qua non for achieving that end. This study looks at people aged 65 and over living in an urban area in Spain who retain the ability to attend Social Centers providing recreational facilities. The aim of this paper is to quantify independence and identify the risk factors involved in its deterioration.

Methods

The sample size was calculated using the equation for proportions in finite populations based on a random proportional sample type, absolute error (e) = 0.05, α = 0.05, β = 0.1, p = q = 0.5. Two-stage sampling was used. In the first place, the population was stratified by residence and a Social Center was randomly chosen for each district. In the second stage, individuals were selected in a simple random sample without replacement in proportion to the number of members at each social center.

A multivariate logistical regression analysis takes functional ADL capacity as the dependent variable. The choice of predictive variables was made using a bivariate correlation matrix. Among the estimators obtained, Nagelkerke's R2 coefficient, and the Odds ratio (CI 95%) were considered. Sensitivity and 1-specificity were adopted to present the results in graphic form.

Results

Out of this sample, 63.7% were fully capable of carrying out ADLs, while the main factors contributing to deterioration, identified on the basis of a logistic regression model, are in order of importance, poor physical health, poor mental health, age (above 75 years) and gender (female). The model employed has a predictive value of 88% and 92% (depending on the age range considered) with regard to the independence in ADLs.

Conclusion

A review of the few Spanish works using similar methodology shows that the percentage of non-institutionalised persons who are independent enough to carry out ADLs is considerably lower than that found in this study of socially-active persons. Participation in recreational activities as part of a community may delay the onset of the dependence associated with ageing.

Background

Life expectancy has increased remarkably in almost all European countries in recent decades. However, it is not clear whether the years gained are years in good health or years lived with disability and in need of help [1]. In this light, the quality of the extra years of life is more important than overall life expectancy [2]. Indeed, 'adding life to the years' has become a key concept of 'successful ageing' [3], itself a prominent theme in contemporary applied gerontology [4].

The concept of successful ageing, introduced by Rowe and Kahn, implies that those ageing successfully would present the combination of low probability of disease and of deficiencies related to diseases, maintenance or strengthening of physical and cognitive functions and full engagement in life, including productive activities and interpersonal relationships [5-7]. These interdependent factors differentiate "normal ageing" from "successful ageing."

Another psychological perspective interprets successful ageing as a general process of adaptation, described as selective optimisation with compensation (SOC). This model presupposes that the three elements—selection of goals because not all opportunities can be pursued, optimisation or enrichment of reserves and resources, and compensation or utilisation of alternative means to reach the same goal—constitute the basic component processes for changes regarding ageing and adaptive capacity, and that the three always interact [8].

Meanwhile, the World Health Organisation has introduced the close concept of "active ageing" [9], which it defines as "the process of optimising opportunities for health, participation, and security in order to enhance the quality of life as people age." This definition refers not only to staying physically active but also "to continuing participation in social, economic, cultural, spiritual and civil affairs" [10]

While the notion of successful ageing is an attractive one, it has proven difficult to operationalise [11]. Some authors define subjects that are ageing successfully as those who score highest on a combined scale of physical function and exercise [12], such as those with minimal interruption of usual functioning, although minimal signs and symptoms of a chronic condition may be present [11]; with good physical and mental function (with a reduced number of chronic conditions), or good mobility, good cognitive function and absence of depression, and capacity to live an independent life [13]; or having a superior quality of life under a subjective perspective [14].

There appears to be no agreement on how to measure successful ageing. Given that as people age their quality of life is largely determined by their ability to maintain autonomy and independence [15,16], the absence of limitation in performing ADLs plays an important role [4,14,17]. In this context, the capacity to perform the activities of daily living appears to be a sine qua non for successful ageing. Thus, being fully functional in the activities of daily living, or having "no disability in performing ADLs," has been adopted as the objective indicator of successful ageing.

There is growing recognition today that social engagement is relevant for successful ageing [18]. Participation in social activities promotes physical health [19], mental health [20], intellectual functioning [21] and survival [22-24]. On the other hand, social activity is associated with better physical functioning, a lower risk of future dependence and functional recovery in order to perform ADLs [25-27].

Social engagement can be interpreted as community involvement, for example, in terms of membership of neighbourhood associations, religious groups or non-governmental organisations. Community involvement is occasionally referred to as formal social relations, compared to informal social relations, a term covering ties to family and friends. Additionally, social networks refer to structural aspects of social relations, comprising both the fabric of individuals with whom one maintains interpersonal relations and the characteristics of the ties (in terms of the number of ties, proximity of relationship, frequency of contact, reciprocity or duration) [28]. This definition permits an approach to be made to social relations on the whole, although there is no consensus on the ideal instrument to measure these relations. An excellent review of the beneficial effects of social relations on several health issues can be seen in the work of Zunzunegui et al. [27].

This paper looks at elderly people living at home in an urban area in Spain. One feature of the sample population is that these people habitually use local recreational facilities. They therefore do participate in social life in the sense of "time spent in social interaction" [24], as the social centers concerned organise activities such as music and theatre groups, card games, and introductory computer courses, among others. To the best of our knowledge, the target population selected for the study is novel in the existing literature. We would argue this because it is made-up of individuals with a high level of independence, given that all participants are still living at home and they use the social resources already present in the community by frequenting social centers. These centers were set up with the treble purpose of encouraging preventive activity that would allow members' physical, psychological and social conditions to be maintained; to create useful support services and to provide channels for community involvement.

Based on this sample population, the aim of this study is to quantify the absence of dependence in performing the basic and instrumental activities of daily life and to identify the factors contributing to functional dependence for older people.

Methods

Population and Sample

This study, which forms part of a wider research project entitled 'Old Age and Dependency in Aragon (Spain),' looks at a target population of people aged 65 or over who live at home and regularly attend social centers. This population comprises 53,632 people in Zaragoza, a city in northeastern Spain with close to 700,000 inhabitants.

The sample size was calculated using the equation for proportions in finite populations, based on a random proportional sample type, absolute error of e = 0.05, α = 0.05, β = 0.1, p = q = 0.5, requiring a degree of precision of over 90%. Having excluded questionnaires containing seven or more errors (i.e. questionnaires in which no response was given to a number of essential questions or containing inconsistencies in the responses given to similar questions) in a preliminary stage, the sample size was 380. Given that the smallest volume of responses to each individual question in the questionnaire was 348, the minimum level of precision is 91.3%.

Two-stage sampling was used. In a first step, the population was stratified by area of residence and a social center was chosen at random for districts with more than one social center. Each social center was assigned a sample size in proportion to the number of members. The second step consisted of stratified sampling for each social center with proportionate allocation by sex and age, and morning or afternoon/evening visits. Sample units were chosen by simple random sampling. If a selected individual expressed unwillingness to form part of the study, he/she was replaced by another of similar characteristics.

Data Gathering

The questionnaire used was the validated and adapted Spanish language version [29,30] of the OARS-MAFQ (Older Americans Resources and Services Program-Multidimensional Functional Assessment Questionnaire) or OARS [31,32]. The OARS questionnaire was chosen because it provides functional assessment of the non-institutionalised elderly persons in five dimensions: social resources, economic resources, mental health, physical health and the capacity to perform ADLs. It therefore allowed a large amount of information to be gathered, although not all was used in this study. The survey was carried out through face-to-face interviews. For the purposes of this study, the variables selected from the exhaustive OARS consisted, in addition to descriptive variables for the population, on those suitable to assess functional capacity in the area of ADLs, personal finances, and mental and physical health. With respect to perceived health, this variable has been classified as good and normal-poor.

Socio-Demographic Variables

The age of the interviewees was recoded in two categories and in two different ways. Thus, the sample was classified into persons aged under 75 years or 75 and older, and into persons aged under 80 years or 80 and older. In the first place, this choice reflects the fact that the percentage of individuals in the oldest age groups is small, given the characteristics of the population studied. Secondly, a more detailed examination of the survey reveals that it is precisely in this 5-year period where the point of inflexion is located with respect to the use of social and health services and, therefore, we may infer that this is also the case in respect to dependency. The level of education was classified as incomplete primary schooling or primary or higher schooling. The three groups established for marital status were, unmarried-separated-divorced, married (including those living in a stable partnership) and widowed. The latter were not included in the first group because we observed that their behavior was different and that they required more assistance.

Measurement of Personal Finances

Personal finances were measured by scoring 16 questions. These referred directly to the pension actually received and whether the participant had any other sources of income. Other indirect questions referred to the individuals' occupation, the characteristics of their dwelling and whether they could afford occasional treats or not.

Measurement of Mental and Physical Health

Mental health was assessed through six questions, one of them being the Short Psychiatry Evaluation Schedule [33,34]. The participants' general concerns were also considered, along with their appraisal of life as routine, boring or interesting.

Nineteen questions were asked to measure physical health. These included the number of visits made to doctors' surgeries or hospitals in the previous six months, as well as all illnesses diagnosed, the number and type of treatments usually taken and the support or prosthetic devices used in daily life.

Based on the responses obtained, the questionnaire allowed the variables of personal finances, mental health and physical health to be scored on a scale of 1 to 6: excellent capacity, good capacity, slight deterioration, moderate deterioration, strong deterioration and total deterioration. For operational reasons, these categories were initially classified as good (excellent and good capacity), normal (slight and moderate deterioration) and poor (strong and total deterioration).

Given the scant frequency and lack of significance of the category "poor," however, only two categories, "good" and "normal-poor," were considered in the multivariate analysis.

Measurement of Independence

In the area of ADLs, the OARS questionnaire contains 7 items referring to basic or personal care activities (BADLs). These are eating and drinking, dressing/undressing, using toilets, walking unassisted (except with a stick), going to bed/getting up and bathing/showering. The questionnaire contains a further 7 items referring to instrumental activities (IADLs), namely using the telephone, mobility (travel), shopping, meal preparation, housework, medication management and money management.

After evaluating 14 items referring to ADLs, functional capacity was classified into 6 categories: excellent, good, slightly unsatisfactory, moderately unsatisfactory, seriously unsatisfactory and completely unsatisfactory. Our interest in this study was to study independent individuals against those presenting some degree of dependence, thus functional situation is classified into two categories. A score of one indicates good functional ADL ability, or independence to perform ADLs. A score of two is indicative of moderate-severe dependence in individuals who require assistance with some activities or need some help from another person every day to carry out ADLs.

Statistical Analysis

In the first place, the variables were described to permit a multivariate logistical regression analysis taking functional ADLs ability as the dependent variable. The choice of independent or predictive variables was made using a bivariate correlation matrix and Kendall's Tau-b correlation coefficient was used. The item exhibiting the highest correlation with the ADLs was considered the principal variable, while the remainder was treated as modifying variables. Where a high correlation was found between two independent variables, the measurement considered least reliable, most difficult to obtain or most subjective, was eliminated from the analysis [35]. In order to control collinearity, models were required not to have high correlation between predictor variables, and to have a Variable Inflation Factor (VIF) lower than 5 [36].

In the second place, we used a logistic regression model with the Backward Stepwise Wald Method. The goodness-of-fit of these models was studied by means of the Hosmer-Lemeshow test, which required > 0.05 in all cases. Among the estimators obtained from this analysis, Nagelkerke's R2 coefficient, the Odds ratio with a confidence interval of 95%, and sensitivity and 1-specificity were employed to present the results in graphic form. Positive and negative predictive values were calculated to quantify the goodness of the model.

Missing information was not considered. The statistical analysis was performed using the SPSS 14.0 application. The study was evaluated and approved by the Ethics Committee at the Regional Government of Aragon, Spain (11/2009 minutes, 17th June, 2009).

Results

Of the population interviewed, 43.2% were men and 56.8% women. By age groups, 49.1% of respondents were between 65 and 74 years old and, 42.5% were between 74 and 84. Only 8.4% of the individuals frequenting social centers who responded to the questionnaire were aged 85 or older. The socio-demographic characteristics of the sample and the remaining variables are shown in Table 1.

Table 1. Socio-demographic characteristics, health and social resources

Characteristics		Number	%
Gender	Male	162	43.2
	Female	213	56.8
Age	Under 75	186	49.1
	75 and older	193	50.9
	Under 80	286	75.5
	80 and older	93	24.5
Education	Incomplete primary schooling	225	61.8
	Primary or higher schooling	139	38.2
Marital status	Unmarried/separated/divorced	45	12.8
	Married/stable partner	152	43.2
	Widowed	155	44
Perceived health	Good	224	63.6
	Normal or poor	128	36.4
Personal finances	Good	216	58.5
	Normal	148	40.1
	Poor	5	1.4
Physical health	Good	216	57.6
	Normal	145	38.7
	Poor	14	3.7
Mental health	Good	280	74.7
	Normal	76	20.3
	Poor	19	5.1
ADLs	Independent	239	63.7
	Moderate/Severely dependent	136	36.3

As reflected in Table 2, given that physical health exhibits the highest correlation with ADLs, this variable was considered the principal variable in the logistic regression analysis, while mental health, age, sex, marital status, education and personal finances were treated as modifying variables. This analysis was performed for two age groups, the first being under 75 and 75 years and older, and the second under 80 and 80 years and older. We shall call the first Model A, and the second Model B. Perceived health was not included in the model given its close linkage to the physical health variable, and in order to avoid collinearity problems.

Table 2. Bivariate correlation matrix for the variables considered in the study

	ADL	SEX	MARITAL STATUS	EDUCATION	PERCEIVED HEALTH	PERSONAL FINANCES	MENTAL HEALTH	PHYSICAL HEALTH	AGE (75 YEARS)	AGE (80 YEARS)
ADL	1.000	.165(**)	.161(**)	-.139(**)	.367(**)	.100	.530(**)	.576(**)	.287(**)	.236(**)
		.001	.002	.000	.000	.000	.000	.000	.000	.000
	375	347	374	347	375	375	375	375	374	374
SEX		1.000	.114(*)	-.115(*)	.105(*)	.102	.137(**)	.095	.005	.065
			.027	.028	.050	.051	.008	.067	.915	.207
		380	352	364	352	369	375	375	379	379
MARITAL STATUS			1.000	-.121(*)	.083	.034	.103(*)	.077	.243(**)	.245(**)
				.018	.106	.516	.046	.136	.000	.000
			352	352	351	346	347	347	351	351
EDUCATION				1.000	-.192(**)	-.084	-.178(**)	-.168(**)	-.138(**)	-.110
					.000	.112	.001	.001	.009	.037
				364	352	358	359	359	363	363
PERCEIVED HEALTH					1.000	.238(**)	.186(**)	.470(**)	.042	.007
						.000	.001	.000	.427	.893
					352	346	347	347	351	351
PERSONAL FINANCES						1.000	.150(**)	.129(*)	.011	-.012
							.004	.013	.833	.816
						369	369	369	368	368
MENTAL HEALTH							1.000	.456(**)	.240(**)	.275(**)
								.000	.000	.000
							375	375	374	374
PHYSICAL HEALTH								1.000	.251(**)	.247(**)
									.000	.000
								375	374	374

* p < 0.05 ** p < 0.01

Model A

In the first step, all the independent variables were taken into consideration, as well as first order interactions between physical health and the remaining variables (Nagelkerke's R^2 = 0.481). Initially, all interactions were eliminated as they were not significant (R^2 = 0.466). Meanwhile, the variables education, marital status and personal finances (R2 = 0.529) were eliminated in a three-step sequential process. This left physical health, mental health, age and sex as the variables in the final model (Table 3).

Table 3. Odds ratio for the predictor variables for functional deterioration in ADLs

	MODEL A		MODEL B	
	OR (CI 95%)	Sig.	OR (CI 95%)	Sig.
Physical health[a]	8.82 (4.96–15.70)	0.0001	9.21 (5.20 – 16.31)	0.0001
Mental health[b]	6.38 (3.31–12.31)	0.0001	6.70 (3.48 – 12.89)	0.0001
Sex[c]	1.95 (1.09–3.51)	0.026	1.82 (1.02 – 3.23)	0.042
Age[d]	2.24 (1.26 – 4.00)	0.006	1.38 (0.72 – 2.64)	0.33
Constant	-3.01		-2.64	
R^2	0.529		0.514	
VIF	1.39		1.36	

[a] Reference category: Good physical health.
[b] Reference category: Good mental health.
[c] Reference category: Male.
[d] Reference category: under 75 (Model A), under 80 (Model B).

Model B

After following an identical process (with the exception of the age-cut off, in this case 80 years), the correlation coefficient initially obtained was similar to the first model (Nagelkerke's R^2 = 0.470). After eliminating all non-significant interactions, the coefficient was 0.457. Once again, a sequential process was applied to eliminate all non-significant variables from the model. This left physical health, mental health, sex and age as the variables in the final model (R^2 = 0.514).

Predictors of Functional Deterioration for ADLs

As shown in Table 3, the possession of normal/poor physical health means the likelihood of some degree of dependence is at least 4.96 higher than in individuals whose physical health is good. In the case of mental health, the likelihood is at least 3.31 higher. The age variable appears in third place as an explanatory factor. Thus, as age increases from under to over 75 years old, the risk of ADLs dependence rises by at least 1.26. With regard to sex, women are at slightly more risk of dependence than men.

In Model B, the risk of becoming to some degree dependent increases at least 5.2 times when physical health deteriorates, while the figures for mental health and sex are similar to those in Model A. Finally, the age of the individuals forming part of the sample ceases to be significant in this model.

Figure 1 shows the ROC (Receiver Operating Characteristic) curves for the two models described. As can be seen, both exhibit high values (over 75%) in terms of sensitivity and specificity. Considering the sensitivity and specificity values for the different cut-off points, it may be observed that the optimum point in both models is 0.4. This cut-off point represents the highest value and it is located to the left on the ROC curve (Chart 1). Note that both curves are very similar for Models A and B. In fact, the area below the Model A curve is 0.880 (95% CI = 0.843—0.916) and that of Model B is 0.866 (95% CI = 0.826—0.906).

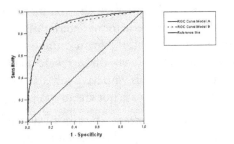

Figure 1. ROC for ADL independence in models A and B.

In addition, the negative predictive value for Model A is 0.88 whilst for Model B it is 0.92 (see Table 4). Thus, where the model indicates that the subject is independent, the prediction will only be erroneous 12% of the time in model A and 8% in model B. Thus, the second model represents an improvement on the first.

Table 4. Predictive values for the model used to determine functional dependence in ADLs.

	Sensitivity	Specificity	1-Sensit.	1-Specif.	Positive predictive value	Negative predictive value
Model A	0.822	0.808	0.178	0.192	0.724	0.88
Model B	0.800	0.833	0.200	0.176	0.68	0.92

Model A: Age two age group: under 75 and 75 years and older.
Model B: Age two age group: under 80 and 80 years and older

Discussion

In this study we sought to obtain empirical evidence of successful ageing in individuals who maintain formal social relations on the assumption that the capacity to carry out the activities of daily living is a prerequisite for successful ageing. Furthermore, to the best of our knowledge, the target population selected is novel in the existing literature, insofar as all participants are still living at home and use the social resources already present in the community. In this context, 63.7% of the population exhibit complete independence in carrying out ADLs defined as basic and instrumental activities.

In view of the magnitude of this figure, it is important to identify the variables that determine, or help explain, independence among the population participating in formal social activity. As expected, the results of the logistic regression analysis performed suggest deterioration of physical and mental health, gender and age as predictors of disability or disability risk factors.

Physical health is the main predictor of disability in both models. In fact, deteriorating physical health was found to increase the risk of ADLs disability by at least 4.9 times. Guralnik and Simonsick showed that the presence of a single chronic condition is a significant predictor of deterioration in functional capacity, since such conditions are, in general, the main cause of difficulty in performing ADLs [37]. In Ho et al. five chronic conditions were analyzed, allowing the researchers to conclude that individuals who have suffered a heart attack are at high risk of becoming dependent, if they are not already so [38].

Deteriorating mental health increases the risk of dependence by at least 3.3 times in our study. Similar results have been obtained by other authors [39-41], who found a strong association between levels of anxiety and functional capacity. Meanwhile, a significant association between the symptoms of depression and functional capacity has been noted [42].

Age and female gender are among the potential risk factor predictors for disability or associated with ADLs dependence [38,43-45]. Age is probably the key factor [46], increasing the risk of functional deterioration by 2.0 every ten years [37]. In our study, age is, indeed, a determining factor for disability, but above the age of 80 it loses significance as a predictor of deterioration, as may be observed in Model B. This is probably because the majority of the population of this age suffers from some level of dependence.

In terms of the sex variable, it is women who have the highest risk of developing disability problems, since gender appears as a risk factor in both models with significant OR values. Furthermore, these results are found regardless of age, given that there are no statistically significant differences between the mean age of men (74.4; CI = 73.4–75.4) and women (75.6; CI = 74.7–76.5). These results are close to the findings of other authors [1,38], who shows that ADLs independence at any given age was lower among women than among men, which reflects the superior longevity of women and the higher mortality of men at all ages. An explanation to these results could be that as long as men are fully independent, they live longer than women, but once their health begins to deteriorate, the progression of disability and the onset of death are faster in men [2]. This is because men suffer more fatal diseases like cardiovascular disease and strokes, whilst women tend to suffer from non-fatal but chronic disabling conditions like arthritis and osteoporosis. However, other scholars have argued that this difference disappears if other factors such as chronic diseases are also controlled [46], and it has been suggested the incidence of dependence is in fact similar taking into consideration the longer life expectancy of the female population [47].

The marital status, education and personal finances variables did not turn out to be significant disability risk predictors. Nevertheless, other authors have found that married people tend to function better than single or widowed individuals [48]; that there is a direct relationship between the number of years people can live independently and their level of education [1,37,49]; and that the likelihood of functional deterioration increases almost three times among individuals with low incomes [49].

One of our most interesting findings is that the model employed has considerable predictive power with regard to independence. Thus, it is relatively unlikely that an independent individual will not be identified as such by the model. In fact, only between 12% and 8% of individuals (depending on the age range considered) would be in this situation. In this light, the model may be interpreted as explaining a significant higher part of ADL disability than the models employed in other, similar studies. However, the advantages of the questionnaire and the model employed need to be considered in light of the difficulties inherent in the data gathering procedure.

The proportion of the population with independence in performing ADLs in our work (63.7%) is higher than the percentages obtained in studies performed in other Spanish cities using identical methodology. Martínez de la Iglesia et al. obtained a figure of 49% [50]; Eiroa et al. found that 57.6% of the population was sufficiently independent to carry out ADLs [51]; and Azpiazu et al. found that this percentage was 58% [52]. All four studies used the multidimensional OARS to measure the functional capacity of non-institutionalised old people, but only our study features social participation by all of the respondents in the survey.

Comparison with results of these national studies allows the conclusion to be reached that, in terms of the Verbrugge and Jette disability model, participation by the elderly population in social center activities is an external or environmental factor that acts as a delaying mechanism for functional dependence [53].

This statement seems to agree with results from the literature concerned with the relation between social networks of the elderly population and the functional capacity of this group to perform ADLs.

Two studies have recently been published in Spain regarding the relation between social networks of the elderly population, whether informal or formal relationships, and ADL disability [27,54]. By means of an index of social engagement in community and family activities in the older population of the city of Leganés, the first study concludes that social engagement is associated with a lower prevalence of ADL disability. [27]. The second analyses the effect of social networks on the level of autonomy (instrumental and basic disability) at the initial stages of old age, in two metropolitan areas. By means of two indicators, a general one for social network and another for involvement, it is found that social engagement has strong influence in delaying the onset of disability, for both BADLs and IADLs, although to a greater degree for basic activities [54].

Among international studies, Mendes de Leon et al. in North America concluded that leisure activities and contact with other persons reduce the risk of disability [55]. Similarly, in the Avlund et al. study for the Nordic countries, diversity in social contacts and high social participation predicted the maintenance of basic activities of daily living [56]. More recently, evidence that diversity of social ties are beneficial with respect to prevalence and recovery from ADL disability among older people in three European countries (CLESA Project), and with respect to four basic activities of daily living [27].

A different analysis is that used for the socially-active elderly (with participation in groups of elderly people or engaged in activities outside the nuclear family) in the metropolitan region of Porto Alegre [14]. Among independent predictive factors of successful ageing are independence in performing daily life activities and autonomy. Consequently, it seems that the adoption of independence in ADLs as a proxy for successful ageing would not be an absurd notion.

The described works appear to agree that social relations of elderly people, considered on the whole, are beneficial for the maintenance of independence and functional capacity for performing ADLs. However, not all factors covered by social relations contribute in the same degree [26,54]; rather, formal social engagement, more than the family network, has a greater protective effect.

Another reading of the results of this work shows that there is a significant proportion (36.3%) of elderly people who regularly frequent these social centers despite presenting a degree of ADL dependency. Dependency, at least in its least severe manifestations, is no obstacle for maintaining formal social relations and participating in both recreational and educational activities.

Although our results are in line with the described published studies, there are two fundamental weaknesses: the first is the adoption of a transversal design that does not allow a relation of causality be established between social engagement and dependence for ADLs, and the second is that this work was only made with socially-active older people.

However, in this regard, the study presented here may be considered as a starting point. To take this research further, it would interesting to apply this method to non-institutionalised subjects who do not take part in social centers in order to examine differences in the results obtained for both populations. Also, a longitudinal analysis of the population aged 65 and over, with and without participation in social centers, would allow clarification of the risk factors explaining the loss of independence in both populations. This would provide a better understanding of the interrelationship between social engagement and successful ageing.

Conclusion

Given the high percentage of independent elderly persons in this study, the high predictive power of the model with regard to independence, the fact that community participation in social activities seems to be beneficial as a protective mechanism against functional dependence, and that deterioration in physical and mental health is the main controllable cause of functional disability, public policies should foster this type of center and develop health promotion programs.

Competing Interests

The authors declare that they have no competing interests.

Authors' Contributions

ER conceived the study, performed the statistical analysis, interpreted findings and drafted the manuscript. AL conceived the study, interpreted findings and drafted the manuscript. ASS interpreted findings and drafted the manuscript.

Acknowledgements

The authors would like to express their thanks to the Zaragoza City Council and to the Management of the Social Centers of that city for providing the information and the facilities necessary in order to carry out the surveys essential to this study.

This study, which forms part of a wider research project entitled 'Old Age and Dependency in Aragon (Spain),' is funded by Regional Government of Aragon (Spain), Project S42.

References

1. Minicuci N, Noale M, Pluijm SMF, Zunzunegi MV, Blums tein T, Deeg DJH, Bardage C, Jylha M: Disability-free life expectancy: a cross-national comparison

of six longitudinal studies on aging. The CLESA project. Eur J Ageing 2004, 1(1):37–44.

2. Pérès K, Jagger C, Liévre A, Barberger-Gateau P: Disability-free life expectancy of older French people: gender an education differentials from the PAQUID cohort. Eur J Ageing 2005, 2(3):225–233.

3. Havighurst RJ, Albrecht R: Older people. New York: Longmans; 1953.

4. Hsu HC: Exploring elderly people's perspectives on successful ageing in Taiwan. Ageing Soc 2007, 27:87–102.

5. Rowe JW, Kahn RL: Human aging: usual and successful. Science 1987, 237(4811):143–149.

6. Rowe JW: The new gerontology. Science 1997, 278(5337):367.

7. Rowe JW, Kahn RL: Successful aging. Gerontologist 1998, 38(2):151.

8. Baltes PB, Baltes MM: Psychocological perspectives on successful aging: The model of selective optimization with compensation. In Successful Aging: Perspectives from behavioral sciences. Edited by: Baltes PB, Baltes MM. Cambridge New York: University Press; 1990:1–34.

9. World Health Organization: WHOQOL_Bref-Introduction, administration, scoring and generic version of the assessment. Field trial version. In World Health Organization Programme on Mental Health. Geneva: World Health Organization; 1996.

10. World Health Organization: Active Ageing: a Policy Framework. WHO/NMH/NPH 02.8. Geneva: World Health Organization; 2002.

11. Strawbridge WJ, Cohen RD, Shema SJ, Kaplan GA: Successful Aging: Predictors and Associated Activities. Am J Epidemiol 1996, 144(2):135–141.

12. Guralnik J, Kapplan GK: Predictors of healthy aging: prospective evidence from the Alameda Country Study. Am J Public Health 1989, 79:703–708.

13. Seeman TE, Bruce ML, Mcavay GJ: Social Networks Characteristics and Onset of ADL Disability: McArthur Studies of Successful Aging. J Gerontol B Psychol Sci Soc Sci 1996, 51(4):S191–200.

14. de Moraes JF, de Azevedo e Souza VB: Factors associated with the successful aging of the socially-active elderly in the metropolitan region of Porto Alegre. Rev Bras Psiquiatr 2005, 27(4):302–208.

15. World Health Organization: Statement development by WHO Quality of Life Working Group. In Published in the WHO Health Promotion Glossary 1998. WHO/HPR/HEP 98.1. Geneva: World Health Organization; 1994.

16. Smith JA, Braunack-Mayer A, Wittert G, Warin M: I've been independent for so damn long!: Independence, masculinity and aging in a help seeking context. J Aging Stud 2007, 21(4):325–335.

17. Wiener JM, Hanley RJ, Clark R, Van Nostrand JF: Measuring the activities of daily living: Comparisons across national surveys. J Gerontol 1990, 45(6):S229–S237.

18. Mendes de Leon CF: Social engagement and successful aging. Eur J Ageing 2005, 2(1):64–66.

19. Bennett KM: Social engagement as a longitudinal predictor of objective and subjective health. Eur J Ageing 2005, 2(1):48–55.

20. Wilson RS, Mendes de Leon CF, Barnes LL, Schneider JA, Bienias JL, Evans DA, Bennett DA: Participation in cognitively stimulating activities and risk of Alzheimer disease. JAMA 2002, 287(6):742–748.

21. Schooler CM, Mulatu MS: The reciprocal effects of leisure time activities and intellectual functioning in older people: a longitudinal analysis. Psychol Aging 2001, 16(3):466–482.

22. Nakanishi N, Tatara K: Correlates and prognosis in relation to participation in social activities among older people living in a community in Osaka, Japan. J Clin Geropsychol 2000, 6(4):299–307.

23. Nakanishi N, Fukuda H, Tatara K: Changes in psychosocial conditions and eventual mortality in community-residing elderly people. J Epidemiol 2003, 13(3):72–79.

24. Maier H, Klumb PL: Social participation and survival at older ages: is the effect driven by activity content or context? Eur J Ageing 2005, 2(1):31–39.

25. Unger JB, McAvay G, Bruce ML, Berkman LF, Seeman T: Variation in the impact of social network characteristics on physical functioning in elderly persons: MacArthur studies of successful aging. J Gerontol B Psychol Sci Soc Sci 1999, 54(5):S245–251.

26. Mendes de Leon CF, Glass TA, Berkman LF: Disability as function of social networks and support in elderly African Americans and whites: the Duke EPESE 1896–1992. J Gerontol B Psychol Sci Soc Sci 2001, 56:S179–190.

27. Zunzunegui MV, Rodriguez-Laso A, Otero A, Pluijm SMF, Nikula S, Blumstein T, Jylhä M, Minicuci N: Disability and social ties: comparative findings of the CLESA study. Eur J Ageing 2005, 2(1):40–47.

28. Berkman LF, Glass T, Brissette I, Seeman TE: From social integration to health: Durkheim in the new millennium. Soc Sci Med 2000, 51:843–857.

29. Grau G: Valoración funcional multidimensional de los adultos de edad avanzada. Versión española del OARS Multidimensional Functional Assessment Questionnaire. Documento técnico de la Consejería de Salud de Andalucía. Dirección General de Coordinación, Docencia e Investigación. Sevilla 1993.

30. Grau G, Eiroa P, Cayuela A: Versión española del OARS Multidimensional Functional Assessment Questionnaire: adaptación transcultural y medida de la validez. Aten Primaria 1996, 17(8):486–95.

31. Duke OARS: Multidimensional functional assessment: the OARS methodology. In Center for the Study of Aging and Human Development, Durham. 2nd edition. New Jersey: Duke University; 1978.

32. Fillembaun GG: Multidimensional functional assessment of older adults: the Duke Older American Resources and services procedures. In Center for the Study of Aging and Human Development. Hillsdale: Lawrence Elrbaum Associates. New Jersey: Duke Universitty; 1988.

33. Haug M, Belgrave LL, Gratton B: Mental health and the elderly: factors in stability and change overtime, J Health Soc Behav 1984, 25(2):100–15.

34. Liang J, Levin JS, Krause NM: Dimensions of the OARS Mental Health Measures. J Gerontol 1989, 44(5):P127–P138.

35. Hosmer DW, Lemeshow S: Applied logistic regression. New York: John Wiley & Sons, Inc; 1989.

36. Keimbaum DG, Kupper LL, Muller KE: Applied Regression Analysis and other Multivariable Methods. 2nd edition. Belmont California: Duxbury Press; 1988.

37. Guralnik JM, Simonsick EM: Physical disability in older Americans. J Gerontol 1993, 48:3–10.

38. Ho HK, Matsubayashi K, Wada T, Kimura M, Kita T, Saijoh K: Factors associated with ADL dependence: A comparative study of residential care home and community-dwelling elderly in Japan. Geriatr Gerontol Int 2002, 2(2):80–86.

39. Leveille SG, LaCroix AZ, Hecht JA, Grothaus LC, Wagner EH: The cost of disability in older women and opportunities for prevention. J Women's Health 1992, 1(1):53–61.

40. Bruce ML, Seeman TE, Merrill SS, Blazer DG: The impact of depressive symptomatology on physical disability: MacArthur studies of successful aging. Am J Public Health 1994, 84(11):1796–1799.

41. Bruce ML: Depression and disability in late life: direction of futures research. Am J Geriatr Psychiatry 2001, 9:102–12.

42. Penninx BW, Guralnik JM, Ferrucci L, Simonsick EM, Deeg DJ, Wallace RB: Depressive symptoms and physical decline in community-dwelling older persons. JAMA 1998, 279(21):1720–1726.

43. Jitapunkul S, Krungkraipetch N, Kamolratanakul P, Dhanamun B: Dependence and active life expectancy of the elderly population living in the central region of Thailand. J Med Assoc Thai 2001, 84(3):349–356.

44. Jang Y, Haley WE, Mortimer JA, Small BJ: Moderating effects of psychosocial attributes on the association between risk factors and disability in later life. Aging Ment Health 2003, 7(3):163–170.

45. Menéndez J, Guevara A, Arcia N, Leon EM, Marín C, Alfonso J C: Chronic diseases and functional limitation in older adults: a comparative study in seven cities of Latin America and the Caribbean. Rev Panam Salud Publica 2005, 17(5/6):353–361.

46. Stuck AE, Walthert JM, Nikolaus T, Büla CJ, Hohmann G, Beck JC: Risk factors for functional status decline in community-living elderly people: a systematic literature review. Soc Sci Med 1999, 48(4):445–469.

47. Strawbridge WJ, Kaplan GA, Camacho T, Cohen RD: The dynamics of disability and functional change in an elderly cohort: results from the Alameda Country study. J Am Geriatr Soc 1992, 40(8):799–806.

48. Goldman N, Korenman S, Weinstein R: Marital status and health among the elderly. Soc Sci Med 1995, 40(12):1717–1730.

49. Berkman LF, Seeman TE, Albert M, Blazer D, Kahn R, Mohs R: High, usual and impaired functioning in community-dwelling older men and women: findings from the MacArthur Foundation Research Network on Sucessful Aging. J Clin Epidemiol 1993, 46(10):1129–1140.

50. Martínez de la Iglesia J, Espejo J, Rubio V, Enciso B, Zunzunegui MV, Aranda JM: Valoración funcional de personas mayores de 60 años que viven en una comunidad urbana. Proyecto ANCO. Aten Primaria 1997, 20:475–484.

51. Eiroa P, Vázquez-Vizoso FL, Veras R: Discapacidades y necesidades de servicios en las personas mayores detectadas en la encuesta de salud OARS-Vigo. Med Clin(Barc) 1996, 106:641–648.

52. Aspiazu Garrido M, Cruz Jentoft A, Villagrasa Ferrer JR, Abanades Herranz C, García Marín N, Alvear Valero de Bernabé F: Factores asociados al mal estado de salud percibido o a la mala calidad de vida en personas mayores de 65 años. Rev Esp Salud Pública 2002, 76(2):683–699.

53. Verbrugge LM, Jette AM: The disablement process. Soc Sci Med 1994, 38(1):1–14.

54. Escobar Bravo MA, Puga D, Martín M: Asociaciones entre la red social y la discapacidad al comienzo de la vejez en las ciudades de Madrid y Barcelona en 2005. Rev Esp Salud Pública 2008, 82(6):637–651.

55. Mendes de Leon CF, Glass TA, Berkman LF: Social engagement and disability in a community population of older people. The New Heaven EPESE. Am J Epidemiol 2003, 157(7):633–642.

56. Avlund K, Lund R, Holstein BJ, Due P, Sakari-Rantala R, Heikkinen RL: The impact of structural and functional characteristics of social relations as determinants of functional decline. J Gerontol B Psychol Sci Soc Sci 2004, 59(1):S44–51.

CITATION

Originally published under the Creative Commons Attribution License. Rubio E, Lázaro A, Sánchez-Sánchez A. "Social participation and independence in activities of daily living: a cross sectional study," in BMC Geriatrics 2009, 9:26. © 2009 Rubio et al; licensee BioMed Central Ltd. doi:10.1186/1471-2318-9-26.

Age-Related Attenuation of Dominant Hand Superiority

Tobias Kalisch, Claudia Wilimzig, Nadine Kleibel,
Martin Tegenthoff and Hubert R. Dinse

ABSTRACT

Background

The decline of motor performance of the human hand-arm system with age is well-documented. While dominant hand performance is superior to that of the non-dominant hand in young individuals, little is known of possible age-related changes in hand dominance. We investigated age-related alterations of hand dominance in 20 to 90 year old subjects. All subjects were unambiguously right-handed according to the Edinburgh Handedness Inventory. In Experiment 1, motor performance for aiming, postural tremor, precision of arm-hand movement, speed of arm-hand movement, and wrist-finger speed tasks were tested. In Experiment 2, accelerometer-sensors were used to obtain objective records of hand use in everyday activities.

Principal Findings

Our data confirm previous findings of a general task-dependent decline in motor performance with age. Analysis of the relationship between right/left-hand performances using a laterality index showed a loss of right hand dominance with advancing age. The clear right-hand advantage present at younger ages changed to a more balanced performance in advanced age. This shift was due to a more pronounced age-related decline of right hand performance. Accelerometer-sensor measurements supported these findings by demonstrating that the frequency of hand use also shifted from a clear right hand preference in young adults to a more balanced usage of both hands in old age. Despite these age-related changes in the relative level of performance in defined motor tasks and in the frequency of hand use, elderly subjects continued to rate themselves as unambiguous right-handers.

Conclusion

The discrepancy between hand-specific practical performance in controlled motor tests as well as under everyday conditions and the results of questionnaires concerning hand use and hand dominance suggests that most elderly subjects are unaware of the changes in hand dominance that occur over their lifespan, i.e., a shift to ambidexterity.

Introduction

The hand-arm system is the most active part of the human upper extremities. Over the lifespan, hands undergo many physiological and anatomical changes [1] where both intrinsic and extrinsic factors contribute to age-related alterations. For example, muscle mass and strength decrease, especially after the age of sixty years [2]. Other age-related changes include decreased abilities to maintain steady forces [3], [4], an increase in time required to manipulate small objects [5], and a clear decrease in finger-pinch strength [6]. With increasing age, a decline of hand movement coordination occurs [7] which can lead to an impaired ability to perform everyday activities [8].

A decline in hand function can result from changes in the peripheral nervous system such as decreased nerve conduction velocity, sensory perception, or excitation-contraction coupling of motor units [9], [10]. It has been suggested that the higher muscle fatigue resistance typically found in the elderly was attributable to differences in both the muscle and the central nervous system [11]. Moreover, impairment of sensory perception is thought to be a key component of decreased fine motor functioning [12]. Besides age-related changes in the sensory motor system, it remains unclear how environmental factors such as declining physical activity associated with aging [13], [14] and sedentary lifestyles contribute to impaired hand function [15], [16].

While a general age-related decline in hand performance and hand function is undisputed, little is known about possible changes in hand dominance, i.e. about asymmetries of hand use that develop with advancing age. Several questionnaires like the "Edinburgh Handedness Questionnaire" [17], "Revised Waterloo Handedness Questionnaire" [18], "Annett handedness questionnaire" [19] as well as practical tests such as the "WatHand Box Test" [20], "Jebsen Test of Hand Function" [21], tapping-tasks and pegboard tests are available to assess hand dominance. These two approaches differ in that questionnaires only detect subjective preferences towards the use of the dominant or non-dominant hand in specified situations, but not necessarily the level of hand performance itself [22]. However, there is agreement that in young healthy subjects self-rated hand dominance and the level of motor performance is highly correlated [23], [24], [25].

We addressed the question of possible changes in hand dominance with advancing age in elderly subjects 65 to 90 years of age. Handedness can be defined as the preference or hand-difference in task performance [22]. To analyze both factors, we combined self-rating with questionnaires with objective measurements of hand dominance using a conventional fine motor test-series (Experiment 1). This provided insight into how the performance of each hand is differentially affected by age. Additionally, sensors were used to record the frequency of hand use during everyday activities which was compared with self-rated hand dominance (Experiment 2). We found that the superior performance of the dominant hand present at younger ages is progressively lost with advancing age due to a more pronounced age-related decline of hand function in the dominant hand. Interestingly, according to the subjective self-rated questionnaire, the older subjects were mostly unaware of these changes.

Methods

Elderly subjects in senior residences were recruited by poster announcements or by word of mouth. All subjects were tested by a clinical neurologist to determine that they were without neurological symptoms and in good physical condition. Eligibility criteria were lucidity, independence in everyday activities, and the absence of motor handicaps such as functional impairment due to arthritis or other causes of joint immobility. Subjects with significant visual or hearing loss, cerebro-vascular or spinal diseases, pathological tremor, or any functional limitations of the upper limbs as a consequence of stroke or Parkinson's disease were excluded from the study. Medication taken by the subjects was documented to prevent the influence of drugs that may affect the central nervous system. An assessment of cognitive abilities was made using the "Mini Mental State Examination" [26]. Only persons with scores of 27 to 30 out of 30 possible points (indicative of "no dementia") participated in the study. Accordingly, the subjects included in our study represent a subpopulation clearly biased towards mental and physical fitness. Young subjects were recruited by poster announcements from the university community. These individuals reported no known neurological disorders.

Hand preference was determined throughout the study with the "Edinburgh Handedness Inventory" (EHI) [17] which classifies handedness on the basis of a short interview on hand preference in the performance of routine practical tasks. The questionnaire evaluates handedness values from −100 for extreme left hand use to +100 for extreme right hand use. Only persons with unambiguous right hand dominance (≥ +70 points) and without a history of hand switching during their lifetime were included. The study was performed in accordance with the Declaration of Helsinki. Subjects gave written informed consent, and the protocol was approved by the local ethics committee of the Ruhr-University Bochum.

Experiment 1

Sixty healthy volunteers (34 females and 26 males) participated in this study. Their self-rated handedness (EHI) was compared to the computer-based assessment of their dexterity. Subjects were divided into four age groups designated "25," "50," "70," and "80" in accordance with the average age of the group. Group 25 included 14 subjects (9 females and 5 males) with a mean age of 24.8±3.1 years; group 50 included 14 subjects (8 females and 6 males) with a mean age of 51.8±3.2 years; group 70 included 18 subjects (9 females and 5 males) with a mean age of 70.9±2.7 years, and group 80 included 14 subjects (8 female and 6 males) with a mean age of 80.7±4.7 years.

Self-rated hand dominance revealed no significant differences between the four groups (Oneway ANOVA, $F(3,59) = 0.042$, $p = 0.989$). The EHI scores were calculated as 85.00±7.60 for group 25, 83.08±9.58 for group 50, 83.89±9.48 for group 70, and 83.93±11.63 for group 80.

Motor Performance Test-Series

According to Fleishman [27], fine motor movements can be factorized with regard to speed, accuracy, and maintenance of upper limb positions. We investigated these aspects during execution of fine motor movements of the arms, hands, and fingers using the four separate tests described below.

"Steadiness" (Fig. 1a) describes the ability to obtain a prescribed arm-hand position and to maintain it for a defined time period. "Line tracing" (Fig. 1b) describes the ability to fulfill precise, simultaneous arm-hand movements. "Aiming" (Fig. 1c) describes the ability to accomplish fast arm-hand movements for small targets. "Tapping" (Fig. 1d) describes the ability to perform very fast, repetitive wrist-finger movements with little emphasis on precision of movement.

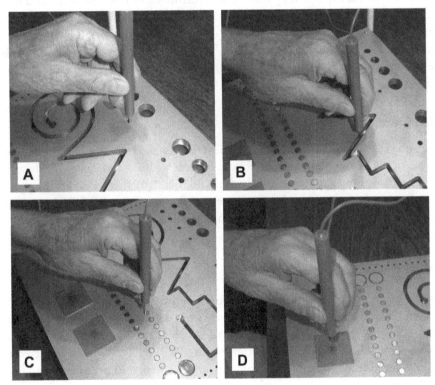

Figure 1. Examination of fine motor performance
A commercial test-series was used to measure the fine motor performance of both hands.
    The "steadiness" task describes the ability to maintain prescribed arm-hand positions (a).
    "Line tracing" describes the ability to fulfil precise arm-hand movements (b).
    "Aiming" describes the ability to accomplish fast movements directed at small targets (c).
    Finally, "tapping" describes the ability to perform very fast, repetitive hand movements (d).

All tests are available in the commercial test-series "MLS" (Dr. G. Schuhfried GmbH, Austria). The MLS is a computerized device for the accurate analysis of fine motor performance. Data registration was performed with the "Vienna-test-system" software, Version 5.05 (Dr. G. Schuhfried GmbH, Austria). We conducted a short form of the tests (10–15 minutes) in order to create a convenient test-situation for the elderly subjects.

The test board for the MLS can be used in both horizontal and vertical orientations. Two contact pencils are connected to the sides. The number and duration of contacts between pencils and test board are measured by closing electrical circuits (5V, 20mA). Data are transferred via an interface to a computer for analysis. The plain surface of the test board contains holes of different diameters, two rows of small contact plates, two large square contact plates, and a long groove.

Each task was explained by reading a standardized instruction sheet, and then the task was demonstrated to ensure that the subjects fully understood what they had to do. While the subjects sat in front of the board, support of the test arm was not permitted. All tests were performed with both the right and the left hands. To prevent systematic errors, subjects were randomly allocated to use the right or the left hand first.

Steadiness

The subject's task was to place the pencil into a small circular hole (5.8 mm) of the vertically positioned board, and hold it there without touching the edges for 32 seconds without support to steady the hand (Fig. 1a). This tested for the ability to hold a steady position, and for the absence of postural tremor [28]. Dependent variables were the number of errors, meaning the number of contacts the pencil made with the circumference the hole.

Line Tracing

Subjects were instructed to insert the pencil perpendicular to the groove in the horizontally positioned board and follow its course without touching the edges (Fig. 1b). This tested ataxia and action tremor by assessing the ability to make visually-controlled, steady, guided movements [28]. Subjects were instructed to make as few errors as possible. Dependent variables were number of errors and the total time required to complete the task. Arm movements were carried out from the periphery to midline for each respective hand.

Aiming

Subjects had to consecutively hit each of a row of 20 linearly arranged small contact fields (diameter 5 mm, midpoint separation 9 mm) with the test pencil (Fig. 1c). This test assessed the degree of ataxia and the speed of movement by the ability to make rapid repeated aimed movements [28]. Again, the dependent variables were the number of errors (missed contact fields) and the total time needed to complete the task.

Tapping

Subjects were required to hit a square contact plate (40 by 40 mm) on the test board with the test pencil as frequently as possible (Fig. 1d). The measured parameter was number of hits achieved in a time interval of 32 seconds and thereby the speed of antagonistic oscillation [28]. Because, in this task, support of the forearm was allowed, the repetitive contacts had to be accomplished by wrist movements.

Experiment 2

Another group of 36 healthy volunteers (16 females and 20 males) participated in the second experiment. This time, the self-rated handedness (EHI) was compared with the sensory-based assessment of hand use in everyday activities. As in the first experiment, the groups were designated according to the average age of the group. Group 25 included 13 subjects (6 females and 7 males) with a mean age of 27.3±4.8 years, group 50 had 9 subjects (3 females and 6 males) with a mean age of 52.4±3.1 years, and group 70 had 14 subjects (7 females and 7 males) with a mean age 72.9±3.6 years.

Self-rated hand dominance revealed no significant difference between the three groups (Oneway ANOVA, $F(2,35) = 0.051$, $p = 0.950$). The EHI scores were 88.46±24.44 for group 25, 87.78±6.67 for group 50, and 88.87±10.27 for group 70.

Assessment of Hand Movements in Everyday Activities

In order to obtain an objective measure of the use of the dominant and non-dominant hands in everyday activities, two ActiTrac® monitors (IM Systems Inc, USA) containing ceramic biaxial piezoelectric accelerometer sensors were used to record physical motion in two planes (vertical and front-to-back axes). The devices were fixed on the wrist of each hand, using belt-clips to allow unrestricted mobility of the subjects during recording, for several hours. The ActiTrac monitors measured acceleration at a rate of 40 Hz and accumulated the acceleration signals every 2 s resulting in 30 epochs per minute in units of mG (sensitivity 1.25 mG), which were stored for off-line analysis.

Statistics

Data obtained for dominant and non-dominant hands in both experiments were analyzed using ANOVAs for the factors "GENDER" and "AGE-GROUP," and repeated measures ANOVA designed for the factor "HAND." In order to detect possible relationships between performance and age, single parameters were correlated with age (Pearson-correlation). To allow a direct comparison between the extent of age-dependency, correlation coefficients were Fisher-transformed and listed as Z-values. In order to discover possible changes in hand dominance, a laterality index (l–r)/(l+r) [l = left hand performance, r = right hand performance] was calculated based on the results obtained in the practical tests (Experiment 1), or hand use in everyday activities (Experiment 2). The indices describe the extent of hand dominance for a given task within a continuum ranging from 1 to –1 (left hand dominance). All indices were aligned so that positive values indicate right hand dominance within a given task. These indices were also correlated with age via Pearson correlations. All statistical analyses were calculated using SPSS version 12.0 (SPSS Inc, USA). A p value of <0.05 was considered significant.

Results

Experiment 1

Age-Dependence of Fine Motor Performance.

For the majority of tasks tested, we found a clear decline in performance with increasing age. For the subtest "steadiness," the number of contacts with the circumference of the hole increased significantly with age for the right-hand ($r = 0.596$, $p<0.001$; $F(3,52) = 14.421$, $p\leq0.001$) and left-hand ($r = 0.414$, $p<0.001$; $F(3,52) = 4.809$, $p = 0.005$) executions of the task (Fig. 2a).

For the subtest "line tracing" measuring the precision of hand movements, the number of errors showed a significant increase with age for the right-hand ($r = 0.625$, $p<0.001$; $F(3,52) = 16.831$, $p\leq0.001$) and left-hand performances ($r = 0.539$, $p<0.001$; $F(3,52) = 8.030$, $p\leq0.001$) (Fig. 2b). Differences in the total time needed to fulfill the task reached significance for right hand performance ($r = 0.275$, $p = 0.034$; $F(3,52) = 1.738$, $p = 0.171$), but not for left hand performance ($r = 0.040$, $p = 0.761$; $F(3,52) = 0.262$, $p = 0.853$) (Fig. 2c).

For the subtest "aiming," measuring the precision of target directed movements, the total time increased with age both for the right-hand ($r = 0.640$, $p<0.001$; $F(3,52) = 14.463$, $p\leq0.001$) and left-hand performances ($r = 0.598$, $p\leq0.001$; $F(3,52) = 11.46$, $p\leq0.001$) (Fig. 2d). The number of errors failed to reach significance for right ($r = 0.209$, $p = 0.109$; $F(3,52) = 0.718$, $p = 0.546$) and left hand performances ($r = 0.236$, $p = 0.069$; $F(3,52) = 3.603$, $p = 0.019$) (Fig. 2e).

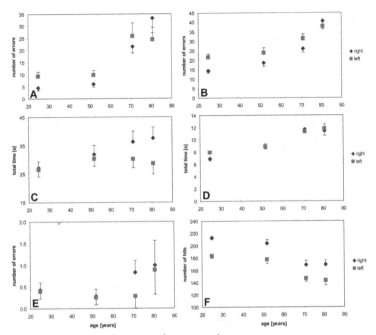

Figure 2. Age-related change of performance in fine motor tasks

Group data illustrating right and left hand performance (±SEM) in "steadiness," "line tracing," "aiming" and "tapping" tasks.

Linear Pearson's correlations revealed a significant influence of age on almost all parameters: steadiness (number of errors, a), line tracing (number of errors, b, total time right hand, c), aiming (total time, d) and tapping (number of hits, f) (p≤0.001).

Only for line tracing (total time for the left hand, c), and aiming (number of errors, e) were a lack of age related influences found (p≥0.069).

The performance of males and females differed only in the tapping task. In general, women made fewer contacts in 32 seconds irrespective of whether the task was performed with the right (F(1,52) = 15.597, p≤0.001) or the left hand (F(1,52) = 7.747, p = 0.007). This gender-specific difference in tapping performance was independent of the subject's age (non-significant interaction AGE*GENDER p≥0.277)

Laterality Indices and Age

To obtain a quantitative measure for the degree of right- or left-hand dominance, we calculated laterality indices ranging from –1 for left-hand superiority to +1 for right-hand superiority for each task and parameter. To find out if laterality changes with age, we performed linear correlation analyses between individual laterality indices averaged over all measured MLS parameters. We found a significant negative correlation (r = –0.406, p≤0.001) indicating a clear shift from right-hand superiority at the younger age towards a balanced performance at the older ages. In the younger age group, only 2 of 14 (14.3%) subjects showed left-hand superiority in average task performance, and 3 of 14 (21.4%) in group 50. However, among the individuals in groups 70 and 80, 13 of 32 subjects (41.6%) showed left-hand superiority. Accordingly, the difference between right- and left-hand performances, i.e. the dominance of the right hand, is largely abolished in the higher age groups. This results in an equalization of the level of performance of both hands, although, according to the Edinburgh questionnaire (EHI), all subjects were unambiguously right-handed (score ≥ +70).

To find out which tasks and parameters are most sensitive to age-related alterations in hand dominance, we analyzed the laterality indices separately (Tab. 1). According to this analysis, 4 of 6 laterality indices showed a significant correlation with age (Fig. 3): The total time for the aiming task ($r = -0.286$, $p = 0.027$) (Fig. 3a), the number of errors in the steadiness task ($r = -0.317$, $p = 0.014$) (Fig. 3b), and the number of errors ($r = -0.313$, $p = 0.015$) (Fig. 3c) and total time ($r = -0.324$, $p = 0.012$) for the line tracing task (Fig. 3d) showed significant correlations with age. Tapping was the only task that showed no indication of change in hand superiority with age. For all tasks and conditions that showed changes in hand superiority, we did not observe any gender-specific differences of the laterality indices ($F(1,59)≤2.234$, $p≥0.140$).

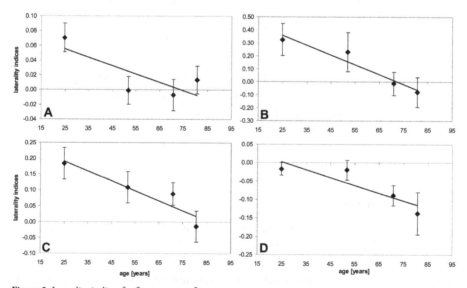

Figure 3. Laterality indices for fine motor performance
Averaged laterality indices (±SEM) for the total time in the aiming task (a), number of errors in the steadiness task (b), the line tracing task (c), and the total time needed for the line tracing task (d).

Table 1. Laterality indices of fine motor performance in different age groups

EXPERIMENT 1						Pearson-Correlation with age		
Task:	Parameter:	Group 25	Group 50	Group 70	Group 80	r	p	Fisher-Z
Aiming	Number of errors	0.38±0.19	0.16±0.14	0.08±0.14	0.41±0.16	−0.065	0.621	−0.065
	Total time	0.07±0.02	0.00±0.02	−0.01±0.02	0.01±0.02	−0.286	0.027*	−0.295
Steadiness	Number of errors	0.33±0.12	0.23±0.15	−0.01±0.09	−0.08±0.11	−0.317	0.014*	−0.328
Line tracing	Number of errors	0.18±0.05	0.11±0.05	0.09±0.04	−0.01±0.05	−0.313	0.015*	−0.324
	Total time	−0.02±0.02	−0.02±0.03	−0.09±0.03	−0.14±0.06	−0.324	0.012*	−0.336
Tapping	Number of hits	−0.07±0.01	−0.07±0.01	−0.07±0.01	−0.08±0.02	0.001	0.991	0.001

Average laterality indices (±SEM) for all parameters obtained for the aiming, steadiness, line tracing, and tapping tasks for the four age groups tested. Individual data (n = 60) were used to calculate the linear (Pearson) correlation coefficients between laterality indices and age for each parameter. Significance levels are given for $p < 0.05$ (*). To obtain comparability of linear (Pearson) correlation coefficients, Fisher-Z values are given after Fisher-transformation.

Time Course of Hand Dominance Changes

While there was a general loss of right-hand dominance with increasing age, the average age at which changes became evident was dependent on the individual task and parameter. For example, the time course for change of the laterality indices for the total time in the aiming task confirmed that subjects in group 25 performed better with their right hand than with their left hand. The right-hand advantage was completely absent in the group 50 subjects. For this group, and for the subjects in groups 70 and 80, we observed a balanced performance between the left and right hands, implying that a shift of laterality is already present in middle age. In contrast, the laterality indices for the number of errors in the steadiness task showed a more gradual change with small shifts in laterality for the subjects in group 50. Only subjects in groups 70 and 80 showed a loss of right hand dominance.

The line-tracing task requires the subject to perform both accurately and as quickly as possible. Since there is a trade-off between speed and error rate, subjects were instructed to make as few errors as possible. The laterality indices for the number of errors and total time both showed a continuous decline with age beginning with group 50, indicating that right hand dominance progressively decreases. However, both parameters differed in that for the total time, a balanced performance with both hands was already evident in group 25 subjects. It is possible that subjects change their strategy with age, shifting from speed-oriented to error-minimizing behavior. However, we found that the reduction in errors with the left hand compared to the right was not achieved by older subjects taking more time to perform the task. On the contrary, the average time needed to complete the task was also shorter for the left hand than for the right hand. This behavior began to emerge in group 70 and continued in group 80. As a result, in the line tracing task, the progressively increasing dominance of the left hand with increasing age was reflected in both the error rates and the time needed for completion.

Less drastic impairment of non-dominant hand performance causes dexterity equalization.

To identify possible contributing factors to shifts in hand dominance, we analyzed the age-related decline of the right and left hand performances separately for each task and parameter by normalizing the correlation coefficients with a Fisher-transformation and subtracted the Z-values (right-left). For all parameters that showed a strong equalization in performance with increasing age, the result of the subtraction was calculated by positive amounts (total time for the aiming task = 0.07, number of errors in the steadiness task = 0.025, number of errors and total time for the line tracing task = 0.013 and 0.024). This indicates that right-hand performance declines more with age than left-hand performance. Accordingly, the loss of right hand dominance appears to be largely due to the right hand being more sensitive to age-related alterations than the left hand. For the two factors which showed no equalization of hand performance, the subtraction of Z-values was also carried out (number of errors in the aiming task = −0.03, number of hits in the tapping task = 0.06).

Experiment 2

Frequency of Hand Use in Everyday Tasks

To address possible age related changes in hand use using objective measurements, 36 subjects agreed to wear acceleration sensors for several hours (average duration 3.12 ± 0.41 h), at approximately the same time of day, during normal everyday indoor activities. Use of a computer was not allowed during this time (because more right hand use would be forced by the one-sidedness of the computer-mouse). The sensors were fixed on the wrists to detect acceleration of the hands during movements. As we were interested in left-right differences, we calculated the laterality index as described above for all subjects (Tab. 2). There was a significant correlation between the laterality indices of hand use and age (Pearson correlation, $r = 0.447$, $p = 0.007$). In the younger, group 25 subjects we found superiority in the frequency of dominant right hand use (laterality index: 0.11 ± 0.01). The same observation was made in group 50 subjects (laterality index: 0.11 ± 0.01). However, the group 70 subjects showed a more balanced frequency of dominant and non-dominant hand use (laterality index: 0.06 ± 0.01) confirming a tendency towards equalization of hand use in older age during everyday activities (Fig. 4). There were no gender-specific differences in the laterality indices found for group 25 ($F(1,12) = 0.482$, $p = 0.502$), group 50 ($F(1,8) = 0.293$, $p = 0.605$), and group 70 ($F(1,13) = 0.703$, $p = 0.418$).

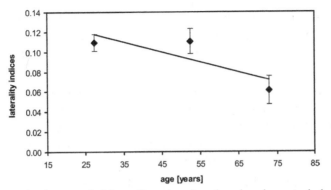

Figure 4. Changing hand use over the lifespan Illustration of age-dependent changes in the laterality indices (±SEM) obtained from acceleration measurements used to objectively assess the frequency of hand use (+1 for right hand superiority to –1 for left hand superiority) for all subjects tested (n = 36).

There is a significant linear correlation of individual laterality indices with age (Pearson, r = –0.447, p = 0.007) indicative of a loss of right (dominant) hand advantage with age.

Table 2. Frequency of hand use during everyday activities

EXPERIMENT 2

group	age [years]	number of subjects	data points [0.5/sec]	mean acceleration [mG]		calculated laterality index	EHI laterality index
				left arm	right arm		
25	27.31±1.51	13	6014±599.28	30.62±3.03	38.61±3.85	0.11±0.01	85.00±8.60
50	52.44±1.02	9	4752±304.63	37.44±4.30	47.53±6.26	0.11±0.01	87.78±2.22
70	72.86±9.97	14	6093±606.42	20.63±2.00	23.78±2.68	0.06±0.01	88.57±2.75

Group data summarizing the assessment of the frequency of hand-use during everyday activities (mean acceleration in mG for the left and right arms). The laterality indices (index) showed a significant reduction with increasing age (Pearson correlation, r = –0.447, p = 0.007), indicating a loss of dominant hand superiority in everyday activities. In contrast, the EHI-scores (Edinburgh Handedness Inventory) remained unaffected by age (Pearson correlation, r = 0.009, p = 0.960).

Discussion

Age-Related Discrepancy between Self-Rated Hand Dominance and Active Performance

We obtained the EHI handedness-scores for the 96 subjects participating in experiments 1 and 2. There was clearly no correlation between the score and the age of the subjects, pointing towards an unchanged dominance of the right hand with increasing age (Linear Pearson-correlation, r = –0.041, p = 0.692). This outcome was not surprising, as strong right-hand dominance was part of the selection criteria.

In experiment 1, we compared self-rated hand dominance with the results of a fine motor test-series (MLS). The scores on the EHI and the actual hand performance collided, as 4 parameters indicated an equalization of hand dominance. In most of the remaining parameters, at least a trend towards hand equalization could be demonstrated. A Pearson correlation of the EHI scores and MLS laterality indices confirmed that both measures were unrelated ($r = 0.130$, $p = 0.322$).

In experiment 2, self-rated hand dominance was compared with the sensor-based assessment of hand-use in everyday living. Similar to the results of experiment 1, the calculated laterality indices for hand-use indicate a loss of right hand dominance with increasing age, but also showed no significant correlation with the obtained EHI-scores ($r = 0.050$, $p = 0.771$).

We conclude that although the older subjects were unaware of changes in their handedness, a clear trend towards an equalization of hand performance under controlled (Experiment 1, Fig. 5) and everyday conditions (Experiment 2, Fig. 4) could be measured. Our results indicate that hand dominance is affected during the normal aging process. The significant correlation between subjective hand preference and task performance present in young subjects [23], [24], [25] disappears with aging. In contrast, while older subjects subjectively reported a strong right-hand preference, a variety of motor tasks and measurements of the frequency of hand use during everyday activities indicate a trend towards equalization of left and right hand use, and of the quality of hand performance with increasing age.

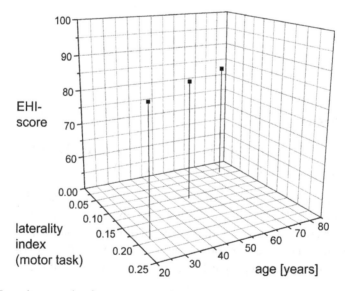

Figure 5. Contradictory results of questionnaires and practical tasks Three dimensional plot illustrating the age-dependence of the sensitivity of the EHI-scores obtained from the Edinburgh Handedness Inventory, and the laterality indices averaged over all motor tasks and parameters (laterality of motor performance) for the 4 age groups investigated.

While subjects of all age groups are characterized by approximately the same handedness score (EHI ≥ 70), there is a distinct reduction in laterality, indicating that the age-related loss of dominant hand advantage remains largely unrealized by the subjects.

General Influence of Aging

As expected, we found a decline of the level of fine motor performance with age, regardless of whether the tests were performed with the right or the left hand. The effect of aging on the capacity to perform fine motor movements is well-established in the experimental literature [29], and the extent of age-related slowing of hand-movements is positively correlated with task-difficulty (coarse versus fine motor performance) [30], [31].

The age related-decline may be the result of a general slowing down of central cognitive processes [29], [32], [33] which is assumed to affect the motor performance of less coordinated repetitive movements, like those in a tapping task [34], [35]. Whether cognitive processing is a leading global factor of the nonspecific slowing down, or if it acts selectively on different aspects of the movement process, is still under debate. Experimental evidence points towards the necessity of separating different movement stages, such as planning and execution, and provide evidence that these are differentially sensitive to age. Specifically, the movement-planning stage is prolonged, and the start of movements is delayed relative to the go signal [36]. But the actual movement seems affected by age as well, as motion trajectories by elderly people in aiming tasks seem to be less linear and more irregular than in younger subjects [36], and compensatory movements in pursuit-tracking experiments, at least at high speeds of the target, are slower and more variable [37].

Changes in Hand Dominance with Age

Apart from replicating the recognized age-related decline, we provide evidence that the decline is greater for the right hand, which shifts the manual dexterity towards a more balanced performance with both hands. These results were obtained under highly controlled experimental conditions, and are supported by recording hand use during everyday activities. This demonstrates that, in contrast to younger subjects, the elderly employ both hands with equal frequency. Interestingly, this completely contradicted the results of the common EHI-questionnaire, which showed no change in self-rated handedness over the lifespan. This was a somewhat surprising result, because previous studies in younger subjects revealed that self-rated handedness and the level of motor performance show significant coherence [23], [24], [25].

Cortical Correlates of Hand Dominance

Handedness is known to be correlated with specific lateral asymmetries; for example, higher cortical excitability in the dominant hemisphere for right-handers that are absent in left-handed-subjects [38], and anatomic asymmetries such as deeper left central sulci that are highly correlated with the degree of handedness in male subjects [39]. If handedness changes with age, as our results indicate, this may be due to asymmetric changes in the hemispheres rather than reflecting long-time plastic changes.

The number of brain areas and the intensity of their activation in motor tasks of the upper extremities changes during the human aging process; for example, greater cortical and more pronounced bilateral activation of sensorimotor regions in EEG recordings in elderly subjects [40], or a greater activation in the contralateral sensorimotor cortex, lateral premotor area, and ipsilateral cerebellum in a fMRI study [41], [42]. The increase in cortical activation with aging is not specific for movement tasks, but is also commonly found in studies addressing visual and mnemonic abilities [43], [44], [45]. These are usually interpreted in terms of recruitment of additional brain areas to provide compensatory processes, or regarded as evidence for reorganization and redistribution of functional networks thereby compensating for age-related structural and neurochemical changes [41]. The common finding of increased bilateral activation and additional activation in the frontocentral cortex (supplementary motor area region) lead to the conclusion that there is an increased need for cognitive control of simple motor tasks in old age [40].

Frontal activity during cognitive performance seems to be less lateralized in the elderly (hemispheric asymmetry reduction in older adults, HAROLD; for a detailed review, see [46], [47] compared to young adults which is supported by functional neuroimaging studies in the domains of working memory [48], perception [49], and inhibitory control [50] and may reflect functional dedifferentiation or more likely, a form of compensation [45]. With the exception of frontal activity, these common findings of age-related changes may provide an explanation for changes in the level of performance with age—like the decrease in movement performance replicated in our data—but do not help to resolve the puzzling tendency toward ambidexterity. Is there any evidence for asymmetric aging of the different hemispheres?

Historically, changes in lateral asymmetry with aging were often reported, and proposed to constitute a basic phenomenon reflecting compensational adaptations in response to age-related changes in neural processing. Some authors assume that the right hemisphere shows a larger age-related decline (called the right hemi-aging model) than the left hemisphere [51], [52]. This assumption was promoted by behavioral studies in the domains of cognitive and sensorimotor processing [53], [54] and interpreted assuming a faster age-related decline in the right hemisphere in terms of blood flow, neuron-death, and other physiological changes [55]. In contrast, other authors have argued the opposite [56], as left-hemispheric functions were believed to be better preserved because they are practiced more intensively than right-hemispheric functions [57]. Recent anatomical studies investigating structural changes that occur during the normal aging of primate cerebral hemispheres (for a review see [58]) disproved assumptions such as a major neuron loss with aging, and showed no asymmetrical developments at all [59].

If neurophysiological evidence does not support asymmetric aging of the hemispheres, a possible explanation for the move towards ambidexterity in upper limb movements with age comes from the concept of use-dependent plasticity. The advantage of the dominant hand is determined early in life, and is intensified by practice through everyday activities. When these activities decrease after retirement, or by the limitations in older age and sedentary lifestyles [16], [60], [61], it is conceivable that the practice-based superior performance of the right hand is no longer maintained, thus approaching the performance level of the left hand. This is in line with the more balanced use of both hands in everyday-life of aged subjects we found. In particular, fine motor abilities, which undergo extreme changes during the aging process, seem to depend on intact physical fitness [1]. That the age-related lateral equalization in motor performance depends on physical abilities is also in accordance with the comparable equalization in cognitive performance, as many studies indicated a correlation between age-related changes in physical fitness and cognitive performance [13], [14], [62].

Questionnaires Versus Practical Performance

In contrast to the findings presented here, several older studies have suggested that right-handedness might increase with age [63], [64], [65]. This view was largely based on the theory of asymmetrical aging of the cortical hemispheres ([51], [53]; for a sex-specific account of age-related changes of visual field asymmetries, see [66]), especially a more rapid aging of the right hemisphere [67], which is not supported by more recent neurophysiological evidence [58]. It should be emphasized that the empirical evidence for an increase in handedness stems, at least partly, from the use of questionnaires [64]. Our data clearly show that questionnaires and self-rating do not estimate actual performance in elderly subjects. As self-ratings provide a reliable tool for the assessment of handedness in young subjects [68], the discrepancy between self-rated handedness and movement data for older subjects may seem puzzling. On the other hand, it has been hypothesized that the elderly are prone to extreme decisions [69], as they show a response bias toward choosing the extreme end of a scale continuum. This bias could provide an explanation for why self-rated handedness increases with age, both in the literature and in our data. However, data from actual movement tasks in our study do not support a shift towards increasing right-handedness with age, as we found severe discrepancies between the results from the Edinburgh Handedness Inventory and practical tasks assessing motor performance with increasing age.

Summary

The use of practical tasks and the recording of hand use during everyday activities shows that the differences in the performance of the dominant and non-dominant hands which are clearly present at young age, diminish with increasing age. Apart from replicating the common finding of a decline in fine motor performance, our results provide evidence that handedness may become more balanced between early and late adulthood.

Self-ratings (EHI) indicate that the elderly subjects were unaware of these changes, i.e. the fact that hand performance had become more balanced. While at younger ages the outcomes of the self-rating and performance measures yielded a high coherence, the lack of awareness of their own performance level imposes severe restrictions on the use of questionnaires for handedness assessments in advanced age groups. As neurophysiological studies do not support a clear neural basis for this finding of changes in laterality, our results provide evidence for the importance of the concept of use-dependent plasticity.

Acknowledgements

We acknowledge expert editing of an earlier version of the MS by San Francisco Edit, Mill Valley, USA.

Authors' Contributions

Conceived and designed the experiments: MT HD TK. Performed the experiments: TK NK. Analyzed the data: HD TK CW NK. Contributed reagents/materials/analysis tools: MT HD TK. Wrote the paper: MT HD TK.

References

1. Carmeli E, Patish H, Coleman R (2003) The aging hand. J Gerontol A Biol Sci Med Sci 58: 146–152.

2. Murray P (1980) Strengh of isometric and isokinetic contractions in knee muscles of man aged 20–86. Phys Ther 60: 412–419.

3. Galganski ME, Fuglevand AJ, Enoka RM (1993) Reduced control of motor output in a human hand muscle of elderly subjects during submaximal contractions. J Neurophysiol 69: 2108–2115.

4. Keen DA, Yue GH, Enoka RM (1994) Training-related enhancement in the control of motor output in elderly humans. J Appl Physiol 77: 2648–2658.

5. Cole KJ (1991) Grasp force control in older adults. J Mot Behav 23: 251–258.

6. Sperling L (1980) Evaluation of upper extremity function in 70-year-old men and women. Scand J Rehabil Med 12: 139–144.

7. Grimby G, Danneskiold-Samsoe B, Hvid K, Saltin B (1982) Morphology and enzymatic capacity in arm and leg muscles in 78–81 year old men and women. Acta Physiol Scand 115: 125–134.

8. Kinoshita H, Francis PR (1996) A comparison of prehension force control in young and elderly individuals. Eur J Appl Physiol Occup Physiol 74: 450–460.

9. Laidlaw DH, Bilodeau M, Enoka RM (2000) Steadiness is reduced and motor unit discharge is more variable in old adults. Muscle Nerve 23: 600–612.

10. Frolkis VV, Martynenko OA, Zamostyan VP (1976) Aging of the neuromuscular apparatus. Gerontology 22: 244–279.

11. Chan KM, Raja AJ, Strohschein FJ, Lechelt K (2000) Age-related changes in muscle fatigue resistance in humans. Can J Neurol Sci 27: 220–228.

12. Warabi T, Noda H, Kato T (1986) Effect of aging on sensorimotor functions of eye and hand movements. Exp Neurol 92: 686–697.

13. Dik M, Deeg DJ, Visser M, Jonker C (2003) Early Life Physical Activity and Cognition at Old Age. J Clin Exp Neuropsychol 25: 643–653.

14. Weuve J, Kang JH, Manson JE, Breteler MM, Ware JH, et al. (2004) Physical activity, including walking, and cognitive function in older women. Jama 292: 1454–1461.

15. Carmelli D, Reed T (2000) Stability and change in genetic and environmental influences on hand-grip strength in older male twins. J Appl Physiol 89: 1879–1883.

16. Ranganathan VK, Siemionow V, Sahgal V, Yue GH (2001) Effects of aging on hand function. J Am Geriatr Soc 49: 1478–1484.

17. Oldfield RC (1971) The assessment and analysis of handedness: the Edinburgh inventory. Neuropsychologia 9: 97–113.

18. Elias LJ, Bryden MP, Bulman-Fleming MB (1998) Footedness is a better predictor than is handedness of emotional lateralization. Neuropsychologia 36: 37–43.

19. Annett J (1970) A classification of handedness. Brit J Psychol 61: 303–332.

20. Cavill S, Bryden P (2003) Development of handedness: comparison of questionnaire and performance-based measures of preference. Brain Cogn 53: 149–151.

21. Jebsen RH, Taylor N, Trieschmann RB, Trotter MJ, Howard LA (1969) An objective and standardized test of hand function. Arch Phys Med Rehabil 50: 311–319.

22. Triggs WJ, Calvanio R, Levine M, Heaton RK, Heilman KM (2000) Predicting hand preference with performance on motor tasks. Cortex 36: 679–689.

23. Bishop DV, Ross VA, Daniels MS, Bright P (1996) The measurement of hand preference: a validation study comparing three groups of right-handers. Br J Psychol 87: (Pt 2)269–285.

24. Corey DM, Hurley MM, Foundas AL (2001) Right and left handedness defined: a multivariate approach using hand preference and hand performance measures. Neuropsychiatry Neuropsychol Behav Neurol 14: 144–152.

25. Henkel V, Mergl R, Juckel G, Rujescu D, Mavrogiorgou P, et al. (2001) Assessment of handedness using a digitizing tablet: A new method. Neuropsychologia 39: 1158–1166.

26. Folstein MF, Folstein SE, McHugh PR (1975) "Mini-mental state." A practical method for grading the cognitive state of patients for the clinician. J Psychiatr Res 12: 189–198.

27. Fleishman EA (1972) Structure and measurement of psychomotor abilities. In: Singer RN, editor. The psychomotor domain. Philadelphia: Lea&Febiger. pp. 78–196.

28. Kraus PH, Przuntek H, Kegelmann A, Klotz P (2000) Motor performance: normative data, age dependence and handedness. J Neural Transm 107: 73–85.

29. Krampe RT (2002) Aging, expertise and fine motor movement. Neurosci Biobehav Rev 26: 769–776.

30. Smith CD, Umberger GH, Manning EL, Slevin JT, Wekstein DR, et al. (1999) Critical decline in fine motor hand movements in human aging. Neurology 53: 1458–1461.

31. Francis KL, Spirduso WW (2000) Age differences in the expression of manual asymmetry. Exp Aging Res 26: 169–180.

32. Carella J (1990) Aging and information processing rates in the elderly. In: Birren JE, Schaie KW, editors. Handbook of the psychology of aging. San Diego: Academic Press. pp. 201–221.

33. Salthouse TA (1996) The processing-speed theory of adult age differences in cognition. Psychol Rev 103: 403–428.

34. Denner D, Wapner S, Werner H (1964) Rhythmic activity and discrimination of stimuli in time. Percept Mot Skills 19: 723–729.

35. Boltz MG (1994) Changes in internal tempo and effects on the learning and remembering of event durations. J Exp Psychol 20: 1154–1171.

36. Liao MJ, Jagacinski RJ, Greenberg N (1997) Quantifying the performance limitations of older and younger adults in a target acquisition task. J Exp Psychol Hum Percept Perform 23: 1644–1664.

37. Jagacinski RJ, Liao MJ, Fayyad EA (1995) Generalized slowing in sinusoidal tracking by older adults. Psychol Aging 10: 8–19.

38. Civardi C, Cavalli A, Naldi P, Varrasi C, Cantello R (2000) Hemispheric asymmetries of cortico-cortical connections in human hand motor areas. Clin Neurophysiol 111: 624–629.

39. Amunts K, Jancke L, Mohlberg H, Steinmetz H, Zilles K (2000) Interhemispheric asymmetry of the human motor cortex related to handedness and gender. Neuropsychologia 38: 304–312.

40. Sailer A, Dichgans J, Gerloff C (2000) The influence of normal aging on the cortical processing of a simple motor task. Neurology 55: 979–985.

41. Mattay VS, Fera F, Tessitore A, Hariri AR, Das S, et al. (2002) Neurophysiological correlates of age-related changes in human motor function. Neurology 58: 630–635.

42. Hutchinson S, Kobayashi M, Horkan CM, Pascual-Leone A, Alexander MP, et al. (2002) Age-related differences in movement representation. Neuroimage 17: 1720–1728.

43. Grady CL, McIntosh AR, Horwitz B, Maisog JM, Ungerleider LG, et al. (1995) Age-related reduction in human recognition memory due to impaired encoding. Science 269: 218–221.

44. Grady CL, McIntosh AR, Bookstein F, Horwitz B, Rapoport SI, et al. (1998) Age-related changes in regional cerebral blood flow during working memory for faces. Neuroimage 8: 409–425.

45. Cabeza R, Grady CL, Nyberg L, McIntosh AR, Tulving E, et al. (1997) Age-related differences in neural activity during memory encoding and retrieval: a positron emission tomography study. J Neurosci 17: 391–400.

46. Dolcos F, Rice HJ, Cabeza R (2002) Hemispheric asymmetry and aging: right hemisphere decline or asymmetry reduction. Neurosci Biobehav Rev 26: 819–825.

47. Cabeza R (2002) Hemispheric asymmetry reduction in older adults: the HAROLD model. Psychol Aging 17: 85–100.

48. Dixit NK, Gerton BK, Dohn P, Meyer-Lindenberg A, Berman KF (2000) Age-related changes in rCBF activation during an N-back working memory paradigm occur prior to age 50. Neuroimage 5: 94.

49. Grady CL, Craik FI (2000) Changes in memory processing with age. Curr Opin Neurobiol 10: 224–231.

50. Nielson KA, Langenecker SA, Garavan H (2002) Differences in the functional neuroanatomy of inhibitory control across the adult life span. Psychol Aging 17: 56–71.

51. Brown JW, Jaffe J (1975) Hypothesis on cerebral dominance. Neuropsychologia 13: 107–110.

52. Albert MS (1988) Geriatric neuropsychology. New York: Guilford Press.

53. McFie J (1975) Assessment of organic intellectual impairment. London: Academic Press.

54. Wechsler D (1958) The measurement and appraisal of adult intelligence. Baltimore: Willliams and Wilkins.

55. Horn JL, Donaldson G (1977) Faith is not enough: A response to the Balter-Schaie claim that intelligence does not wane. Am Psychologist 31: 269–273.

56. Meudell PR, Greenhalgh M (1987) Age related differences in left and right hand skill and in visuo-spatial performance: their possible relationships to the hypothesis that the right hemisphere ages more rapidly than the left. Cortex 23: 431–445.

57. Kocel KM (1980) Age-related changes in cognitive abilities and hemispheric specialization. In: Heron J, editor. The neuropsychology of left-handedness. New-York: Academic Press. pp. 293–302.

58. Peters A (2002) Structural changes that occur during normal aging of primate cerebral hemispheres. Neurosci Biobehav Rev 26: 733–741.

59. Morrison JH, Hof PR (1997) Life and death of neurons in the aging brain. Science 278: 412–419.

60. Schut LJ (1998) Motor system changes in the aging brain: what is normal and what is not. Geriatrics 53: Suppl 1S16–19.

61. Hughes S, Gibbs J, Dunlop D, Edelman P, Singer R, et al. (1997) Predictors of decline in manual performance in older adults. J Am Geriatr Soc 45: 905–910.

62. Chodzko-Zajko WJ (1991) Physical fitness, cognitive performance, and aging. Med Sci Sports Exerc 23: 868–872.

63. Sand PL, Taylor N (1973) Handedness: evaluation of binomial distribution hypothesis in children and adults. Percept Mot Skills 36: 1343–1346.

64. Fleminger JJ, Dalton R, Standage KF (1977) Age as a factor of handedness in adults. Neuropsychologia 15: 471–473.

65. Weller MP, Latimer-Sayer DT (1985) Increasing right hand dominance with age on a motor skill task. Psychol Med 15: 867–872.

66. Hausmann M, Gunturkun O, Corballis M (2003) Age-related changes in hemispheric asymmetry depend on sex. Laterality 8: 277–290.

67. Goldstein G, Shelly C (1981) Does the right hemisphere age more rapidly than the left? J Clin Neuropsychol (Netherlands) 3: 65–78.

68. Peters M (1998) Description and validation of a flexible and broadly usable handedness questionnaire. Laterality 3: 77–96.

69. Porac C (1993) Are age trends in adult hand preference best explained by developmental shifts or generational differences? Can J Exp Psychol 47: 697–713.

CITATION

Originally published under the Creative Commons Attribution License. Kalisch T, Wilimzig C, Kleibel N, Tegenthoff M, Dinse HR. "Age-Related Attenuation of Dominant Hand Superiority," in PLoS ONE 1(1): e90. © 2006 Kalisch et al. doi:10.1371/journal.pone.0000090.

Are Sedatives and Hypnotics Associated with Increased Risk of Suicide in the Elderly?

Anders Carlsten and Margda Waern

ABSTRACT

Background

While antidepressant-induced suicidality is a concern in younger age groups, there is mounting evidence that these drugs may reduce suicidality in the elderly. Regarding a possible association between other types of psychoactive drugs and suicide, results are inconclusive. Sedatives and hypnotics are widely prescribed to elderly persons with symptoms of depression, anxiety, and sleep disturbance. The aim of this case-control study was to determine whether specific types of psychoactive drugs were associated with suicide risk in late life, after controlling for appropriate indications.

Methods

The study area included the city of Gothenburg and two adjacent counties (total 65+ population 210 703 at the start of the study). A case controlled study of elderly (65+) suicides was performed and close informants for 85 suicide cases (46 men, 39 women mean age 75 years) were interviewed by a psychiatrist. A population based comparison group (n = 153) was created and interviewed face-to-face. Primary care and psychiatric records were reviewed for both suicide cases and comparison subjects. All available information was used to determine past-month mental disorders in accordance with DSM-IV.

Results

Antidepressants, antipsychotics, sedatives and hypnotics were associated with increased suicide risk in the crude analysis. After adjustment for affective and anxiety disorders neither antidepressants in general nor SSRIs showed an association with suicide. Antipsychotics had no association with suicide after adjustment for psychotic disorders. Sedative treatment was associated with an almost fourteen-fold increase of suicide risk in the crude analyses and remained an independent risk factor for suicide even after adjustment for any DSM-IV disorder. Having a current prescription for a hypnotic was associated with a four-fold increase in suicide risk in the adjusted model.

Conclusion

Sedatives and hypnotics were both associated with increased risk for suicide after adjustment for appropriate indications. Given the extremely high prescription rates, a careful evaluation of the suicide risk should always precede prescribing a sedative or hypnotic to an elderly individual.

Background

The use of psychotropic drugs among the elderly is high [1-3] and health risks associated with these drugs are a topic of public health concern. Increased risk for fall accidents [4,5], adverse drug reactions [6], drug related morbidity [7] and unfavourable interactions with other medication [8] have been emphasized in the literature. While the induction of suicidality is a concern in younger age groups [9], there is increasing evidence that antidepressants may be beneficial in the prevention of suicide late in life [10-15]. Regarding a possible association between other types of psychoactive drugs (sedatives, hypnotics, antipsychotics) and suicide, results are inconclusive [16-18]. Sedatives and hypnotics are widely prescribed to elderly persons with symptoms of depression, anxiety, and sleep disturbance. They are regularly detected in post-mortem analyses of elderly suicide victims, and often implicated in lethal overdoses in Sweden [19].

The elderly consume more psychotropics than any other age group in Sweden [20]. There have been substantial changes in the sales of psychotropic drugs to persons aged 65 and above over the last fifteen years. While sales of antipsychotics (Figure 1a) and sedatives (Figure 1b) decreased by almost 50% during this time period, sales of hypnotics remained extremely high (Figure 1c). Sales rates are particularly high in the older elderly, with Defined Daily Dosages (DDD)/1000 inhabitants and day as high as 165 in women and 120 in men (Figure 1c). SSRIs were introduced in Sweden in 1990 and antidepressant sales have seen a near five-fold increase since that time (Figure 1d.)

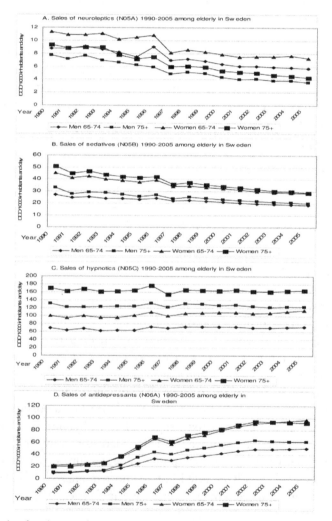

Figure 1. The sales of psychotropic drugs 1990–2005 among elderly in Sweden.

Given the high psychotropic prescription rates and the high suicide rates in this age group it would be appropriate to examine whether different types of psychotropic drugs are associated with increased suicide risk. When testing for a possible association, it is important to rule out confounding by indication. However, studies that include detailed evaluation of psychiatric symptoms in elderly suicides and population-based controls are lacking. The aim of the current study was thus to determine whether different types of psychotropic drugs were associated with increased risk of suicide in persons aged 65 years and above after adjustment for appropriate indications.

Methods

One hundred consecutive cases of suicide among persons aged 65 years and above who underwent necropsy at the Gothenburg Institute of Forensic Medicine were reviewed. Close informants for 85 suicide cases (46 men, 39 women, mean age 75 years) accepted to participate in an interview with a psychiatrist (MW). We have previously shown that the study cases were representative of all suicides among persons age 65 and above in the catchment area during the study period; antidepressants and/or lithium were detected at post-mortem analysis in 38% of the study cases and 40% of the 100 suicides (65+) evaluated at the Forensic Institute during the study period [21]. The study area included the city of Gothenburg and two adjacent counties (total 65+ population 210 703 at the start of the study).

In order to create a population comparison group, two people with the same sex, year of birth and zip code as each suicide case were randomly selected from the tax register. If a potential comparison person declined participation, another was invited (up to eight per case). In all, 240 people were invited to take part in the study and 153 participated (84 men, 69 women). Comparison subjects were interviewed face-to-face, using the same questionnaire. Primary care and psychiatric records were reviewed for suicide cases and comparison subjects. All available information was used to determine past-month mental disorders in accordance with DSM-IV [21].

Ongoing prescriptions were classified according to the Anatomical Therapeutic Chemical (ATC) classification [22]. For the purpose of this study the following drugs were classified as sedatives: diazepam, alprazolam, buspirone, hydroxizine and dixyrazine. These were classified as hypnotics: flunitrazepam, nitrazepam, zopiclone, zolpidem, oxazepam, levomepromazine, propiomazine and alimemazine. Dixyrazine is a neuroleptic, but here classified as sedative in accordance with its main clinical use. Levomepromazine, a neuroleptic and alimemazin, an antihistamine were both included in the hypnotic group because they are prescribed in Sweden to elderly with sleep disturbance.

Ethics

The study was approved by the Ethics Committee at the Faculty of Medicine, Gothenburg University.

Statistics

Crude odds ratios for suicide were calculated for the different classes of psychotropic drugs with bivariate logistic regression. In a second set of analyses, odds ratios were adjusted for age, sex and appropriate indications (see footnote Table 1) using multivariate logistic regression.

Table 1. Psychotropic drugs prescribed to elderly suicide cases (n = 85) and population controls (N = 153).

Type of drug prescribed	Cases n (%)	Controls n (%)	OR(95% CI)[a]	OR(95% CI)[b]
Any antidepressant	34 (40)	9 (6)	10.7 (4.8–23.8)	0.9 (0.2–3.2)
SSRIs	23 (27)	8 (5)	6.7 (2.9–15.9)	0.8 (0.2–2.9)
Antipsychotics	9 (11)	4 (3)	4.4 (1.3–14.8)	2.7 (0.8–10.1)
Sedatives	31 (36)	6 (4)	14.1 (5.6–35.6)	4.4 (1.3–15.2)
Hypnotics	48 (56)	16 (10)	10.8 (5.4–21.5)	4.2 (1.6–11.0)

a Unadjusted
b Adjusted for age, sex and indication. Antidepressants adjusted for affective and anxiety disorders, antipsychotics adjusted for psychotic disorders, sedatives and hypnotics adjusted for any DSM IV Axis I disorder.

Results

Psychotropic drugs were widely prescribed to the suicide cases and all drug types were associated with suicide in the unadjusted analyses (Table 1). Antidepressants were prescribed to 40% of the cases at the time of the suicide. Antidepressant medication was associated with a ten-fold increase in suicide risk in the initial analysis. However, after adjustment for affective disorders and anxiety disorders, neither antidepressants in general nor SSRIs in particular showed an association with suicide.

Antipsychotics were prescribed to one tenth of the suicide group at the time of death. Only 3% of those in the comparison group had a prescription for an antipsychotic. Antipsychotic medication was associated with increased suicide risk in the initial analyses, but the association did not remain after adjustment for psychotic disorder.

While sedative use was uncommon among persons in the control group, over a third of the suicide cases were prescribed sedatives at the time of death. A prescription for a sedative was associated with an almost twelve-fold increase of suicide risk after adjustment for anxiety disorders (OR 11.8 C.I. 4.6–30.6). As sedatives may be indicated for anxious elderly with other types of mental disorder (depression, psychosis) a second model was constructed which adjusted for any DSM IV disorder. Sedative treatment remained an independent four-fold risk factor for suicide even in the fully adjusted model (Table 1).

Hypnotics were the most widely prescribed drug type in both cases and controls. Half of the suicide cases had a prescription for a hypnotic at the time of the suicide. Hypnotics were prescribed to one tenth of the comparison group. A prescription for a hypnotic drug was associated with a ten-fold increase in suicide risk in the crude analysis. As sleep disturbance may occur in most psychiatric disorders, the final model was adjusted for any Axis I disorder. Having a current hypnotic prescription was associated with a four-fold increase in suicide risk in this adjusted model.

As interactions between psychoactive drugs and alcohol may trigger impulsive behavior, we wanted to test whether suicide victims who used sedatives/hypnotics were more likely to have a positive post-mortem test for alcohol. For hypnotic users, 11 out of 47 cases had a positive test for alcohol compared to 14 out of 35 for non-users (p = 0.146, Fisher's exact test). For sedative users, 8 out of 31 had positive registration for alcohol compared to 17 out of 51 for non-users (p = 0.622, Fisher's exact test).

Discussion

We found a four-fold increased suicide risk among elderly using sedatives and/or hypnotics after adjustment for appropriate indications. A recent Canadian register-based study showed a similar odds ratio [23], but it should be noted that those results pertained specifically to use of benzodiazepines, and diagnoses were based on physicians claims files. Our finding is at odds with that of Barak and colleagues [18] who compared rates of benzodiazepine prescription in elderly patients with major depression who did/did not attempt suicide. The proportion with benzodiazepines was larger in the group that did not attempt suicide. Several methodological differences may help to explain the conflicting results. First, the outcome variable in the study by Barak and colleagues was attempted rather than completed suicide. Second, controls in that study were psychiatric patients; our study employed face-to-face interviews with individuals randomly selected from the underlying population. Third, the proportion of women was greater in the Barak study. Women tend to have higher rates of sedative use than men, and men higher suicide rates than women.

One possible explanation for the observed increase in suicide risk associated with sedatives and hypnotics in our study may be that these drugs trigger aggressive behavior [24]. Further, interactions between benzodiazepines and alcohol may intensify impulsive tendencies, thereby increasing risk of suicide. We have previously reported that 29% of the suicide cases in the current study had a positive post-mortem test for alcohol [25]. However, proportions a positive post-mortem test for alcohol were similar in suicide cases both with and without sedatives/hypnotics, indicating that interaction with alcohol cannot fully explain the observed increase in risk. Another partial explanation might be that persons with prescriptions for these drugs have ready access to a suicide method. Availability of suicide methods increases suicide risk [26]. The current study design does not allow us to tease out the contribution of availability of suicide means on suicide risk. Suicide methods vary widely in different cultural settings. It would be of interest to test whether sedatives and hypnotics are associated with increased late-life suicide risk even in settings where other suicide methods, such as hanging or shooting are more common.

The finding that sedatives and hypnotics were associated with increased suicide risk does not in itself imply causality. It is possible that these drugs are merely markers for some other factor related to suicide risk, such as somatic illness, functional disability, alcohol use disorder, interpersonal problems, lack of social network [27] and sleep disturbance [28]. Persons with these problems might be more likely to seek health care and perhaps more likely to receive prescriptions for psychotropic drugs.

To the best of our knowledge, this is the first study to examine use of psychotropic drugs in elderly suicides and matched population controls who have been subjected to a detailed evaluation of psychiatric symptoms. We did not find support for the hypothesis that SSRIs increase suicide risk in the elderly, after controlling for indication. Taken together with previous observations from ecological studies [10-13], from retrospective analyses of patient records [18] and register studies [14], this is important information for the clinician, who might be less inclined to prescribe antidepressants in the aftermath of the "black box warning" that was issued regarding risks in adolescents [29].

Strengths and Limitations

A major strength of this study was the inclusion of detailed psychopathological data which eliminated issues of confounding by indication. Study cases were representative of all suicides in the catchment area [21]. Major limitations include small numbers and the fact that diagnoses of the suicide victims are based on data accrued by proxy interviews. However, there is evidence that suicide studies based on proxy interviews with next of kin (so-called "psychological autopsies") provide reliable diagnoses [30]. Further, medical records were readily available for cases and controls, and these were authored by physicians who were "blind" to the suicide outcome. Another methodological concern was that the participation rate for the comparison subjects was lower (64%) than that of the suicide informants (85%). Potential comparison persons with mental illness and psychotropic drug treatment might be less likely to participate, which would result in inflated odds ratios.

Implications for the Clinician

While antidepressant prescription was not an independent predictor of suicide risk, it is noteworthy that 40% of the elderly cases committed suicide despite prescribed treatment for affective illness. Treatment failure cannot be ascribed to lack of adherence to antidepressants, as we have previously shown that antidepressants were detected at post-mortem screening to a large degree [21]. As pointed out by Szanto and colleagues [31], there is a need for intensive treatment and follow-up of the depressed and suicidal elderly. In the present study, sedatives and hypnotics were related to increased risk for late life suicide. Clinicians need to be aware of this as these drugs are widely prescribed to the elderly. A careful evaluation of the suicide risk should be carried out when an elderly person presents with symptoms of anxiety and sleep disturbance.

Conclusion

In conclusion sedatives and hypnotics were both associated with increased risk for suicide after adjustment for appropriate indications. Given the extremely high prescription rates, a careful evaluation of the suicide risk should always precede prescribing a sedative or hypnotic to an elderly individual.

Competing Interests

The authors declare that they have no competing interests.

Authors' Contributions

MW was responsible for the design of the study and the study interviews. AC was responsible for collection and analysis of the pharmacological data and performed the statistical analyses. Both AC and MW prepared the manuscript. Both authors read and approved the final manuscript.

Acknowledgements

Supported by grants from the Swedish Foundation for Health Care Science and Allergy Research (V98 226), the Swedish Council for Social Research (F0042 and 0914) and the Swedish Research Council (K2004-21PD-15102-01A).

References

1. NOMESCO: Medicines Consumption in the Nordic Countries 1999–2003. NOMESCO, Copenhagen; 2004.

2. Linjakumpu T, Hartikainen S, Klaukka T, Koponen H, Kivelä SL, Isoaho R: Psychotropics among the home-dwelling—elderly increasing trend. Int J Geriatr Psychiatry 2002, 17:874–883.

3. Linjakumpu TA, Hartikainen SA, Klaukka TJ, Koponen HJ, Hakko HH, Viilo KM, Haapea M, Kivelä SL, Isoaho RE: Sedative drug use in home-dwelling elderly. Ann Pharmacother 2004, 38:2017–2022.

4. Panneman MJ, Goettsch WG, Kramarz P, Herings RM: The costs of benzodiazepine associated hospital treated fall injuries in the EU: a Pharmo study. Drugs Aging 2003, 20:833–839.

5. Leipzig RM, Cumming RG, Tinetti ME: Drugs and falls in older people: a systematic review and metaanalysis: I. Psychotropic drugs. J Am Geriatr Soc 1999, 47:30–39.

6. Routledge PA, O'Mahony MS, Woodhouse KW: Adverse drug reactions in elderly patients. Br J Clin Pharmaco 2004, 57:121–126.

7. Klarin I, Wimo A, Fastbom J: The association of inappropriate drug use with hospitalization and mortality: a population based study of the very old. Drugs Aging 2005, 22:69–82.

8. Bjorkman IK, Fastbom J, Schmidt IK, Bernsten CB: Pharmaceutical Care of the Elderly in Europe Research (PEER) Group. Drug-drug interactions in the elderly. Ann Pharmacother 2002, 36:1675–1681.

9. Cipriani A, Barbiui C, Geddes JR: Suicide, depression and antidepressants. BMJ 2005, 330:373–374.

10. Hall WD, Mant A, Mitchell PB, Rendle VA, Hickie IB, McManus P: Association between antidepressant prescribing and suicide in Australia, 1991–2000: trend analysis. BMJ 2003, 10:1008.

11. Kelly CB, Ansari T, Rafferty T, Stevenson M: Antidepressant prescribing and suicide rate in Northern Ireland. Eur Psychiatry 2003, 18:325–328.

12. Grunebaum MF, Ellis SP, Li S, Oquendo MA, Mann JJ: Antidepressants and suicide risk in the United States, 1985–1999. J Clin Psychiatry 2004, 65:1456–1462.

13. Barak Y, Aizenberg D: Association between antidepressant prescribing and suicide in Israel. Int Clin Psychopharmacol 2006, 21:281–284.

14. Rahme E, Dasgupta K, Turecki G, Nedjar H, Galbaud du Fort G: Risks of suicide and poisoning among elderly patients prescribed selective serotonin reuptake inhibitors: a retrospective cohort study. J Clin Psychiatry 2008, 69:349–357.

15. Barbui C, Esposito E, Cipriani A: Selective serotonin reuptake inhibitors and risk of suicide: a systematic review of observational studies. CMAJ 2009, 180:291–297.

16. Juurlink D, Herrmann N, Szalai J, Kopp A, Redelmeier D: Medical Illness and the Risk of Suicide in the Elderly. Arch Intern Med 2004, 164:1179–1184.

17. Youssef NA, Rich CL: Does acute treatment with sedatives/hypnotics for anxiety in depressed patients affect suicide risk? A literature review. Ann Clin Psychiatry 2008, 20:57–169.

18. Barak Y, Olmer A, Aizenberg D: Antidepressants reduce the risk of suicide among elderly depressed patients. Neuropsychopharmacology 2006, 31(1):178–181.

19. Carlsten A, Waern M, Holmgren P, Allebeck P: The role of benzodiazepines in elderly suicides. Scand J Public Health 2003, 31:224–228.

20. National Corporation of Swedish Pharmacies (Apoteket AB): Sales statistic. Apoteket AB, Stockholm 2006.

21. Waern M, Runeson BS, Allebeck P, Beskow J, Rubenowitz E, Skoog I, Wilhelmsson K: Mental disorder in elderly suicides: a case-control study. Am J Psychiatry 2002, 159:450–455.

22. WHO: The WHO Collaborating Center for Drug Statistics Methodology. Norwegian Institute of Public Health. WHO, Oslo.

23. Voaklander DC, Rowe BH, Dryden DM, Pahal J, Saar P, Kelly KD: Medical illness, medication use and suicide in seniors: a population-based case-control study. J Epidemiol Community Health 2008, 62:138–146.

24. Waern M: Alcohol dependence and misuse in elderly suicides. Alcohol Alcohol 2003, 38:249–54.

25. Ben-Porath DD, Taylor SP: The effects of diazepam (valium) and aggressive dispostion on human aggression: an experimental investigation. Addict Behav 2002, 27:167–177.

26. Gunnell D, Middleton N, Frankel S: Method availability and the prevention of suicide—a re-analysis of secular trends in England and Wales 1950–1975. Soc Psychiatry Psychiatr Epidemiol 2000, 35:437–443.

27. O'Connell H, Chin A-V, Cunningham C, Lawlor BA: Recent developments: Suicide in older people. BMJ 2004, 329:895–899.

28. Singareddy RK, Balon R: Sleep and suicide in psychiatric patients. Ann Clin Psychiatry 2001, 13:93–101.

29. Valuck RJ, Libby AM, Orton HD, Morrato EH, Allen R, Baldessarini RJ: Spillover Effects on Treatment of Adult Depression in Primary Care After FDA Advisory on Risk of Pediatric Suicidality with SSRIs. Am J Psychiatry 2007, 164:1198–1205.

30. Conner KR, Duberstein PR, Conwell Y: The validity of proxy-based data in suicide research: a study of patients 50 years of age and older who attempted suicide. I. Psychiatric diagnoses. Acta Psych Scand 2001, 104:204–209.

31. Szanto K, Mulsant BH, Houck PR, Miller MD, Mazumdar S, Reynolds CF 3rd: Treatment outcome in suicidal vs. non-suicidal elderly patients. Am J Geriatr Psychiatry 2001, 9:261–268.

CITATION

Mental Rotation of Faces in Healthy Aging and Alzheimer's Disease

Cassandra A. Adduri and Jonathan J. Marotta

ABSTRACT

Background

Previous research has shown that individuals with Alzheimer's disease (AD) develop visuospatial difficulties that affect their ability to mentally rotate objects. Surprisingly, the existing literature has generally ignored the impact of this mental rotation deficit on the ability of AD patients to recognize faces from different angles. Instead, the devastating loss of the ability to recognize friends and family members in AD has primarily been attributed to memory loss and agnosia in later stages of the disorder. The impact of AD on areas of the brain important for mental rotation should not be overlooked by face processing investigations—even in early stages of the disorder.

Methodology/Principal Findings

This study investigated the sensitivity of face processing in AD, young controls and older non-neurological controls to two changes of the stimuli—a rotation in depth and an inversion. The control groups showed a systematic effect of depth rotation, with errors increasing with the angle of rotation, and with inversion. The majority of the AD group was not impaired when faces were presented upright and no transformation in depth was required, and were most accurate when all faces were presented in frontal views, but accuracy was severely impaired with any rotation or inversion.

Conclusions/Significance

These results suggest that with the onset of AD, mental rotation difficulties arise that affect the ability to recognize faces presented at different angles. The finding that a frontal view is "preferred" by these patients provides a valuable communication strategy for health care workers.

Introduction

Imagine trying to find a familiar face as you walk into a crowded party. As you scan the room, you recognize your friend fairly easily even though she may not be looking directly at you. When you finally make your way towards her and engage in a conversation, she may turn away for a second. Even though the act of your friend turning results in different retinal input, you are not led to believe that you are now speaking to a different person. Similarly, at this same party, you may put your glass down on a table, and despite looking at it from a different angle when you pick it up, you still recognize it as your glass. This success in recognizing people and objects from different viewpoints relies on robust image representations that are resilient to large changes in retinal inputs. How one derives these invariant representations is one of the crucial questions in vision science. Despite the large amount of research in this field, the question of whether or not the strength of these representations changes during our lifetime, or is affected by neurodegenerative disorders like Alzheimer's disease (AD), has been largely unstudied.

Previous research has shown that even though we are able to recognize objects and faces from different vantage points, it is not done without cost. Viewpoint-dependent theories of recognition suggest that the viewer must mentally rotate the image to a canonical orientation and the further the presentation angle is from this canonical view, the more recognition times and errors increase [1]–[5]. For face recognition, there is strong evidence that a three-quarter view is the canonical view, as it produces the fastest and most accurate responses [6]–[9]. However it should be noted that the three-quarter view advantage is still being debated, as some have found that a frontal view of a face can show the greatest advantage [10].

In addition to rotations in depth, planar transformations have also been shown to have significant effects on face recognition. Yin (1969) [11] was the first to find that faces were more difficult to recognize when they were inverted, the inversion effect, leading him to conclude that faces are not represented in a face-centered or view-invariant way. Yin also found that face recognition was disproportionately impaired by stimulus inversion when compared to recognition of other objects. This result has been replicated many times and is a standard in the literature [for review, see 12].

The cost of mentally rotating objects appears to be fundamentally affected by aging. Although older viewers consistently show longer reaction times than younger viewers when mentally rotating objects and shapes [13]–[16], it has also been argued that different strategies may be utilized. While young adults utilize a holistic approach when mentally rotating simple shapes, and a piecemeal/parts-based approach when mentally rotating complex shapes, older adults use a holistic approach for both tasks. This strategy may serve to reduce cognitive load with the older adults [16].

In contrast to object rotation, there has been little investigation of the effects of normal and pathological aging on our ability to mentally rotate faces. Only recently has the ability to rotate faces in healthy older individuals been investigated [17]. In a behavioral study using synthetic face stimuli to measure thresholds for detecting differences between similar faces, older and younger adults were not found to differ for faces in identical views but there was a marked impairment in healthy older individuals making matches across a 20° rotation from full-face [17]. This exciting finding highlights the need to extend this research using real images of faces over longer duration periods. Further, the effect that pathological disorders like AD have on the ability to mentally rotate faces has been less studied.

AD is a neurodegenerative disease that progressively destroys a person's memory, and his or her ability to learn and reason, make judgments, communicate, and carry out daily activities. While memory is often what is affected first, it is typically followed by a progressive decline of executive functions, language, perception and visuospatial skills [18], [19]. As the disease progresses towards a more moderate stage, patients may require assistance in everyday tasks. At this stage, patients may be unable to recall names of family members, or their addresses—this memory impairment will continue to worsen as the late stage of the disease begins [20].

A widespread cortical network has been implicated in mental rotation. Cortical activation areas have been found spanning the parietal, prefrontal, occipital, and temporal areas [21]–[24]. Given that AD is associated with extensive damage to many of these same areas [25], [26], it may not be surprising that many people with AD have difficulty mentally rotating shapes and objects [27]–[29]. This is true despite the fact that, at least in the early to moderate stages of the disease, they perform as well as healthy elderly controls on matching tasks that do not require mental rotation [28], [29].

An investigation by Murphy, Kohler, Black and Evans (2000) revealed that AD patients had great difficulty matching identical, non-symmetrical shapes (called Blake shapes) when they were presented at different orientations to one another. In fact, research has shown that as the angular disparity increases between objects, the AD patients' performance is more impaired than that of age-matched controls [27], [29]. Furthermore, participants with mild to moderate AD perform poorly on tasks in which they have to visually identify objects rotated at various rotation angles, provide the canonical orientation, and mentally rotate these same objects [30]. This deficit, called orientation agnosia, can also be seen when AD patients copy simple pictures—their drawings can be rotated, sometimes up to 90° or 180°, when compared to the original [31].

What about the ability of individuals with AD to mentally rotate faces? Is it disproportionately impaired in these individuals? Traditionally, AD studies of face processing have generally concentrated on famous face tasks [32] and those that test new retrieval methods to associate specific faces with names [33]. Although these kinds of studies have revealed that people with AD are impaired at face recognition tasks, they assume that the impairment is primarily the result of a memory deficit. More recently, a study by Lee, Buckley, Gaffan, et al. (2006) revealed that AD participants in early stage of the disease were not impaired when required to choose the odd-one-out of four faces, which could be presented at the same or different angle to the other 3 faces, but were impaired when scenes were presented this way. Although mental rotation difficulties with scenes were observed in this study, the AD participants were not impaired when selecting the "odd" rotated face [34]. It is worth noting, however, that this sample of AD participants was fairly young (mean age 70.14, S.D. = 5.55), and had high MMSE scores (mean = 23.57, S.D. = 3.91), which suggests that these AD patients may have been in very early stages of the disease. Additionally, AD patients had an unlimited amount of time to complete the task and reaction times were not reported in the manuscript. It is quite possible then that a speed-accuracy trade-off may have occurred, and with a time-limited face mental rotation task, difficulties with faces may have been observed.

As a first step in our investigation, we tested the effects of aging on the ability of young adults and older non-neurological controls to complete a time-limited face-matching task in which stimuli could be rotated in depth and/or inverted. We wanted to determine if aging on its own hampers the ability of individuals to mentally rotate faces; if orientation affects young and old participants in the same way and if there was a "preferred" view of the face in which performance was better for each of the groups. Following this, the performance of a subgroup of our "oldest" aged controls was compared to a group of AD patients in order to determine if individuals with AD are disproportionately impaired in their ability to mentally rotate a face.

Materials and Methods

Ethics Statement

Procedures involving experiments on human subjects were done in accord with the ethical standards of the University of Manitoba's Research Ethics Board.

Experiment #1 — The Effects of Aging on Face Processing

Participants

All participants gave informed, written consent before beginning the study. Fifteen young adults (7 males, 8 females, mean age = 23.0 years old, S.D. = 4) were recruited from an Introduction to Psychology course at the University of Manitoba and received credit toward a course requirement for their participation. These individuals were right-handed with normal or corrected-to-normal vision. Twelve healthy elderly adults (5 males, 7 females, mean age = 74.9 years old, S.D. = 9) were recruited from the Center on Aging at the University of Manitoba. All participants were right-handed, had no known neurological problems, did not self-report a depression diagnosis, and had their near visual acuity tested under binocular viewing conditions (see Table 1). No participants self-reported having glaucoma or macular degeneration.

Prior to testing on the computerized face-matching task, all participants over 60 years of age completed two cognitive screening tools: the Mini Mental State Examination [35], and the Dementia Rating Scale [36]. On the MMSE, the healthy elderly group received a mean score of 28.8 (S.D. = 2.8), or 1.4 standard deviations above normative data [37], and a DRS mean score of 140.3 (S.D. = 2.2), or 0.8 standard deviations above the norm [36]. All MMSE and DRS scores were converted into T scores (M = 50, SD = 10). On both measures, all participants in the healthy elderly group had T scores above the mean of 50, which indicates that all participants were above normative data (see Table 1).

Table 1. Information on age, education, MMSE, DRS, and visual acuity scores for the older control group.

Subject	Age	Education (years)	MMSE (raw score)	MMSE (T-score)	DRS (raw score)	DRS (T-score)	Visual acuity	Cataracts
1	63	18	30	57.7	142	59.5	20/30	No
2	75	21	29	56.3	140	56.0	20/30	Yes (right)
3	86	12	29	65.0	138	56.0	20/20	No
4	73	12	30	68.8	144	66.0	20/20	No
5	78	10	29	63.3	140	56.0	20/30	No
6	68	10	30	64.3	139	50.0	20/20	No
7	64	21	30	57.7	141	56.0	20/20	No
8	84	12	29	67.4	138	56.0	20/25	Yes (both)
9	81	11	30	71.7	137	52.0	20/20	No
10	62	12	29	55.9	143	62.5	20/25	No
11	78	9	30	70.0	140	56.0	20/25	Yes (both)
12	87	9	29	65.0	142	66.0	20/30	No

Apparatus

The experiment was conducted on a Macintosh PowerBook G4 computer with subjects making their responses on a USB keypad. Stimuli were presented on a 15.4-inch color monitor, positioned approximately 50 cm from the participant, using PsyScope experimental software version 1.2.1 [38]. The stimuli consisted of colour pictures of male and female faces obtained from the Max-Planck Face Database, which contains three–dimensional (3D) models of real faces. Three presentation angles around a vertical axis were selected for the experiments: frontal (F), right three-quarter (T), and right profile (P) (see Figure 1) [8]. The face database was provided by the Max-Planck Institute for Biological Cybernetics in Tuebingen, German. The faces were collected as 3D models and colour maps using a Cyberware™ 3D laser-scanner. Hair was trimmed from the images, leaving the face area alone [39]. Each face was positioned on a black square background (7.5 cm×7.5 cm). A total of 97 faces were used from the database to produce the experimental trials.

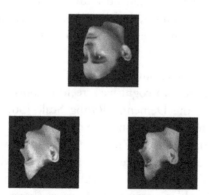

Figure 1. Sample of stimuli used (from Max-Planck Face Database) inverted and rotated in depth around the vertical axis.

Three stimulus faces were presented on each trial. A target face appeared 16.5 cm from the left side of the screen, and 5.5 cm from the top of the screen. Below this face, two choice faces appeared; one was presented on the lower left side of the screen (9.5 cm from left, 16 cm from top), and the other was presented on the lower right side of the screen (22.5 cm from left, 16 cm from top). All three stimuli appeared at once on a gray background. In one set of trials, all stimuli (target and choices) were presented upright; in the other set of trials, all stimuli were inverted. Although the rotation angle of the target face could differ from the rotation angle of the choice faces by 0°, 45°, or 90°, the two choice faces were always rotated to the same presentation angle within a trial (see Figure 1). This resulted in a total of nine possible target×choice face combinations. In three of the combinations (FF, TT, PP) the rotation angle of the target and choice faces did not differ (0°). In four of the combinations the rotation angle between target and choice faces differed by 45° (FT, TF, TP, PT). In two of the combinations, the rotation angle between the target and choice faces differed by 90° (FP, and PF). Participants were presented with 4 blocks (2 blocks of upright, and 2 blocks of inverted trials), consisting of 72 trials each, which were counterbalanced across all subjects.

Design and Procedure

Each trial began with a fixation cross appearing on the computer screen for 250 ms, followed by the three faces (one on top and two on the bottom). Participants were instructed that their task was to determine which of the two bottom choice faces (left or right) was the same as the top (target) face, regardless of how the pictures were rotated. The stimuli appeared on the screen until a participant made a response or a 10 s time limit had passed, at which point the faces were replaced by a gray screen. After a response had been logged, the experimenter initiated the next trial. The 10 s time limit was chosen because in a pilot study comparing a young and older non-neurological control group, the exposure time for the faces was unlimited and participants in the older group often took an exceedingly long time to respond, without a noticeable benefit in accuracy.

All participants were told to make their responses as quickly, but as accurately as possible. Any responses made after the 10 s time limit, at which point the faces disappeared, were coded as incorrect in order to avoid a reliance on memory. This resulted in 0.53% of the trials for the young participants, and 2.83% of trials for the healthy elderly participants, being coded as incorrect on these grounds.

Analysis

Since there were nine possible target×choice face combinations, with three of the combinations (FF, TT, PP) the rotation angle differing by 0°, four of the combinations (FT, TF, TP, PT) differing by 45°, and two of the combinations (FP, PF) differing by 90°, the number of trials was not standard across the different levels of angular disparity. Therefore, the mean percentage of trials that were answered correctly was computed at each level of this variable, for both upright and inverted displays. These mean values were entered into a 2×2×3 [group (older and younger)×planar orientation (upright and inverted)×angular disparity between target and choice face (0°, 45°, 90°)] repeated measures analysis of variance (ANOVA). For all post-hoc comparisons, an alpha level of .05 was adjusted for multiple comparisons using a Bonferroni correction.

Experiment #2 — The Effects of Alzheimer's Disease on Face Processing

Participants

Nine right-handed participants who were diagnosed with AD by a qualified medical professional (7 males, 2 females, mean age = 85.9, S.D. = 4.9, mean education = 11.7 years, S.D. = 2.7) were recruited from the Alzheimer Society of Manitoba and from personal care homes in Winnipeg, MB. On the MMSE, this AD group received a mean score of 18.3 (S.D. = 4.5), or −4.8 standard deviations below normative data [37], and a mean score of 95.8 (S.D. = 13.2) on the DRS, or −2.4 standard deviations below the norm [36] (see Table 2).

Table 2. Information on age, education, MMSE, DRS, and visual acuity scores for the AD group.

Subject	Age	Education (years)	MMSE (raw score)	MMSE (T-score)	DRS (raw score)	DRS (T-score)	Visual acuity	Cataracts
1	78	11	15	−30	86	24	20/30	Yes (both)
2	81	16	22	−5.6	110	28	20/50	Yes (both)
3	80	8	17	7.9	78	24	20/30	No
4	87	12	23	35	93	24	20/30	Yes (both)
5	91	10	18	10	90	24	20/50	Yes (both)
6	90	16	22	11.5	106	28	20/25	No
7	90	10	23	35	111	30.5	20/50	Yes (both)
8	88	10	10	−30	80	24	20/70	Yes (both)
9	88	12	15	−5	108	28	20/25	No

Because our AD group was significantly older than the full sample of older individuals who participated in Experiment 1, their performance was compared to the subset of healthy controls who were age 78 or over (the age of our youngest AD participant). This was deemed necessary, as within the full sample of elderly participants there was a significant negative correlation between age and overall accuracy [r = −0.63, p<0.05]. This subgroup, hereafter referred to as the "oldest-old" comparison group, consisted of 1 male and 5 females, with a mean age of 82.3 years (SD = 3.9), and a mean of 10.5 years of education (SD = 1.4). An independent samples t-test revealed that the oldest-old control group and the AD group were matched on age [$t_{(13)}$ = –1.49, p>0.05], and education [$t_{(13)}$ = –0.68, p>0.05].

Design and Procedure

The face-matching task of Experiment 2 generally replicated the design and procedure of Experiment 1. The exception was that all but two of the AD participants (Cases 3 and 4) indicated their responses by pointing to the choice face that they felt matched the target face on the screen, rather than entering their responses on the computer keyboard directly. This was done due to motor difficulties using the USB keypad. Once a choice was made, the experimenter entered the response. As in Experiment 1, responses made after the 10 s limit were coded as incorrect; 22.19% of AD participants' responses were coded as incorrect by this criterion. Although we coded responses made after 10 seconds as incorrect, analysis revealed that of these "incorrect" responses, 55.43% of those responses would be correct, with an average response time of 13.04 seconds. From the responses given after the 10 s time limit—36% of that data required a 45° rotation, 45% required a 90° rotation, while only 19% required no transformation in depth. Since the experimenter responded for most of the AD patients, we felt an analysis of reaction times would not be meaningful.

Analysis

Most AD participants were only able to comfortably complete two blocks of trials (one upright and one inverted). This meant that, often, the oldest-old participants had completed twice as many trials as the AD participants. To compensate for this, in the analyses described below the scores obtained by the AD group were compared to the scores obtained by the oldest-old participants' performance on the first two blocks of trials only (n = 144 trials, in total).

Results

Experiment #1 — The Effects of Aging on Face Processing

Effects of Inversion on Face Recognition

The older group was less accurate than the younger group, regardless of planar orientation [$F_{(1,25)}$ = 13.70, p<0.001]. Despite this, both groups exhibited an effect of stimulus inversion, producing more errors when faces were inverted [$F_{(1,25)}$ =

36.15, p<0.001]. This effect, however, was more pronounced in the older participants [planar orientation×group, $F_{(1,25)}$ = 5.91, p<0.05] (see Figure 2).

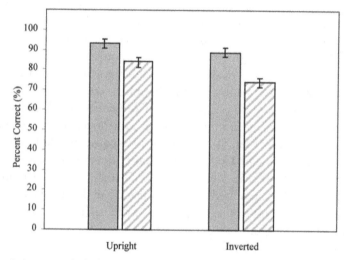

Figure 2. Accuracy under both upright and inverted conditions for both the younger and older viewers (errors bars: SEM's).

Effects of Increasing Angular Disparity between Target and Choice Faces

As the angular disparity between the target and choice face increased, participants both showed a systematic decrease in accuracy [$F_{(2,50)}$ = 47.61, p<0.001]. The most accurate responses occurred when no transformation in depth was required (0°), more errors were produced during a 45° rotation, and the most errors were generated when a 90° rotation was required. No interaction was present between rotation angle and group [$F_{(2,50)}$ = 0.14, p>0.05].

Is there a "Better" View?

The mean percent correct for each combination of target and choice faces, in each orientation, were submitted to a 2×2×3×3 [group×planar orientation×target face (frontal, three-quarter, profile)×choice face (frontal, three-quarter, profile)] repeated measures ANOVA. A significant interaction between the target face and choice faces was observed [$F_{(4,100)}$ = 31.80, p<0.001], and follow-up tests confirmed that the most accurate responses occurred when all faces were presented in the three-quarter view. A significant interaction between planar orientation and target face was also found [$F_{(2,50)}$ = 5.00, p<0.01]. Follow-up tests on this interaction showed that, while inverting the faces did not impair accuracy for profile views, accuracy was significantly impaired for frontal and three-quarter views (see Figure 3). There were no significant main effects or interactions involving Group.

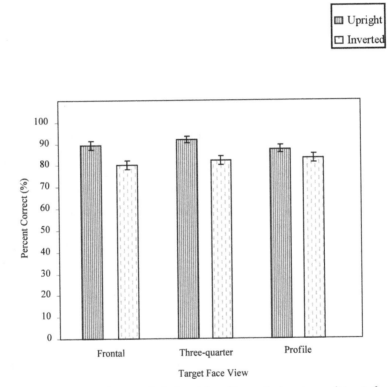

Figure 3. The effects of target face views for both upright and inverted orientation conditions in frontal, three-quarter, and profile views, on older and young adults' ability to match faces (error bars: SEM's).

Experiment #2—The Effects of Alzheimer's Disease on Face Processing

Face Matching Task

When faces were upright and required no transformation in depth (0°), the majority of AD participants performed within the 95% CI of the oldest-old controls (see Figure 4A, 4B represents inverted faces). However, three participants (1, 6, and 8) were below chance levels in this simple matching condition, which suggests that either they did not understand the task, or that they have a basic perceptual problem that affects their ability to discriminate between faces. For these reasons, they were excluded from subsequent analyses. Group comparisons confirmed that the six remaining AD participants continued to be matched in age and education with the six oldest-old controls.

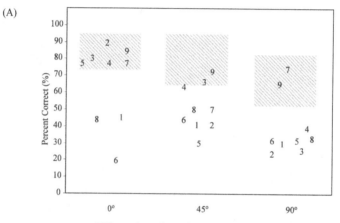

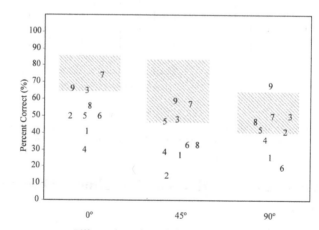

Figure 4. Comparing the impact of increasing the angular disparity between target and choice face angle (0°, 45°, and 90°) for each AD participant (number) in both (a) upright and (b) inverted orientation conditions. Hatched regions represent 95% CIs for the oldest-old control participants (n = 6).

Effects of Inversion and Rotation Angle

The AD participants were significantly less accurate than the oldest-old control group [$F_{(1,10)}$ = 5.95, p<0.05]. Both groups exhibited the inversion effect, producing more errors when faces were inverted compared to when upright [$F_{(1,10)}$ = 30.51, p<0.001]. However, a significant three-way interaction was observed between planar orientation, rotation angle, and group [$F_{(2,20)}$ = 3.92, p<0.05]. In the oldest-old control group, while inverting a face did not impair accuracy for a 0° or a 45° difference, accuracy was significantly impaired when faces were presented at a 90° difference (see Figure 5A). In contrast, for the six AD participants, inverting a face significantly impaired accuracy even in the 0° condition (see Figure 5B). Performance was equally poor with upright and inverted faces when the target and choice faces differed by either 45° or 90°.

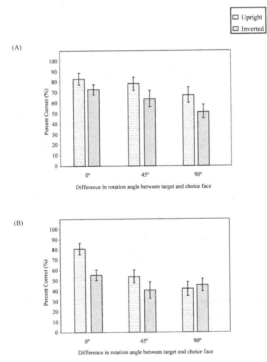

Figure 5. The effects of planar orientation and angular disparity between target and choice face in accuracy for the (A) oldest-old control participants, and (B) for the six AD participants (error bars: SEM's).

What "View" Works Best for AD Patients?

The mean percent correct was computed for each combination of target and choice faces, in each orientation. For both groups, a significant interaction between target and choice face was found [$F_{(4,40)}$ = 10.40, p<0.001], with the most accurate responses occurring when all faces were presented in frontal views. A significant interaction between choice face angle and group was found [$F_{(2,20)}$ = 5.92, p<0.01]. While for the oldest-old controls, accuracy was not affected by the orientation of the choice faces, the AD group performed significantly better when choice faces were presented in the frontal view (see Figure 6).

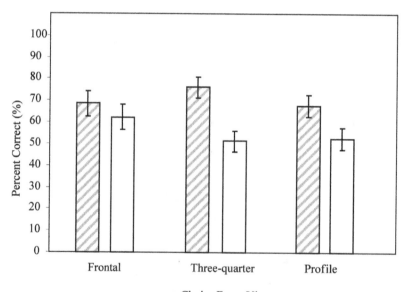

Figure 6. The impact of choice face orientation on matching accuracy for both the oldest-old control participants, and the six AD participants (error bars: SEM's).

Correlations

For all nine participants in the AD group, Pearson's bivariate correlations were computed between overall accuracy on the face mental rotation task, the visual acuity scores, and T scores for both the MMSE and the DRS. No significant correlations were found between overall accuracy and visual acuity [r = –0.25, p = 0.52], between T scores on the DRS and overall accuracy [r = 0.04, p = 0.93], or between T scores on the MMSE and overall accuracy [r = 0.43, p = 0.25].

Since the AD participants did poorly with any mental rotation, or inversion, and therefore this could be a reason why no correlations were observed, the same correlations as above were computed with the exception of correlating performance on the upright 0° condition with all specified scales. Once again, no significant correlations were found between accuracy and visual acuity [r = −0.19, p = 0.62], and between T scores on the DRS and accuracy [r = 0.01, p = 0.80], between T scores on the MMSE and accuracy [r = 0.27, p = 0.48].

For the six AD participants that completed the face-matching task, no significant correlations were found between accuracy and visual acuity [r = 0.21, p = 0.69], between T scores on the DRS and accuracy [r = 0.26, p = 0.63], or between T scores on the MMSE and accuracy [r = −0.35, p = 0.55].

Discussion

The experiments presented here examined the ability of young adults, older non-neurological controls and patients with Alzheimer's disease to complete a time-limited face-matching task in which stimuli could be rotated in depth and/or inverted. We wanted to determine if aging on its own hampers the ability of individuals to mentally rotate faces; if AD patients are disproportionately impaired; if orientation affects young and old participants in the same way and if there was a "preferred" view of the face in which performance was better for each of the groups.

Normal Aging

Even though healthy older controls were less accurate on the face-matching task than young adults, both groups exhibited a systematic effect of rotating a face in depth—as the angular disparity between the target and choice faces increased, accuracy decreased. This result has extended the recent finding that healthy older individuals are impaired at making matches with synthetic faces across a 20° rotation from full-face [17]. Here we show that this impairment is present even with real 3D images of faces.

When faces were inverted, the healthy elderly viewers were significantly more impaired than younger controls. Inverting a face disrupts the ability to process it holistically [11], [12], and triggers the adoption of a parts-based analysis. It is possible that the older group had more difficulty switching to a parts-based strategy. This finding coincides with those of Dror et al. [16], who showed that when older adults rotated simple and complex images, a holistic approach was used for both tasks to reduce cognitive load—a pattern not seen with younger viewers. Application of a holistic/configural processing approach may not always serve older viewers well. In fact, recent research has shown that aging results in an impairment in the ability to encode configural information during facial expression recognition [40], [41]. Therefore, when examining age-related changes in holistic processing, it may be very important to consider the particular task demands.

Alzheimer's Disease

When faces were upright and no mental rotation was involved, the majority of AD participants performed within the 95% CI of the oldest-old control participants. However, participants 1, 6, and 8 performed below chance levels on this simple matching task. It may be that these individuals did not fully understand the task, or they have a basic perceptual problem that affects their ability to discriminate between faces.

For the rest of the AD patients, the fact that they performed comparably with the oldest-old group when no mental rotation was required does support previous findings [28], [29] and suggests that their basic face perception abilities were intact. However, when the target and choice faces differed in orientation or when the stimuli were inverted, the AD group was severely impaired. This lack of a systematic effect of orientation suggests that the underlying difficulties impairing their ability to mentally rotate objects carries over even to "special" stimuli like faces.

Of course, it is also possible that the damage that occurs in Alzheimer's disease to areas like the ventral temporal cortex, which is important in face processing [42], [43], may cause these individuals to rely on processing approaches that do not work well with changes in viewpoint, or face inversion. In short, a form of viewpoint-dependency, where any rotation in depth results in a reduction in accuracy, could be contributing to the AD group's difficulty with matching faces presented at different angles. These results are intriguingly similar to those seen when prosopagnosic patients are run on the same task [44].

Interestingly, in contrast to the young adults and the younger-elderly participants, who both showed a three-quarter view preference, both the oldest-old subgroup and the AD group performed most accurately with frontal views of the target face. It may be that as we age, an increased reliance on holistic processing [16], makes the frontal view more useful for face processing—a valuable piece of information for anyone with an AD patient in their care.

Limitations

Even though it is difficult to make strong conclusions based on the small number of AD patients presented in the current investigation, the findings do suggest that it is not memory alone that contributes to AD patients' difficulty with face recognition. Instead, an underlying problem with mental rotation may affect patients ability to process or match any object or face.

Although there was a 10 second time limit for viewing these faces (which was followed by a blank screen), it may be the case that with more time, the AD group could have done better at this task. However, it is also possible that the AD participants equated the blank screen that appeared after 10 seconds as a reminder to respond to the task. Future studies are required to view if an increase in time for viewing the faces would benefit the AD participants.

Conclusions

Although face recognition deficits in people with AD have traditionally been associated with memory impairments, our results suggest that an underlying problem with mental rotation may compound these recognition difficulties. This difficulty forming representations that are robust to a variety of spatial transformations appears to occur early in the disease process, so it seems surprising that the literature does not typically report problems with face perception/recognition until later stages of AD [45], [46]. Of course, these reports are often from caregivers who are not conducting the kind of detailed face perception/mental rotation experiments carried out in the current study. It may be that the AD patients' ability to use contextual cues hides this deficit until later stages of AD when the damage has spread to more frontal regions of the brain and they are no longer able to piece together the contextual cues.

Acknowledgements

We thank the Center on Aging at the University of Manitoba, Deer Lodge Center, and the Alzheimer Society of Manitoba for their aid in recruitment. We also thank Lee Baugh and Loni Desanghere for their comments on the manuscript.

Authors' Contributions

Conceived and designed the experiments: JJM. Performed the experiments: CAA. Analyzed the data: CAA. Wrote the paper: CAA JJM.

References

1. Bülthoff H, Edelman S (1992) Psychophysical support for a two-dimensional view theory of object recognition. Proceedings of the National Academy of Science 89: 60–64.

2. Shepard RN, Metzler J (1971) Mental rotation of three-dimensional objects. Science 171: 701–703.

3. Tarr M, Pinker S (1989) Mental rotation and orientation dependence in shape recognition. Cognitive Psychology 21: 233–282.

4. Bruce V, Humphreys GW (1994) Recognizing objects and faces. Visual Cognition 1: 141–180.

5. Troje NF, Kersten D (1999) Viewpoint-dependent recognition of familiar faces. Perception 28: 483–487.

6. Baddeley A, Woodhead M (1983) Improving face recognition ability. In: Llooyd-Bostock S, Clifford B, editors. Evaluating witness evidence. Chichester: Wiley. pp. 125–136.

7. Bruce V, Valentine T, Baddeley A (1987) The basis of the ¾ view advantage in face recognition. Applied Cognitive Psychology 1109–1120.

8. Troje NF, Bulthoff HH (1996) Face recognition under varying poses: the role of texture and shape. Vision Research 36: 1761–1771.

9. O'Toole AJ, Edelman S, Bulthoff HH (1998) Stimulus-specific effects in face recognition over changes in viewpoint. Vision Research 38: 2351–2363.

10. Liu CH, Chaudhuri A (2002) Reassessing the ¾ view effect in face recognition. Cognition 83: 31–48.

11. Yin RK (1969) Looking at upside-down faces. Journal of Experimental Psychology 81: 141–145.

12. Valentine T (1988) Upside down faces: A review of the effect of inversion upon face recognition. British Journal of Psychology 79: 471–491.

13. Craik FI, Dirkx E (1992) Age-related differences in three tests of visual imagery. Psychol Aging 7: 661–665.

14. Dror IE, Koslyn SM (1994) Mental imagery and aging. Psychology and Aging 9: 90–102.

15. Sharps MJ, Nunes MA (2002) Gestalt and feature-intensive processing: Toward a Unified Model of Human Information Processing. Current Psychology 21: 68–84.

16. Dror IE, Schmitz-Williams IC, Smith W (2005) Older adults use mental representations that reduce cognitive load: mental rotation utilizes holistic representations and processing. Exp Aging Res 31: 409–420.

17. Habak C, Wilkinson F, Wilson HR (2008) Aging disrupts the neural transformations that link facial identity across views. Vision Res 48: 9–15.

18. Ball MJ (1977) Neuronal loss, neurofibrillary tangles and granulovacuolar degeneration in the hippocampus with ageing and dementia. A quantitative study. Acta Neuropathol 37: 111–118.

19. Von Gunten A, Bouras C, Kovari E, Ginnakopoulous P, Hof PR (2006) Neural substrates of cognitive and behavioral deficits in atypical Alzheimer's disease. Brain Research Reviews 51.

20. Steele C, Rovner B, Chase GA, Folstein M (1990) Psychiatric symptoms and nursing home placement of patients with Alzheimer's disease. Am J Psychiatry 147: 1049–1051.

21. Alivisatos B, Petrides M (1997) Functional activation of the human brain during mental rotation. Neuropsychologia 35: 111–118.

22. Carpenter PA, Just MA, Keller TA, Eddy W, Thulborn K (1999) Graded functional activation in the visuospatial system with the amount of task demand. J Cogn Neurosci 11: 9–24.

23. Jordan K, Heinze HJ, Lutz K, Kanowski M, Jancke L (2001) Cortical activations during the mental rotation of different visual objects. Neuroimage 13: 143–152.

24. Vingerhoets G, Santens P, Van Laere K, Lahorte P, Dierckx RA, et al. (2001) Regional brain activity during different paradigms of mental rotation in healthy volunteers: a positron emission tomography study. Neuroimage 13: 381–391.

25. Jack CR Jr, Shiung MM, Gunter JL, O'Brien PC, Weigand SD, et al. (2004) Comparison of different MRI brain atrophy rate measures with clinical disease progression in AD. Neurology 62: 591–600.

26. Visser PJ, Verhey FR, Hofman PA, Scheltens P, Jolles J (2002) Medial temporal lobe atrophy predicts Alzheimer's disease in patients with minor cognitive impairment. J Neurol Neurosurg Psychiatry 72: 491–497.

27. Lineweaver TT, Salmon DP, Bondi MW, Corey-Bloom J (2005) Differential effects of Alzheimer's disease and Huntington's disease on the performance of mental rotation. J Int Neuropsychol Soc 11: 30–39.

28. Murphy KJ, Kohler S, Black SE, Evans M (2000) Visual object perception, space perception, and visuomotor control in Alzheimer's disease. Retrieved from http://cognet.mit.edu/library/conferences/paper?paper_id=47310.

29. Tippett LJ, Blackwood K, Farah MJ (2003) Visual object and face processing in mild-to-moderate Alzheimer's disease: from segmentation to imagination. Neuropsychologia 41: 453–468.

30. Caterini F, Della Sala S, Spinnler H, Stangalino C, et al. (2002) Object recognition and object orientation in Alzheimer's disease. Neuropsychology 16: 146–155.

31. Della Sala S, Muggia S, Spinnler H, Zuffi M (1995) Cognitive modeling of face processing: evidence from Alzheimer patients. Neuropsychologia 33: 675–687.

32. Fahlander K, Wahlin A, Almkvist O, Backman L (2002) Cognitive functioning in Alzheimer's disease and vascular dementia: further evidence for similar patterns of deficits. J Clin Exp Neuropsychol 24: 734–744.

33. Hawley KS, Cherry KE (2004) Spaced-retrieval effects on name-face recognition in older adults with probable Alzheimer's disease. Behav Modif 28: 276–296.

34. Lee ACH, Buckley MJ, Gaffan D, Emery T, Hodges JR, et al. (2006) Differentiating the roles of the hippocampus and perirhinal cortex in processes beyond long term declarative memory: A double dissociation in dementia. Journal of Neuroscience 26(19): 5198–5203.

35. Folstein MF, Folstein SE, McHugh PR (1975) "Mini-mental state." A practical method for grading the cognitive state of patients for the clinician. J Psychiatr Res 12: 189–198.

36. Mattis S (1973) Dementia rating scale professional manual. Odessa, FL.: Psychological Assessment Resources.

37. Fields RB (1998) The Dementias. In: Snyder PJ, Nussbaum PD, editors. Clinical Neuropsychology: A pocket handbook for assessment. Washington, D.C.: APA.

38. Cohen J, MacWhinney B, Flatt M, Provost J (1993) PsyScope: A new graphic interactive environment for designing psychology experiments. Behavioral Research Methods Instruments & Computers 25: 257–271.

39. Blanz V, Vetter T (1999) A morphable model for the synthesis of 3D faces. SIGGRAPH'99 Conference Proceedings 187–194.

40. Calder AJ, Young AW, Keane J, Dean M (2000) Configural information in facial expression perception. J Exp Psychol Hum Percept Perform 26: 527–551.

41. Murray J, Ruffman T, Halberstadt J (2008) Age-related changes in face processing. Journal of Vision 8: 196.

42. Grill-Spector K (2003) The functional organization of the ventral visual pathway and its relationship to object recognition. In: Kanwisher N, Duncan J, editors. Attention and Performance: Functional Brain Imaging of Visual Cognition. London: Oxford University Press.

43. Kanwisher N, McDermott J, Chun MM (1997) The fusiform face area: A module in human extrastriate cortex specialized for face perception. The Journal of Neuroscience 17: 4302–4311.

44. Marotta JJ, McKeeff TJ, Behrmann M (2002) The effects of rotation and inversion on face processing in prosopagnosia. Cognitive Psychology 19: 31–47.

45. Bourgeois MS, Schulz R, Burgio L (1996) Interventions for caregivers of patients with Alzheimer's disease: a review and analysis of content, process, and outcomes. Int J Aging Hum Dev 43: 35–92.

46. Selwood A, Johnston K, Katona C, Lyketsos C, Livingston G (2007) Systematic review of the effect of psychological interventions on family caregivers of people with dementia. J Affect Disord 101: 75–89.

CITATION

Originally published under the Creative Commons Attribution License. Adduri CA, Marotta JJ. "Mental Rotation of Faces in Healthy Aging and Alzheimer's Disease," in PLoS ONE 4(7): e6120. © 2009 Adduri, Marotta. doi:10.1371/journal.pone.0006120.

Preventing Falls in Older Multifocal Glasses Wearers by Providing Single-Lens Distance Glasses: The Protocol for the VISIBLE Randomized Controlled Trial

Mark J. Haran, Stephen R. Lord, Ian D. Cameron,
Rebecca Q. Ivers, Judy M. Simpson, Bonsan B. Lee,
Mamta Porwal, Marcella M. S. Kwan and Connie Severino

ABSTRACT

Background

Recent research has shown that wearing multifocal glasses increases the risk of trips and falls in older people. The aim of this study is to determine whether the provision of single-lens distance glasses to older multifocal glasses wearers, with recommendations for wearing them for walking and outdoor activities, can prevent falls. We will also measure the effect of the intervention on health status, lifestyle activities and fear of falling, as well as the extent of adherence to the program.

Methods/Design

Approximately 580 older people who are regular wearers of multifocal glasses people will be recruited. Participants will be randomly allocated to either an intervention group (provision of single lens glasses, with counseling and advice about appropriate use) or a control group (usual care). The primary outcome measure will be falls (measured with 13 monthly calendars). Secondary measures will be quality of life, falls efficacy, physical activity levels and adverse events.

Discussions

The study will determine the impact of providing single-lens glasses, with advice about appropriate use, on preventing falls in older regular wearers of multifocal glasses. This pragmatic intervention, if found to be effective, will guide practitioners with regard to recommending appropriate glasses for minimising the risk of falls in older people.

Trial Registration

The protocol for this study was registered with the Clinical Trials.gov Protocol Registration System on June 7th 2006 (#350855).

Background

Refractive error due to presbyopia is the most prevalent form of visual impairment in older people [1]. Presbyopia is a visual condition in which the crystalline lens of the eye loses its flexibility, making focusing on close objects difficult [2]. To correct for presbyopia, older people are either prescribed separate single lens glasses for distant and near vision, or for convenience, a single pair of multifocal (bifocal, trifocal or progressive lens) glasses. Multifocal glasses have specific benefits for tasks that require changes in focal length, including everyday tasks of driving, shopping and cooking. However, multifocal glasses also have certain disadvantages as outlined below.

Bifocal glasses have optical defects, such as prismatic jump at the top of the reading segment, which causes an apparent displacement of fixed objects [2,3]. The lower lenses of all types of multifocal glasses blur distant objects in the lower visual field and this factor, in particular, may represent a significant problem for older people [3,4]. Multifocal glasses have been shown to impair distant depth perception and contrast sensitivity [5] and two recent studies have found that multifocal glasses impair step negotiation and the accuracy of foot placement when stepping onto a raised surface in older people [6,7]. These studies reported that when wearing multifocal glasses versus single-lens glasses, older people displayed more variability in minimum vertical toe clearance when stepping up [7] and in foot placement (lead or trail toe-to-platform distance) when negotiating a raised surface [6]. Contacts with the raised platform's edge [7] and trips [6] only occurred in the multifocal glasses conditions.

Other studies have reported significant associations between wearing multifocal glasses and tripping incidents [8,9] and a prospective study of 156 older people found that multifocal glasses wearers were significantly more likely to fall during 12 months of follow-up, after adjusting for age and known physiological risk factors for falls [5]. Findings from this study also indicated that multifocal glasses wearers were more likely to fall when outside their homes, and when walking up or down stairs. The population attributable risks of regular multifocal glasses use were 35% for any falls and 41% for falls outside the home.

To date, there is no evidence that falls can be prevented by restricting the use of multifocal glasses. To address this need we are conducting the Visual Intervention Strategy Incorporating Bifocal & Long-distance Eyewear (VISIBLE) study. The primary aim of this randomized controlled trial is to determine whether the provision of supplementary single-lens distance glasses to elderly multifocal glasses wearers, together with recommendations for wearing them for standing and outdoor activities, can reduce falling rates over a 13-month period. Secondary aims include determining whether, as a result of fewer falls, the intervention has beneficial effects for physical and emotional wellbeing, falls efficacy and activity levels, and whether a range of physical, psychological and socio-demographic factors can predict compliance in the intervention group.

Methods

Participants

Participants will be recruited from the following sources: sampling of older people from the electoral roll in northern Sydney, residents of retirement (self-care apartment) villages in the Central Coast, Blue Mountains and Illawarra regions of New South Wales, outpatients and inpatients discharged from rehabilitation and orthopaedic wards at hospitals within northern and eastern Sydney, and responders to newspaper advertisements and media coverage.

Eligibility will be assessed primarily with a telephone screening questionnaire. Participants will be eligible for the trial if they are at a relatively high risk of falls defined as (a) aged 65–79 years with a Timed Up and Go Test (TUG) of 15+ seconds [10] and/or a history of one or more falls in the last 12 months or b) aged 80 years and over. Other inclusion criteria comprise: a user of bifocal, trifocal or progressive-lens glasses 3+ times/week when walking outdoors; having had a review by an optometrist or ophthalmologist in the last 12 months; not currently using single-lens distance glasses; and indicating that they are at least 'quite confident' of complying with the study recommendations.

People will be ineligible if they: reside in a high-care residential facility, have a cognitive impairment (a Mini-Mental State Examination (MMSE) [11] score of less than 24), a severe visual impairment (Melbourne Edge Test score of <16 dB) [12], insufficient English language skills to understand the assessment and/or intervention procedures, have planned ophthalmic surgery in the next 12 months, or have an unstable medical condition at enrollment.

The University of New South Wales Human Research Ethics Committee (HREC) approved the study protocol (HREC Number 04229) as have ethics committees at the participating hospitals. The participant flow diagram is presented as Figure 1.

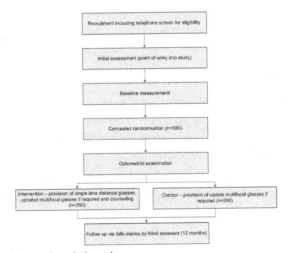

Figure 1. Flow of participants through the trial.

Initial Assessment

Initially the study aims will be explained, and then the participants will be asked if they feel confident that they could manage and wear two pairs of glasses if randomized to the intervention. Only participants who answer 'very or quite confident,' as opposed to 'not confident' or 'very unconfident,' will undergo consent, assessments and randomization. The approach of targeting the intervention to those who consider it to be reasonable and manageable is based on strong evidence from hip protector trials [13].

Demographic details and potential confounders will be recorded, including age, gender, living alone status, MMSE [10], number of falls in past year, number and type of medications, physical activity levels, gait aids, medical conditions relevant to falls (e.g. arthritis, syncope, diabetes and depression), current glasses use and ocular history. Baseline lifestyle activities (using the Adelaide Activities Profile (AAP)) [14], health status (SF-12) [15], and falls efficacy (Falls Efficacy Scale—International (FES-I scale) [16] will be measured.

The following vision and physical assessments will then be carried out. Corrected and uncorrected uniocular and binocular visual acuity will be assessed using a 3-metre logMAR chart [17]. Falls risk will be measured with the short-form Physiological Profile Assessment (PPA) [18] which includes five validated measures of physiological falls risk (visual contrast sensitivity, postural sway, quadriceps strength, reaction time and lower limb proprioception). In multivariate models these variables provide an overall falls risk score, and can predict those at risk of falling with 75% accuracy in community and retirement village settings. Functional mobility will be assessed with the TUG test [10], and the sit-to-stand test [19].

Randomization

After completion of the initial assessment, participants will be formally entered into the study and randomized to intervention or control groups. Participants will be stratified by assessment site (Greenwich or Prince of Wales Hospitals) and recruitment source (hospital or community contact).

Then each stratum will be randomly allocated in permuted blocks of 10 using sequentially-numbered opaque envelopes containing group assignment.

Intervention

Following the initial assessment, participants in the intervention group will have an optometrist examination that will include measurement of objective refraction. These participants will then be prescribed a pair of single-lens distance glasses. For the estimated 80% of intervention participants who will not need an update of multifocal lens prescription (n = 234), the correction for the single-lens glasses will be matched with the prescription of the upper segment of their pretrial multifocal glasses. For the expected 20% of participants (n = 56) who have had a significant change in distance correction since the most recent prescription [defined as a change of +/- 0.5 dioptres for refractive error (using the spherical equivalent: sphere + half cylinder), or more than 0.75 dioptres for astigmatism (measured in minus cylinders) and at least +0.50 dioptres change in the reading addition], updated multifocal lenses will be provided. The distance component of the updated multifocal correction will then be used for the prescription of new single-lens distance glasses, thereby making the transition between the two types of glasses easier.

At a second visit (approximately four weeks following the initial assessment) participants will be fitted with their glasses. An optometrist/counselor will then demonstrate how multifocal glasses can impair the visual abilities required for detecting obstacles and judging depth. This will be achieved by administering the tests of distance edge contrast sensitivity and depth perception with participants viewing the visual stimuli through the upper and lower portion of their multifocal lenses and distance lenses [5]. The difference in upper and lower visual performance will be shown to the participant and the rationale for performing these tests will then be explained. Figure 2 shows the tests for measuring upper and lower lens contrast sensitivity used in the counseling (Figure 2).

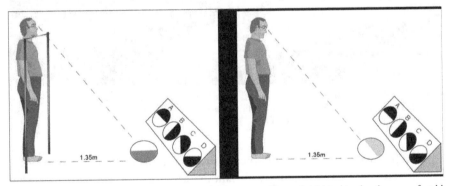

Figure 2. The distance contrast sensitivity tests. Panel A shows a participant (with his chin placed on a comfortable chinrest to prevent head movement) viewing the test plate through the lower (reading) segments of his glasses. Panel B shows the participant looking directly at the test stimulus through the upper (distance) segments of his glasses. Any difference in contrast vision between the two conditions is then shown to the participant as part of the counseling session.

Participants will also be shown photographs of steps and streetscapes with and without simulated lower field blur to further reinforce why multifocal glasses may increase the risk of falls. Figure 3 shows the photographs of a simulated view of a street scene as viewed through single-lens distance (panel a) and bifocal glasses (panel b) (Figure 3). This counseling session will be conducted in line with the health belief model [20], and is considered crucial to convince older wearers of multifocal glasses that they are personally at risk of fall-related injuries and that the benefits of wearing single lens glasses outweighs the inconvenience of dealing with two pairs of glasses.

Figure 3. Simulated view of street scene as viewed through single-lens distance (panel A) and bifocal glasses (panel B). The footpath misalignment (a commonly reported environmental factor involved in outdoor falls) is clearly seen in panel A, but blurred in panel B.

The optometrist/counselor will then instruct participants to use their new single-lens glasses for most walking and standing activities, and in particular when undertaking the following activities: walking up or down stairs outside the home, walking in the street and at shopping centers, walking or standing in other people's homes and unfamiliar buildings, negotiating rough or uneven ground and getting off and on public transport. Multifocal glasses use will not be discouraged for seated tasks, especially those that require frequent changes in focal length such as driving, and indoor standing and walking tasks that require changes in focal depth where there is minimal risk (e.g. cooking and selecting items in the supermarket).

To assist with the swapping of glasses, participants will be provided with a "glasses cord" for wearing around the neck and/or a hard or cloth glasses holder for keeping the second pair of glasses in a pocket. Intervention group participants will also be provided with a booklet reinforcing the advice. Table 1 outlines the strategies that will used to maximize compliance throughout the trial (Table 1). Any perceived barriers to compliance will be recorded.

Table 1. Strategies to be used to maximize compliance

1	A strong evidence-based rationale for the intervention and clear recommendations will be provided.
2	The mechanism by which multifocal glasses can predispose to falls will be demonstrated during counselling using the example of the most commonly reported environmental factor involved in outdoor falls (Figure 2).
3	New glasses will be provided at no cost.
4	Perceived barriers to using two pairs of glasses will be identified and discussed during counselling.
5	An information brochure with recommendations and illustrations on appropriate use of glasses will be provided.
6	A hard/soft case glasses holder (to be worn around the neck and/or placed in the participant's pocket) to aid the swapping of glasses will be provided.
7	Second counselling sessions delivered over the phone or at home visits will address perceived barriers to use of glasses for participants with inadequate compliance, i.e. "never/occasionally" complying with the recommendations.
8	Written reminder cards will be provided with falls diaries and prompts will be made during follow-up telephone calls.

Participants in the control group will also have the same optometric examination as participants in the intervention group. For the estimated 80% of participants who will not need a change of prescription (n = 234), no new glasses will be prescribed. For the expected 20% who will require a prescription change (n = 56), updated multifocal lenses will be prescribed according to the above criteria and fitted at a second visit. Control group participants will be given no specific advice regarding the use of their glasses.

Outcomes

Primary Outcome Measure

The primary outcome measure will be the number of participant falls in the 13-month follow-up period following randomization (this allows for an average time of one month for the provision of single-lens glasses following randomization and 12-month further follow-up). Falls will be defined as 'inadvertently coming to rest on the ground or other lower level with or without loss of consciousness, and other than as a consequence of sudden onset of paralysis, epileptic seizure, excess alcohol intake or overwhelming external force'

[21]. All participants will receive 13 calendars and questionnaires at the time of the baseline assessment. Participants will be asked to record falls (as well as information regarding any fall injuries and need for medical care) on the calendars and return them in pre-paid envelopes to the research center each month. Participants who do not return calendars will be telephoned to ask for the information. Staff who receive falls calendars, make follow-up phone calls and enter data will be unaware of group allocation.

Secondary Outcome Measures

Secondary outcome measures will include physical activity levels, falls efficacy and quality of life assessed at baseline and 12 months. These measures are included to determine whether, as a result of fewer falls, the intervention has beneficial effects on an older person's abilities and quality of life. Physical activity levels will be assessed with the Adelaide Activities Profile, a validated scale which measures lifestyle activities using 21 items which generate four domains (domestic chores, household maintenance, service to others, and social activities) [14]. Physical and emotional wellbeing will be assessed using the SF 12 Version 2 composite physical and mental scores [15]. Falls efficacy will be assessed using the FES-I [16] in which level of concern about falling when carrying out 16 activities is rated on a 4-point scale.

Compliance and Adverse Events

Compliance in the intervention group will be measured monthly by asking how often the participants complied with recommendations about wearing single-lens distance glasses when undertaking activities such as walking outdoors and descending steps etc. Reporting of the type of glasses worn during any fall will be recorded on the same questionnaire. Participants will be asked to return their compliance questionnaires in a separate pre-paid envelope to the falls calendar to the research center each month. Staff who receive compliance questionnaires, make follow-up phone calls and enter data will be unaware of the occurrence of any participant falls. Any perceived barriers to compliance will be recorded at baseline and throughout the trial.

Adverse Effects

Falls occurring as a result of switching from single-lens to multifocal glasses or vice versa while undertaking an activity or any non-fall adverse outcome occurring as a result of the intervention will be recorded on the falls calendars.

Sample Size Calculation

Approximately 580 participants (290 per group) will be recruited. The study will have 80% power to detect as significant at the 5% level a 23% reduction in the rate of falling (i.e. an IRR of 0.77 using negative binomial regression analysis) in the 13-month follow-up period. This represents a reduction from 0.85 falls/person (an established rate of falling in a high-risk control group equivalent to our sample) [22,23] to 0.65 falls/person in the intervention group. This allows for an 8% loss to follow-up defined as withdrawing from the study and/or not completing the falls diaries, 70% compliance in the intervention participants and a rate of 0.57 falls/person among those complying with the intervention

Statistical Analysis

The primary analysis will be by intention to treat. The number of falls per person-year will be analyzed using negative binomial regression to estimate the difference in falls rates between the two groups, adjusted for confounding variables if required. This analysis takes into account all falls during the trial, and the distribution of falls, which is Poisson-like but containing a wider, higher tail in the distribution [24]. Three a priori subgroup analyses will be performed using a test for statistical interaction to assess whether the intervention effect size differs according to: (i) baseline physiological falls risk [18]; (ii) number of participant-reported falls in the year prior to the trial; and (iii) participant-reported outdoor activity levels during the trial as assessed with the AAP [14]. Secondary analyses will assess the effects of the intervention on falls that occur (i) within and (ii) beyond the participants' homes. Longitudinal mixed models will be used to assess the effect of group allocation on the continuously-scored secondary outcome measures. Predictors of adherence will be established using multivariate modeling techniques. Analyses will be conducted using the SPSS [25] and Stata [26] computer programs.

Discussion

Falls are the leading cause of injury-related death and hospitalization in older persons [26]. Falls can also result in disability, restriction of activity and fear of falling—all of which reduce quality of life and independence [27,28]. Furthermore, falls can contribute to the placement of an older person into institutional care [28]. There is mounting evidence that multifocal glasses impair vision, hinder obstacle avoidance and increase the risk of falls in older people.

There are over 2.4 million people over 65 years of age in Australia, one-third of whom fall annually at least once [27]. Given that 52% of older people rely solely on multifocal glasses [1] and only 1% use separate reading and distance glasses [1], over one million Australians could potentially benefit from this trial immediately. Current guidelines recognize the importance of vision in falls and recommend that older people not walk while wearing multifocal glasses [29,30]. It is worth noting, however, that this recommendation also removes the correction of distance refractive error which may increase falls risk in some people. Our intervention, which involves providing multifocal glasses wearers with an additional pair of single-lens distance glasses for high-risk situations that increase postural threats (such as walking on stairs and in unfamiliar outdoor settings), will avoid this problem.

Thus, this program is optimally designed to address the major problem of falls in older people and could be readily translated into clinical practice. The study we are undertaking is designed and adequately powered to assess the effect of the program. There will be far-reaching benefits for older people, their carers and the community if this program can be demonstrated to assist in reducing the number of falls in this high-risk population.

Conclusion

The study will determine the impact of providing single-lens glasses, with advice about appropriate use, on preventing falls in older regular wearers of multifocal glasses.

Competing Interests

The authors declare that they have no competing interests.

Authors' Contributions

This manuscript was drafted by SL and MH. The other authors are also actively involved in the study. Many contributed to the writing of the grant application for this project which was funded by the Australian National Health and Medical Research Council in 2005. The grant application formed the basis for this manuscript. All authors contributed to the manuscript's critical review and approved the final version.

Acknowledgements

This study is funded by an Australian National Health and Medical Research Council (Reference Number: NHMRC ID 350855).

References

1. Attebo K, Ivers RQ, Mitchell P: Refractive errors in an older population: the Blue Mountains Eye Study. Ophthalmology 1999, 106:1066–72.

2. Donahue SP: Loss of accommodation and presbyopia. In Ophthalmology. Edited by: Yanoff M, Duker JS. Mosby, London; 1999.

3. Duke Elder S: The practice of refraction. Churchill, London; 1963.

4. Tinetti ME: Clinical practice. Preventing falls in elderly persons. N Eng J Med 2003, 348:42–9.

5. Lord SR, Dayhew J, Howland A: Multifocal glasses impair edge contrast sensitivity and depth perception and increase the risk of falls in older people. J Am Geriatr Soc 2002, 50:1760–66.

6. Johnson L, Buckley JG, Harley C, Elliott DB: Use of single-vision eyeglasses improves stepping precision and safety when elderly habitual multifocal wearers negotiate a raised surface. J Am Geriatr Soc 2008, 56:178–80.

7. Johnson L, Buckley JG, Scally AJ, Elliott DB: Multifocal spectacles increase variability in toe clearance and risk of tripping in the elderly. Investigative OphthalmolVis Sci 2007, 48:1466–71.

8. Connell BR, Wolf SL: Environmental and behavioral circumstances associated with falls at home among healthy elderly individuals. Atlanta FICSIT Group. Arch Phys Med Rehabil 1997, 78:179–86.

9. Davies JC, Kemp GJ, Stevens G, Frostick SP, Manning DP: Bifocal/varifocal spectacles, lighting and missed-step accidents. Safety Sci 2001, 38:211.

10. Mathias S, Nayak US, Isaacs B: Balance in elderly patients: The 'get-up and go' test. Arch Phys Med Rehab 1986, 67:387–9.

11. Folstein MF, Folstein SE, McHugh PR: Mini-mental state: a practical method for grading cognitive state of patients for the clinician. J Psychiatr Res 1975, 12:189–98.

12. Verbaken JH, Johnston AW: Population norms for edge contrast sensitivity. Am J Opt and Physiol Optics 1986, 63:724–732.

13. Kurrle SE, Cameron ID, Quine S: Predictors of adherence with the recommended use of hip protectors. J Gerontol 2004, 59:M958–61.

14. Clark MS, Bond MJ: The Adelaide Activities Profile: A measure of the lifestyle activities of elderly people. Aging Clin Exp Res 1995, 7:174–84.

15. Gandek B, Ware JE, Aaronson NK, Apolone G, Bjorner JB, Brazier JE, Bullinger M, Kaasa S, Leplege A, Prieto L, Sullivan M: Cross-validation of item selection and scoring for the SF-12 Health Survey in nine countries:results from the IQOLA Project. J Clin Epidemiol 1998, 51:1171–8.

16. Yardley L, Beyer N, Hauer K, Kempen G, Piot-Ziegler C, Todd C: Development and initial validation of the Falls Efficacy Scale International (FES-I). Age Ageing 2005, 34:614–9.

17. Bailey IL, Lovie JE: New design principles for visual acuity letter charts. Am J Optom Physiol Opt 1974, 53:740–5.

18. Lord SR, Menz HB, Tiedemann A: A physiological profile approach to falls risk assessment and prevention. Phys Ther 2003, 83:237–252.

19. Csuka M, McCarty DJ: Simple method for measurement of lower extremity muscle strength. Am J Med 1985, 78:77–81.

20. Rosenstock I: The health belief model and preventive health behavior. Health Education Monographs 1978, 2:354–386.

21. Gibson MJ, Andres RO, Isaacs B, Radebaugh T, Worm-Petersen J: The prevention of falls in later life. Danish Medical Bulletin 1987, 34(Supplement No. 4):1–24.

22. Campbell AJ, Robertson MC, Gardner MM, Norton RN, Tilyard MW, Buchner DM: Randomized controlled trial of a general practice program of home based exercise to prevent falls in elderly women. BMJ 1997, 315:1065–69.

23. Cumming RG, Thomas M, Szonyi G, Salkeld G, O'Neill E, Westbury C, Frampton G: Home visits by an occupational therapist for assessment and modification of environmental hazards: a randomized trial of falls prevention. J Am Geriatr Soc 1999, 47:1397–1402.

24. Glynn RJ, Buring JE: Ways of measuring rates of recurrent events. BMJ 1996, 312:364–6.

25. SPSS for windows: Release 17. SPSS Inc, Chicago, Il; 2008.

26. StataCorp: Stata Statistical Software: Release 7.0. Stata Corporation, College Station, TX; 2001.

27. Lord SR, Sherrington C, Menz H, Close JCT: Falls in Older People: Risk factors and Strategies for Prevention. Second edition. Cambridge University Press, Cambridge, UK; 2007.

28. Tinetti ME, Williams CS: Falls, injuries due to falls, and the risk of admission to a nursing home. New Eng J Med 1997, 337:1279–1284.

29. Kenny RA, Rubenstein LZ, Martin FR, Tinetti ME: Guideline for the prevention of falls in older people. J Am Geriatr Soc 2001, 49:664–672.

30. Tinetti ME: Where is the vision in fall prevention. J Am Geriatr Soc 2001, 49:676–677.

CITATION

Originally published under the Creative Commons Attribution License. Haran MJ, Lord SR, Cameron ID, Ivers RQ, Simpson JM, Lee BB, Porwal M, Kwan MMS, Severino C. "Preventing falls in older multifocal glasses wearers by providing single-lens distance glasses: the protocol for the VISIBLE randomised controlled trial," in BMC Geriatrics 2009, 9:10. © 2009 Haran et al; licensee BioMed Central Ltd. doi:10.1186/1471-2318-9-10.

Emotional Stress as a Trigger of Falls Leading to Hip or Pelvic Fracture. Results From the Tofa Study — A Case-Crossover Study Among Elderly People in Stockholm, Sweden

Jette Möller, Johan Hallqvist, Lucie Laflamme,
Fredrik Mattsson, Sari Ponzer, Siv Sadigh and Karin Engström

ABSTRACT

Background

Sudden emotions may interfere with mechanisms for keeping balance among the elderly. The aim of this study is to analyze if emotional stress and specifically feelings of anger, sadness, worries, anxiety or stress, can trigger falls leading to hip or pelvic fracture among autonomous older people.

Methods

The study applied the case-crossover design and was based on data gathered by face to face interviews carried out in Stockholm between November 2004 and January 2006 at the emergency wards of two hospitals. Cases (n = 137) were defined as persons aged 65 and older admitted for at least one night due to a fall-related hip or pelvic fracture (ICD10: S72 or S32) and meeting a series of selection criteria. Results are presented as relative risks with 95% confidence intervals.

Results

There was an increased risk for fall and subsequent hip or pelvic fracture for up to one hour after emotional stress. For anger there was an increased relative risk of 12.2 (95% CI 2.7–54.7), for sadness of 5.7 (95% CI 1.1–28.7), and for stress 20.6 (95% CI 4.5–93.5) compared to periods with no such feelings.

Conclusion

Emotional stress seems to have the potential to trigger falls and subsequent hip or pelvic fracture among autonomous older people. Further studies are needed to clarify how robust the findings are—as the number of exposed cases is small—and the mechanisms behind them—presumably balance and vision impairment in stress situation.

Background

Falls leading to hip or pelvic fractures among older people is a large public health concern in many high income countries, not least in Sweden and in the other Nordic countries. In Sweden which has one of the oldest populations in the world, the life time risk (at 50 years) of a hip fracture is 28.5% among women and 13.1% among men [1]. Among those afflicted by hip fracture the majority are old (average age 81 years) and of female sex (75%) [2]. In Sweden, the majority of the people aged 65+ live alone in their own house or flat and about 15% of them have home-help in their dwelling or live in special service accommodation [3].

While age-related individual factors increase the vulnerability for fall-related injuries it has been established that environmental factors, both social and physical, also have an impact [4-6]. Among physical environmental factors, slippery floors, bad lightning and loose carpets are common direct reasons for falls among elderly people living at home [6]. Remote factors such as type of housing and marital status account for the social factors susceptible to increase the risk for hip fractures [5,7-9]. Likewise, environmental failures/mishaps, situations and events occurring shortly before a fall and subsequent injury are important dimensions to consider to better understand the causal chain leading to fractures [10,11]. Such factors, called triggers, influence both the occurrence of fall and the consequences of the fall in terms of injuries. In the present research project, called ToFa (Triggers of Fall), emotional stress has been studied as one potential trigger of fall-induced hip or pelvic fractures among the elderly.

Earlier studies, performed among both young people and adults, show that emotional stress influences the risk of being injured. In a study among children aged 10–15 we have established that different kinds of emotional states, caused by peer victimization as well as perceived school failure have the potential to trigger unintentional injuries [12-15]. Another study has shown that although anger is strongly related to intentional injuries among adult men and women it does not seem to have the same impact on fall injuries [16].

To the best of our knowledge, studies on the importance of different emotional stress and the risk for fall-related injuries and hip or pelvic fractures among elderly have not yet been published. The research on older people's emotions, so far, has focused on anxiety or fear of falling and how these factors affect the incidence of falls [17]. The aim of this study is to assess the trigger effects of emotional stress such as feelings of anger, sadness, worries, anxiety and stress, on the risk of falls and subsequent hip or pelvic fracture.

Methods

To study triggers an epidemiological study design, called case-crossover, has proven useful [18]. The case-crossover design has been applied several times in order to study trigger factors of injuries. Alcohol, medication, intellectual exertion, and sleep deprivation are examples of factors that so far have been identified as triggers of different kinds of injuries [19-25].

The ToFa study is a case-crossover study designed to identify triggers of fall-induced hip or pelvic fractures. Cases were defined as patients admitted at least one night with a diagnosis of hip or pelvic fracture (ICD10: S72 or S32) and recruited at two emergency hospitals in Stockholm during November 2004 to January 2006. Screening of admissions to identify eligible cases was performed by one specially assigned research nurse at each hospital. Nine inclusion criteria had to be met. The patient should be (1) aged 65 years or above, (2) resident in Stockholm County, (3) Swedish-speaking, (4) living at home and not in an institution,

(5) able to walk (with or without help of technical aids), (6) mentally healthy and free from addiction, (7) high scoring on the Short Portable Mental Status Questionnaire (SPMSQ) test of the cognitive capability: score ≥ 8 [26,27], (8) exposed to a recent low energy injury, and (9) without a pathologic fracture. Non-Swedish speaking patients (1.3%) were excluded to ensure high information quality in the interviews and cost effectiveness in data collection.

Participation was voluntary and written consent was compulsory. Totally 137 patients participated in the study, the majority (93%) diagnosed as ICD10 S72 (hip fracture) and seven percent with ICD10 S32 (pelvic fracture). Characteristics of the study population are shown in Table 1. The two hospitals have geographically defined catchment areas and together they treat about half of all hip fractures in the Stockholm region. During the study period the organization registered 881 patients, about half of the admitted cases (according to hospital register information). Fifty percent (440 patients) did not meet our inclusion criteria and of these, 40% did not pass the cognitive function test and 43% were too sick to participate. Among the 441 patients that did meet our inclusion criteria 33% were asked for participation. The patients not registered or not asked for participation were missed due to the work schedule and work load of the research nurses or participated in other studies. Among the patients asked, 79% agreed to participate. The 21% declining participation were similar to participants with regard to age and diagnosis, however a somewhat higher proportion was male (33%).

A group of four research nurses were trained to conduct the interviews and also participated in regular meetings to discuss and solve arisen problems. The research nurses interviewed the patients to obtain exposure information and to establish the time of the injury. Ninety-three percent of the interviews were conducted within four days of the injury. The interview was structured and lasted approximately one hour. The questionnaire covered information about the injury, the timing of all main activities of the patient during the two days before the injury, exposure to potential triggers (48 hours prior to the injury and usual level of exposure frequency), general information on health status and health habits, individual information (such as sex, age, length, weight etc), and a score measuring the patient's fear of falling.

Emotional stress was defined as feelings of anger, sadness, worries, anxiety and stress and measured by the question: "Did it happen on the day of your injury or the day before that you felt angry?" If a patient answered 'yes' the patient was asked to specify exactly when they were angry during the 48 hours prior to the injury. Further, all patients were asked "How often during the last six months have you experienced feelings of anger?" The answer could be given as times per day/week/month or year, and was re-calculated to represent a usual annual frequency of exposure. The wording of the questions was similar for all the 5 different emotions.

Table 1. Characteristics (%) of the study population in the ToFa study, Stockholm County, Sweden, 2004–2006, 137 patients.

Characteristic		Percentage
Gender	Men	21.9
	Women	78.1
Age	65–79	35.8
	80+	64.2
Birth country	Sweden	94.8
Civil status	Married	39.4
Housing type	House	17.5
	Apartment	75.2
	Flat in a block of service flats	5.1
	Other	2.2
Go out shopping		77.4
Eat at least 3 meals per day		73.7
Takes care of the home		70.8
Walks in stairs		62.0
Have home-help service		22.6
Get help to go out		16.1
Place of injury	Outdoor	32.3
	Indoor	67.6

A fall and subsequent injury as defined in this study can only occur during time spent walking, standing or in intermediate positions while changing posture. Accordingly, the person-time from which the control windows are sampled should ideally be restricted to times of outcome opportunity [18]. To correctly assess exposure frequencies during the person-time at risk of falling, every episode of walking, standing, sitting, lying and changing posture was thoroughly charted in the interview through a schedule of activities during the two days prior to the injury. To aid the ability to recall, the interviewer first ascertained usual daily activities and situations (for example time for getting out of bed, eating, medication intake, personal hygiene, visits, watching television, significant events etc). Information regarding the level of activity was retrieved for each 15-minute interval but smaller time intervals were registered in case the duration was shorter, such as a visit to the toilet.

Statistical Analyses

The risk estimate was calculated as the ratio between the observed odds of exposure to the expected odds of exposure [28]. The observed odds of exposure (either 1 or 0) were determined by the observed frequency of emotional stress during the hazard-period prior to the injury. Results are presented for hazard-periods of one hour. Hazard-periods of varying length were tested but the number of exposure events outside the one-hour interval prior to the injury was small, limiting the induction time analyses. The expected odds of exposure were defined as the proportion of exposed person-time to the proportion of unexposed person-time, and were assessed on basis of the recorded frequency of exposure during the two types of control periods preceding the fall, i.e., 48 hours or 6 months. In an exact calculation we based the expected odds of exposure on the complete person-time at risk (of falling) during the 24-hour period prior to injury (A in Table 2). Exposed time was determined from reported episodes of trigger exposure and calculated as the sum of the time at risk during each hazard-period following an episode of trigger exposure. Unexposed time was calculated as the total time at risk minus the exposed time. In cases experiencing emotional stress close to the injury this assessment of exposed time might lead to an underestimation of the exposed time since the last period of exposure sometimes was interrupted by the fall. In these rather few instances (n = 4) a more correct estimate of the exposed person time was obtained by adding the remaining time at risk at the same point in time the day before.

To examine the robustness of the exact calculation, we also calculated the expected odds of exposure using self-reported usual frequency of trigger exposure according to the standard case-crossover methodology and applied three different estimations of the time at risk leading to four alternative analytical strategies (B-E in Table 2) [18]. Person-time at risk of falling was defined as (B) the total time during a year equal to 8760 hours, (C) the total time minus individual sleeping time measured by the time between going to bed and waking multiplied by 365, and (D+E) as an average time at risk per year calculated from the proportion of time spent standing or walking during the 48 hours prior to the injury divided by 2 and multiplied by 365. Exposed person-time was estimated as usual annual frequency multiplied with the assumed length of the hazard period (B, C, D) implicitly making the conservative assumption that all trigger exposures occurred during time at risk. In (E) we instead explicitly assumed that trigger exposure occurred independently of whether the person was at risk of falling or not. Hence, exposed person-time was proportionally reduced. In all these calculations the trigger exposures were assumed to be discrete and their effect periods not to overlap.

Table 2. Relative risk (95% CI) of fall-induced hip or pelvic fracture within 1 hour after emotional stress (n = 122).

Analytic strategy	Expected odds based on		Emotion		
			Anger	Sadness	Stress
	Time at risk	Exposed time	Number of exposed cases = 4	Number of exposed cases = 2	Number of exposed case = 6
A	Standing/walking time[1]	Exact calculation[5]	12.2 (2.7–54.7)	5.7 (1.1–28.7)	20.6 (4.5–93.5)
B	All[2]	Usual frequency[6]	22.5 (7.4–68.8)	3.3 (0.7–15.0)	15.1 (3.8–59.8)
C	Time awake[3]	Usual frequency[6]	14.3 (4.7–44.0)	2.4 (0.5–10.8)	8.8 (2.3–34.6)
D	Standing/walking[4]	Usual frequency[6]	3.9 (1.2–12.7)	0.8 (0.2–3.2)	3.7 (1.0–13.8)
E	Standing/walking[4]	Usual frequency corrected for time at risk[7]	14.3 (4.7–44.0)	2.4 (0.5–10.8)	8.8 (2.3–34.6)

Five different analytic strategies varying the bases for the expected odds.
[1] Time spent walking or standing during the 24 hour period prior to injury
[2] Total time per year (8760 hours)
[3] Time awake per year (estimated from time from going to bed until waking up)
[4] Yearly time spent walking or standing (estimated from time spent walking or standing during the 48 hour period prior to injury)
[5] Sum of the time at risk during each hazard period following an episode of trigger exposure. in the 24 hour period prior to injury.
[6] Annual frequency of exposure
[7] Corrected annual frequency of exposure under the assumption that exposure occurs independently of time at risk, correction is made of the frequency of exposure to correspond to the proportion of time at risk (of time awake).

The effect estimates were calculated as standard Mantel-Haenszels estimates with confidence intervals for sparse data [18,29]. Results are presented as relative risks and with 95% confidence intervals. All calculations were performed using the SAS software package version 8.02. The study was approved by the Regional Ethics Committee in Stockholm.

Results

Thirty three of the 137 patients (24.1%) experienced at least one of the emotions during the 24 hour period prior to injury. Among these there were 15 who reported that they had experienced the emotions during the day of the injury but could not state any specific episode with an exact timing, in most cases because the emotion lasted all day. They were excluded from further analyses.

During the hazard period of one hour before the fall 3.3% (4/122) of the cases were exposed to anger, 1.6% (2/122) to sadness and 4.9% (6/122) to stress, in all 10 patients. Only one of the patients experienced more than one type of emotion during the hazard period. The number of cases exposed to feelings of worry and anxiety during the hazard period was too small to be further analyzed.

The analysis based on a complete follow-up of exposed and unexposed person-time during the 24 hours prior to injury (A column in Table 2) shows that the increased relative risk of fall-related hip or pelvic fracture during the first hour after emotional stress is 12.2 (95% CI 2.7–54.7) for anger, 5.7 (95% CI 1.1–28.7) for sadness and 20.6 (95% CI 4.5–93.5) for stress.

The other analytical strategies confirm the results although the risk estimates change slightly. Analysis B, does not take into account that the risk of a fall and subsequent injury is only present when standing, walking or changing posture. Further, the proportion of exposed follow-up time at risk is obtained through the question of usual frequency multiplied with the hazard period. It resulted in relative risks of 22.5 (95% CI 7.4–68.8) for anger and 15.1 (95% CI 3.8–59.8) for stress. If the expected odds of trigger exposure are independent of whether the case is at risk of falling or not the effect estimate would be the same. The slightly higher RR in B than in A suggests that the expected odds of exposure are underestimated in B due to an overestimation of the proportion of unexposed time at risk. In analysis C this problem is reduced by restricting the study base to time awake, correctly assuming that no falls happen while asleep, and that there is also no conscious emotional stress while sleeping. Accordingly, the relative risk estimates were reduced to 14.3 (95% CI 4.7–68.8) for anger and 8.8 (95% CI 2.3–34.6) for stress.

In the results presented in the last two rows, the study base time was restricted to the time at risk estimated by the individual, self-reported time at risk of falling during the 48 hours before the fall. In analysis D we assumed that all emotional stress events in the self-reported usual frequency occur during time at risk of falling. As some of the anger, sadness or stress may have happened during time spent sitting or lying we consciously overestimated the exposed time implying an underestimation of the relative risk [18,28]. Still, these analyses showed elevated risk estimates, 3.9 (95% CI 1.2–12.7) for anger and 3.7 (95% CI 1.0–13.8) for stress. In analysis E we instead assumed independence between emotional events and the risk of falling implying that a proportionate number of emotional events take place during time at risk and during time not at risk. How many of the exposure events that should be assumed to take place during time at risk was decided by the proportion of time at risk of the total time awake. For example, a patient reporting standing or walking approximately 3 hours per day would be at risk of falling 20 percent of the time (given that 9 hours of sleeping time is excluded). If the same patient reported anger occurring five times per day the reported frequency of emotional stress will be reduced to 20%, namely to once a day. Analysis E resulted in elevated relative risk of 14.3 (95% CI 4.7–44.0) for anger and 8.8 (95% CI 2.3–34.6) for stress, similar to the effect estimates reported for analysis A.

Discussion

This study shows that, among autonomous elderly, emotional stress may trigger falls leading to hip or pelvic fractures. The results reveal an increased risk in the period of one hour after anger or perceived stress. This is the first study to report a trigger effect of emotional stress on the risk for fall and subsequent hip or pelvic fracture. In a combined case-control and case-crossover study Vinson et al found an association between state anger and risk of injury. In sub-analyses of different types of injury they reported increased risk estimates for fall injuries however statistically non-significant [16].

There are two possible and closely related mechanisms that may explain the relationship between emotional stress and injurious fall: impoverished postural control and gaze strategy. Aging is frequently accompanied by deterioration in postural control (and preparation) and it has been observed that in order to maintain balance, age-specific compensatory strategies involving the hip (rather than the ankle as in younger subjects) and local muscles (thigh ones) are used to counterbalance a decrease in anticipation [30]. Further, age-related deficits in the neuro-musculoskeletal systems may impede ability to effectively execute "change-in-support" (CIS) balance-recovery reactions that involve rapid stepping or reaching movements that play a critical role in preventing falls [31].

In some groups of frail older people (e.g., women aged 75–86 years with low bone mass), falls-related self-efficacy may be independently associated with both balance and mobility after accounting for age, current physical activity level, and performances in relevant physiological domains [32].

The changing gaze strategy is related to poor vision. It is known that poor visual acuity reduces postural stability and significantly increases the risk of falls and fractures in older people [33]. It is also possible that older people, in stress situation would adopt a gaze strategy detrimental to their balance control, more specifically, a premature gaze transfer which in turn has been associated with decline in stepping accuracy and precision [34]. Studies that have included multiple visual measures have found that reduced contrast sensitivity and depth perception are the most important visual risk factors for falls [34].

Methodological Considerations

A peculiar feature of studies of falls is that person time at risk is not continuous in real time. The fact that there is only an outcome opportunity when walking, standing, sitting or changing posture calls for special measures when applying the case-crossover design. Measuring information of the time at risk is not easy and many studies do not have the possibility to control for it when calculating the effect estimates. In this study we made an effort to collect such information in addition to traditional information for making case-crossover analyses. Therefore we had the opportunity to test different analytic strategies and the robustness of the found effects. Each analytic strategy implies, however, different degrees and aspects of methodological problems.

The first analysis (A) we performed is as close to correct as we think is possible. The analysis was based on an exact calculation of exposed time thoroughly controlled for time at risk. The narrow time interval of 24 hours was chosen to minimize the risk for information bias and to be able to impute risk time information for exposed cases. Ignoring this could lead to a small underestimation of the exposed time among the exposed cases which in turn could lead to an overestimation of the relative risk.

Had we not had information regarding this time at risk we would have been limited to basing the exposed time on the reported usual frequency of trigger exposure. Besides any information bias in such a variable, the analysis based on this has the disadvantage that the reported exposure information is not restricted to times at risk. For this reason we tested to restrict the study base time in three different ways. Analysis B is a straight forward case-crossover analysis falsely assuming that the risk of falling is equal over the whole day. This leads to an overestimation of the proportion of unexposed time and an overestimation of the relative risk. Analysis C restricted the study base time to time spent awake. This reduces the overestimation of the unexposed time but still, depending on the exposure, leads to an overestimation of the relative risk. Analyses D and E was based on the reported person time at risk. The methodological objection to these analyses is due to the fact that the reported usual trigger frequency was not restricted to time at risk. Making assumptions regarding the distribution of exposure over time at risk and time not at risk is difficult. In analysis D, we made the conservative assumption that all episodes of emotional stress occurred during time at risk. Under the premise that it is unlikely that emotional stress always takes place during time at risk, this analysis leads to an overestimation of the exposed time leading to underestimated effect estimates. In analysis E, the overestimation of exposed time was reduced as the frequency of exposure was distributed correspondingly to the distribution of time at risk to time not at risk. However, this assumption might for some individuals not be true, for example if a person becoming angry always starts to move around in order to physically get rid of the anger. On the other hand, if a patient becoming angry always tries to calm down by sitting down this would also be an incorrect assumption. To test this further, additional information has to be measured on an individual level, which we prompt future studies to take into account if not using the exact method.

The restriction to patients with a normal cognitive function and in some analyses to an exposure period of 24 hour or 48 hour prior to the injury, together with an interview after as little delay as possible, would diminish recall problems leading to non-differential misclassification. To further improve recall of exposures, the interview charted activities during the two days prior to the event and there were good possibilities to link the fall and also potential triggers to activities, events and things that had happened during this time period. According to information from the research nurses carrying out the interviews, this group of elderly people lives organized lives with little variations in their daily activities which further aid their memory. Only a small number of patients reported that the 48 hour period prior to injury was "an unusual day." Excluding those from the analysis did not alter our conclusions. It is also reassuring that analyses based on short and long recall periods show similar results.

Additionally, to combat differential misclassification of exposure between the case and control period, neither the patients nor the interviewers were aware of the assumptions regarding the induction times and were instructed to pay equal interest to the whole 48-hour period, hence, an attribution of exposures in those periods seems unlikely. As hip or pelvic fracture is an acute event, exposure in the case period might otherwise be less likely to be missed or forgotten.

Instances of emotional stress were assessed by simple interview questions allowing for a dichotomous classification. No intensity scales were used. Because the case information is compared with control information from the same patient, there will be no differential exposure misclassification as long as each individual patient uses the same definition of every exposure in all questions, which seems a reasonable assumption. Exposure to other triggers of falls in the period prior to onset might imply confounding. However, co-exposure between the five different emotions was uncommon in this study. Only three patients were exposed to two or three emotions simultaneously during the three-hour period prior to the injury. Exclusion of these cases in the analyses decreased power but did not alter the direction of the estimated effects.

Non-participation because the research nurses did not manage to contact all cases would foremost be a result of the nurses' work load and working hours and not selective with regard to the case's exposure status in the case period. The patients were not informed about the questions in the interview and unaware of the fact that emotional stress would be covered. Hence, it is unlikely that exposure status would have affected their participation. Survival bias is no problem as mortality in the early period after a hip or pelvic fracture is low. The restriction in the recruitment of all patients with a hip or pelvic fracture caused by the inclusion criteria does not lead to bias.

If the patients have difficulties in differentiating between emotions prior to the injury and emotions arising due to or post the injury, it would imply bias. However, of the ToFa patients reporting emotions prior to the injury (in the case window) only one reported it to have happened directly in relation to the fall. The rest of the patients reported it to happen longer before the event (such as 25 minutes) which most likely cannot be mistaken for the feelings arising due to the fall.

A further limitation of the study is the small number of exposed cases. Although the effects are statistically significant, the confidence intervals are wide and this challenges the robustness of the estimated effects.

Conclusion

Emotional stress seems to have the potential to trigger falls and subsequent hip or pelvic fracture among autonomous older people. However, to our knowledge, this is the first study to report this and further studies are needed to clarify how robust those findings can be and disentangle the mechanisms behind these findings, likely to be related to balance and vision impairment in stress situation.

Competing Interests

The authors declare that they have no competing interests.

Authors' Contributions

JM contributed to the conception and design of the ToFa study, made the statistical analyses and interpretation of the data, and drafted the manuscript. JH contributed to the conception and design of the ToFa study, interpretation of data, and drafting of the manuscript. LL conceived the ToFa study and contributed to the conception, design, acquisition of data, and interpretation of data, and drafting of the manuscript. KE contributed to the conception, design, and acquisition of data and have revised the manuscript for important intellectual content. FM participated in the statistical analyses and interpretation of the data and has revised the manuscript for important intellectual content. SP made substantial contributions to acquisition of data and has revised the manuscript for important intellectual content. SS made substantial contributions to acquisition of data and has revised the manuscript for important intellectual content. All authors have read and approved the final manuscript.

Acknowledgements

ToFa was financially supported by the Swedish Rescue Services Agency. We thank the research nurses Ingemo Sundberg-Petersson and Anita Söderqvist for their professionalism and valuable contribution during the data collection.

References

1. Kanis JA, Johnell O, De Laet C, Jonsson B, Oden A, Ogelsby AK: International variations in hip fracture probabilities: implications for risk assessment. J Bone Miner Res 2002, 17:1237–44.

2. The national board of health and welfare, Sweden [http:/ / www.socialstyrelsen. se/ NR/ rdonlyres/ 7456A448-9F02-43F3-B776-D9CABCB727A 9/ 6169/ 20051114.pdf].

3. The national board of health and welfare, Sweden [http:/ / www.socialstyrelsen. se/ NR/ rdonlyres/ 4D793100-E677-4E98-997F-4805BDB1FEC C/ 9991/ 20081316.pdf].

4. Nyberg L: Falls in the frail elderly. Incidence, characteristics and prediction with special reference to patients with stroke and hip fracture. PhD thesis. Umeå University, New Series No 483; 1996.

5. Sadigh S, Reimers A, Andersson R, Laflamme L: Falls and fall-related injuries among the elderly: a survey of residential-care facilities in a Swedish municipality. J Community Health 2004, 29:129–40.

6. Morley JE: Falls—where do we stand? Mo Med 2007, 104:63–7.

7. Farahmand B, Persson P-G, Michaelsson K, Baron JA, Parker MG, Ljunghall S: Socioeconomic status, marital status and hip fracture risk. A population-based case-control study. Osteoporosis Int 2000, 11:803–808.

8. Hökby A, Reimers A, Laflamme L: Hip fractures among older people: do marital status and type of residence matter? Public Health 2003, 117:196–201.

9. Peel NM, McClure RJ, Henrikz JK: Psychosocial factors associated with fall-related hip fractures. Age and ageing 2007, 36:145–151.

10. Engström K: Social differences in injury risk in childhood and youth. Exploring the roles of structural and triggering factors. PhD thesis. Karolinska Institutet, Department of Public Health Sciences; 2003.

11. Diderichsen F, Laflamme L, Hallqvist J: Understanding the mechanism of social differences in injuries. In Safety promotion research. A public health approach to accident and injury prevention. Edited by: Laflamme L, Svanström L, Shelp L. Stockholm: Karolinska Institutet, Department of Public Health Sciences; 1999:177–202.

12. Engström K, Hallqvist J, Möller J, Laflamme L: Do episodes of peer victimization trigger physical injuries? A case-crossover study. Scan J Public Health 2005, 33:19–25.

13. Laflamme L, Engstrom K, Moller J, Hallqvist J: Peer victimization during early adolescence: an injury trigger, an injury mechanism and a frequent exposure in school. Int J Adolesc Med Health 2003, 15:267–79.

14. Laflamme L, Engstrom K, Moller J, Hallqvist J: Is perceived failure in school performance a trigger of physical injury? A case-crossover study of children in Stockholm County. J Epidemiol Community Health 2004, 58:407–11.

15. Laflamme L, Möller J, Hallqvist J, Engström K: Peer victimization and intentional injuries: Quantitative and qualitative accounts of injurious physical interactions between students. Int J Adolesc Med Health 2008, 20:201–8.

16. Vinson DC, Arelli V: State anger and the risk of injury: a case-control and case-crossover study. Annals of Family Medicine 2006, 4:63–68.

17. Scheffer AC, Schuurmans MJ, van Dijk N, Hooft T, de Rooij SE: Fear of falling: measurement strategy, prevalence, risk factors and consequences among older persons. Age Ageing 2008, 37:19–24.

18. Maclure M, Mittleman MA: Should we use a case-crossover design? Annu Rev Public Health 2000, 21:193–221.

19. Vinson DC, Mabe N, Leonard LL, Alexander J, Becker J, Boyer J, Moll J: Alcohol and injury. A case-crossover study. Arch Fam Med 1995, 4:505–11.

20. Borges G, Cherpitel C, Orozco R, Bond J, Ye Y, Macdonald S, Rehm J, Poznyak V: Multicenter study of acute alcohol use and non-fatal injuries: data from the WHO collaborative study on alcohol and injuries. Bull World Health Organ 2006, 84:453–60.

21. Neutel CI, Perry S, Maxwell C: Medication use and risk of falls. Pharmacoepidemiol Drug Saf 2002, 11:97–104.

22. Hoffmann F, Glaeske G: New use of benzodiazepines and the risk of hip fracture: A case-crossover study. Z Gerontol Geriatr 2006, 39:143-8. In German.

23. Petridou E, Mittleman MA, Trohanis D, Dessypris N, Karpathios T, Trichopoulos D: Transient exposures and the risk of childhood injury: a case-crossover study in Greece. Epidemiology 1998, 9:622–5.

24. Edmonds JN, Vinson DC: Three measures of sleep, sleepiness, and sleep deprivation and the risk of injury: a case-control and case-crossover study. J Am Board Fam Med 2007, 20:16–22.

25. Fisman DN, Harris AD, Rubin M, Sorock GS, Mittleman MA: Fatigue increases the risk of injury from sharp devices in medical trainees: results from a case-crossover study. Infect Control Hosp Epidemiol 2007, 28:10–7.

26. Pfeiffer E: A short portable mental status questionnaire for the assessment of organic brain deficit in elderly patients. J Am Geriatr Soc 1975, 23:433–41.

27. Erkinjuntti T, Sulkava R, Wikstrom J, Autio L: Short Portable Mental Status Questionnaire as a screening test for dementia and delirium among the elderly. J Am Geriatr Soc 1987, 35:412–6.

28. Maclure M: The case-crossover design: a method for studying transient effects on the risk of acute events. Am J Epidemiol 1991, 133:144–53.

29. Rothman KJ, Greenland S: Modern Epidemiology. Philadelphia: Lippinkott Williams & Wilkins; 1998.

30. Bleuse S, Cassim F, Blatt JL, Labyt E, Derambure P, Guieu JD, Defebvre L: Effect of age on anticipatory postural adjustments in unilateral arm movement. Gait Posture 2006, 24:203–10.

31. Maki BE, Cheng KC, Mansfield A, Scovil CY, Perry SD, Peters AL, McKay S, Lee T, Marquis A, Corbeil P, Fernie GR, Liu B, McIlroy WE: Preventing falls in older adults: New interventions to promote more effective change-in-support balance reactions. J Electromyogr Kinesiol 2008, 18:243–54.

32. Liu-Ambrose T, Khan KM, Donaldson MG, Eng JJ, Lord SR, McKay HA: Falls-related self-efficacy is independently associated with balance and mobility in older women with low bone mass. J Gerontol A Biol Sci Med Sci 2006, 61:832–8.

33. Lord SR: Visual risk factors for falls in older people. Age Ageing 2006, 35(Suppl 2):ii42–ii45.

34. Chapman GJ, Hollands MA: Evidence that older adult fallers prioritise the planning of future stepping actions over the accurate execution of ongoing steps during complex locomotor tasks. Gait Posture 2007, 26:59–67.

CITATION

Health Status Transitions in Community-Living Elderly with Complex Care Needs: A Latent Class Approach

Louise Lafortune, François Béland,
Howard Bergman and Joël Ankri

ABSTRACT

Background

For older persons with complex care needs, accounting for the variability and interdependency in how health dimensions manifest themselves is necessary to understand the dynamic of health status. Our objective is to test the hypothesis that a latent classification can capture this heterogeneity in a population of frail elderly persons living in the community. Based on a person-centered approach, the classification corresponds to substantively meaningful groups of individuals who present with a comparable constellation of health problems.

Methods

Using data collected for the SIPA project, a system of integrated care for frail older people (n = 1164), we performed latent class analyses to identify homogenous categories of health status (i.e. health profiles) based on 17 indicators of prevalent health problems (chronic conditions; depression; cognition; functional and sensory limitations; instrumental, mobility and personal care disability) Then, we conducted latent transition analyses to study change in profile membership over 2 consecutive periods of 12 and 10 months, respectively. We modeled competing risks for mortality and lost to follow-up as absorbing states to avoid attrition biases.

Results

We identified four health profiles that distinguish the physical and cognitive dimensions of health and capture severity along the disability dimension. The profiles are stable over time and robust to mortality and lost to follow-up attrition. The differentiated and gender-specific patterns of transition probabilities demonstrate the profiles' sensitivity to change in health status and unmasked the differential relationship of physical and cognitive domains with progression in disability.

Conclusion

Our approach may prove useful at organization and policy levels where many issues call for classification of individuals into pragmatically meaningful groups. In dealing with attrition biases, our analytical strategy could provide critical information for the planning of longitudinal studies of aging. Combined, these findings address a central challenge in geriatrics by making the multidimensional and dynamic nature of health computationally tractable.

Background

The general approach to studying older people's health has been to look at relationships among measures of chronic conditions, cognition, frailty and steps along the disablement pathway [1-4]; the goals have been to identify determinants and rates of decline and recovery [5-14], and predict mortality [15], institutionalization [16,13], and service utilization [17-21]. Although these measures serve as valid outcomes predictors, they relate differently to various dimensions of health. Trends in any one of them are not evidence of health trends overall [22,23]. Evidence suggests that population subgroups have qualitatively and quantitatively different patterns of change and probabilities of adverse outcomes [24,9,6,26,27,14,25,8]. In fact, elderly populations are highly heterogeneous in their health status owing to the variability and interdependency in how health dimensions manifest themselves overtime. Thus, it becomes increasingly clear that health changes cannot be fully described by any one dimension. Instead, it takes an approach that allows for multiple measures of health to embrace this complexity.

Many valid approaches exist to study the relationship between multiple health indicators and the unobservable (or latent) construct of health [28]. The choice depends, among other considerations, on the distribution of health indicators. Because health indicators commonly used in geriatrics are rarely continuous or normally distributed, factor analysis is unwieldy. Instead, we chose a latent class analytical framework [29], also called finite mixture models [30]. Latent class analysis (LCA) is a "person-centered" approach designed to divide the population under study into latent subpopulations (i.e. classes) that share a distinct interpretation pattern of relationships among indicators [31]. In populations of older people with complex health needs, observed health indicators can serve to identify groups of individuals—unobserved a priori—who present with similar constellations of health problems. The goal is to find the smallest number of health profiles that can describe the association among the set of observed health indicators.

Latent class models present several advantages over classical statistical models (e.g. cluster analysis) [32,33]: the classification is model-based and statistical diagnostic tools exist to assess the quality of the classification, variables may take several forms, there is no need to make parametric assumptions about the relationship between observations, and covariates can be included during model estimation to describe the groups. For the prediction of a dependant variable, LCA offer a parsimonious alternative to models with an unmanageably high number of variables and interaction terms.

Whereas LCA studies class membership using cross-sectional data, latent transition analysis (LTA) studies change in class membership using longitudinal data [31]. LTA combines the cross-sectional measurement of health state profiles and the description of transition over time from one health profile to another [34]. It thus supports the analysis of qualitative change in health status in contrast to continuous models were quantitative change is observed.

Several applications of LCA exist in the social, medical and economic literature, where heterogeneous populations constitute the typical object of analysis. In geriatrics, LCA was used to test criterion validity of physical frailty [35], measure mobility disability [36] and study behavioral syndromes in Alzheimer' patients [37]. We found no example of its use to model heterogeneity in older individuals' health status. Grade of membership (GoM), another latent classification technique [38], was successful in identifying clinically meaningful health profiles in community-living elderly [39-42]. GoM assumes that individuals can be partial members of more than one class of a continuous distribution of latent variables [43]. In contrast, LCA assumes that individuals are full members of one of the class for a discrete latent variable. Practically, LCA classifies individuals in health profiles whereas GoM estimates their degree of proximity to each profile. To our knowledge, these studies did not study transitions in profile membership; though Portrait et al. [41] studied mortality. Conversely, Markov models were used to describe transitions along the disablement pathway [14,9,8,44,45]. Although similar in spirit to LTA [34], these models focused on observable disability indicators as opposed to directly unobservable health status profiles.

The objective of this paper is to identify profiles of health status and study their evolution in a population of frail elderly individuals living in the community. We used LCA to group individuals into homogenous categories of health status based on observed indicators of prevalent health problems. We then applied LTA to study transitions in profile membership over 2 consecutive periods of 12- and 10-months, respectively. The question we address is whether there exists a latent classification that makes substantive sense to study the dynamic of health status in older populations presenting with complex care needs. Many research and policy questions call for a person-centered as opposed to a variable-centered approach [31]. Practical examples include planning of long-term care resources and identification of appropriate sub-groups for prevention studies. We identified and validated a classification that captures the multidimensional and dynamic nature of health, and show how a person-centered approach could prove useful to address these important questions.

Methods

Data Sources

This research uses data collected for the randomized trial of the SIPA program (French acronym for System of Integrated Care for Older Persons), carried out in Montreal, Canada (1999–2001). SIPA aimed at improving continuity of community- and institution-based care for older people with complex health and social needs; its distinguishing features and the results of the experimentation are described elsewhere [46,47]. Eligible people were older than 64 years, competent in French or English (either the individual or the caregiver), had a score of -10 or less on the Functional Autonomy Measurement System (SMAF; -87 represents the worst health state) [48], and no plans for institutionalization within 3 months. Participants were recruited from two public community organizations responsible for home care (with a small proportion from other sources) and randomly assigned to the SIPA program or usual care. Health and sociodemographic data were collected via structured home interviews at baseline (T0), 12-months (T1), and 22-months (T2). Living status (deceased or not) and information on nursing home use come from administrative databases. Given the short intervention period, SIPA had no effect on change in health status or mortality [46]. For our purpose, we thus combined the two groups and included those who had a baseline questionnaire (n = 1164). The research ethics committee of the Montreal Jewish General Hospital granted approval for these secondary analyses.

Health Indicators

LCA uses observed response patterns on a set of health indicators to model heterogeneity and reveal health profiles. We selected 17 indicators based on their high prevalence in older populations. Chronic conditions (no/yes) include self-reported hypertension, stroke, diabetes, cancer as well as circulatory, respiratory, joint and arthritis, stomach and bladder problems. To increase validity, individuals were asked whether a physician had confirmed the diagnosis. Sensory limitations (no/yes) refer to self-declared problems with speech, audition and/or vision. Cognition is measured with the Short Portable Mental Health Questionnaire (scores < 4 indicate no cognitive impairment; scores ≥4 indicate cognitive impairment) [49], and depression with the Geriatric Depression Scale (GDS scores < 3.5 indicate no depression; scores between 3.5–8.5 indicate moderate depression; scores ≥8.5 indicate severe depression) [50]. Functional limitations (no/yes) are defined as difficulty performing upper limbs (≥ 1; raising arms, picking/handling small objects, lifting 5 kg) and lower limbs (≥ 1; pulling/pushing large objects, bending/kneeling; using stairs) movements [51]. Disability is defined as requiring help with mobility activities of daily living (ADL) (getting up from bed/chair; using the toilet; taking a bath/shower; moving around the home, going up/down stairs, walking one block); personal care ADL (eating, drinking, dressing upper and lower body, personal grooming, washing) [52]; and instrumental ADL (IADL; using the phone, transportation, shopping, meal preparation, light housework, taking medication, managing money [53]. Disability measures are coded as three-levels categorical variables. The sociodemographic covariates used in our analyses are gender, age (64–75; 75–84; 85+ years) and living arrangements (whether people live alone or not).

Analysis Strategy

LTA requires a step-wise approach to be accurate in its account of the change process. Using Mplus [54], we first explored alternative LCA models at (T0) to reveal the latent class variable that best captures the health status of our sample (i.e. number and characteristics of health profiles). Then, we fitted LCA models at

(T1) and (T2) to examine the profiles' stability over time. Finally, we fitted LTA models to study transitions in health status.

Baseline Health State Profiles

Under latent class theory [55], individuals are assumed to belong to one of a number of unobserved categories (i.e. latent classes). Here, these categories represent health status profiles. It is assumed that a sufficient number of profiles result in conditional independence among observed health indicators [29]. In Mplus [54], LCA relies on maximum-likelihood methods to estimate the posterior probability p [c/Y] of membership in health state profile c for an individual with an observed health indicator response pattern Y. These posterior probabilities determine the relative prevalence and serve to characterize the profiles [28].

Two types of parameters serve to estimate class membership: health indicator and health profile probabilities. Health indicator probabilities are profile-specific and consist of the probability that a response is associated with the profile (e.g. probability of cognitive problems given membership in health profile 1). Within classes, individuals have comparable health indicator probabilities. Health profile probabilities represent individuals' probability of belonging to each profile; assignment to one profile proceeds on the basis of their highest health profile probability.

Using data for our 17 health indicators, we fitted LCA models successively for 1 through 5 classes. We used multiple start values to avoid convergence on local maxima [34] and assumed a missing at random mechanism (MAR), i.e. missingness depends on the observed components of the complete data and not on missing data [56]. For most health indicators, MAR is justified given the low proportion of missing data. For depression, however, the depression score is missing for all individuals with cognitive impairment (n = 174) and for those who are missing on the cognitive test (n = 101). To test the effect of the MAR hypothesis on the classification, LCA models were fitted with and without the depression indicator for 1) the complete sample (n = 1164), 2) a sub-sample with complete data for depression (n = 890) and 3) a sub-sample that excludes individuals with missing data for the cognition and depression scores (n = 1063). These sensitivity analyses showed that although depression significantly contributes to the classification, assuming a MAR mechanism for missing depression scores did not significantly change how individuals are grouped.

We decided on the best model based on the lowest values observed for the Akaike Information Criteria (AIC), Bayesian Information Criteria (BIC) and adjusted BIC (aBIC), which combine goodness of fit and parsimony [34]. The adjusted Lo-Mendell-Rubin likelihood ratio test (aLMR-LRT) served to decide on the number of classes [57]; it compares improvement in fit (p < 0.001) between sequential class models through an approximation of the LRT distribution. We used an entropy measure to assess how well the model predicts class membership; values range from 0 to 1 and high values are preferred [54]. Bivariate residual statistics served to confirm local independence [34].

Including covariates in LCA models can serve three purposes: describe the formation of health profiles, characterize and validate them. This is accomplished by the concurrent identification of the latent health profiles variable and its multinomial regression on covariates of interest [29,36]. This approach avoids the limits of post-hoc regressions (i.e. performed on a priori classified individuals), which assume that the latent classification is an observed variable measured without error [29]. We found that including gender, age and/or living arrangements as covariates did not significantly influence the formation of the profiles. Therefore, the LCA model without covariates is our basecase. LCA regression models are used for characterization and validation.

To further validate the profiles, we fitted 2 LCA models to estimate the association between profile membership and distal outcomes: mortality and use of nursing home services at 22-months. These associations were estimated by allowing the proportion for each outcome to vary across profiles [54]. Age and gender are included to control for confounding. Differences between classes are reported as odd ratios. Finally, to evaluate construct validity, we measured the relationship between class membership and disability measures (ADL-Personal care, ADL-Mobility, IADL), cognitive status and comorbidity (# chronic conditions) with chi-square and contingency coefficient statistics.

T1 and T2 Health State Profiles

Using available data, we proceeded the same way to identify health status profiles at subsequent time points. To assess whether the health status profiles identified at T1 and T2 have the same substantive meaning as those identified at baseline, we first compared patterns of health indicators probabilities over time. Then, we compared the classifications to those obtained by LCA models constrained to have baseline class-specific health indicator probabilities. The concordance (%) in how individuals are grouped is used to confirm the classification' stability over time. This also allows testing of the classification' robustness to death and lost to follow-up (LTF) attrition.

Latent Transition Analysis

Building from LCA, LTA studies change in class membership using longitudinal data [31,58]. Health indicators are measured repeatedly over time to identify profile membership at each occasion. The transition probability matrices are estimated by a logistic regression for nominal response (e.g. probability of health profile membership at T1, conditional on baseline health profile membership) [33,54]. When covariates are included in LTA, transition probabilities are no longer conditioned only on the previous time(s) membership but also on covariate values [59].

We ran two separate LTA: one to assess transitions in health status between T0 and T1 (LTA-T0T1); one to assess transitions between T1 and T2 (LTA-T1T2). Age and gender are included in our models—a provision that allows us to compare transition probabilities across these covariates. Death and LTF are modeled as absorbing states. Thus, each model adjusts the probability of health status transition for the competing risks of death and being LTF. Finally, we imposed measurement invariance (MI; constraining conditional health indicator probabilities to be the same across time points) to ensure health profiles have the same meaning at each occasion. The plausibility of MI was confirmed by comparing the classifications with and without MI constraints: the differences in classification were not significant ($\chi2$; $p < 0.001$) and information criterion favored the constrained model.

Results

Sample Description

The baseline sample consists of 1164 people between 64 and 104 years old, with 70.9% female and considerable socioeconomic variability (Table 1). Comorbidity ranges from none to seven chronic conditions. Comparisons of health indicators proportions reveal great heterogeneity in health status (Table 2). Given the low proportions of missing values, attribution under the MAR hypothesis had no effect on health indicator proportions, except for depression, cognition and bladder problems. For these indicators, Mplus modeled proportions appear in parentheses.

At 12-months (T1), 11.5% (n = 134) of the sample had died, 7.6% (n = 88) did not complete the interview and 12.4% (n = 144) were lost-to-follow-up (LTF). At 22-months (T2), 20.8% (n = 242) of the sample had died, 38.2% (n = 446) were LTF. Individuals who died were more likely to be male, older; they had worst functional scores and were less likely to live alone. In bivariate analyses, LTF individuals, either at T1 or T2, did not significantly differ from the rest of the sample on sociodemographic characteristics or baseline health status (i.e. IADL, ADL-personal care, cognition, depression, sensory deficits, functional limitations) except for ADL-mobility at T1 (χ^2, $p < 0.05$) and sensory deficits at T2 ($\chi2$, $p < 0.05$). In multivariable models, only ADL-mobility remained significantly associated with LTF status at T1.

Table 1. Sample characteristics at T0; % of deceased and LTF individuals at T1 and T2 *

		Baseline	T1		T2	
		% (n = 1164)	Deceased % (n = 134)	LTF % (n = 232)§	Deceased % (n = 242)	LTF % (n = 446)
Age	Average (± SD)	82.2 (7.2)	**83.7 (7.6)**	82.2 (7.3)	**84.3 (7.2)**	82.1 (7.4)
	64–74 yrs	16.1	**12.7**	15.9	**9.5**	16.1
	75–84 yrs	43.8	**35.8**	41.4	**38.8**	42.0
	>84 yrs	40.2	**51.5**	42.7	**51.7**	41.9
Gender	Female	70.9	**56.0**	73.3	**57.4**	74.0
	Male	29.1	**44.0**	26.7	**42.6**	26.0
Marital status†	Married	33.2	**40.6**	34.3	**40.8**	30.4
	Not married	66.8	**59.4**	65.7	**59.2**	69.6
Living arrangements	Lives alone	43.7	**26.1**	46.8	**30.7**	46.7
	Not alone	56.6	**73.9**	53.2	**69.3**	53.3
Education	Primary	32.3	33.8	32.6	35.2	32.7
	Secondary	48.6	44.9	47.3	47.0	48.8
	Higher	19.1	21.3	20.1	17.8	18.5
Income sufficiency†	Sufficient	62.7	64.9	59.5	62.7	61.4
	Not sufficient	37.3	35.1	40.5	37.3	38.6
Comorbidity	0	29.1	31.6	33.8	28.8	31.2
	1–2	42.8	37.6	41.1	37.9	43.6
	3+	28.1	12.6	25.1	33.3	25.2
SMAF ᶜ	Average (± SD)	-23.5 (12.0)	**-29.0 (13.8)**	-23.3 (11.7)	**-27.9 (12.7)**	-23.5 (12.5)

* Statistical difference calculated using χ² statistics for categorical variables and logistic regression for age and SMAF; values in bold are significantly different at p < 0.001.
§ Includes individuals who did not complete the interview (n = 88) at T1 but who did at T2.
† Married includes having a common law spouse. Income sufficiency: Does your income currently satisfy your needs? Sufficient = very well or adequately; Not sufficient = with some difficulty, not very well or totally inadequate. SMAF: French acronym for Functional Autonomy Measurement System; a minimum score of -87 indicates the worst health state [48].

Table 2. Health indicator proportions (%) at baseline, 12 months and 22 months*

		T0 (n = 1164)	T1 (n = 797)	T2 (n = 475)
Cognitive problems	Yes	28.6 (31.4)	24.1 (25.8)	33.3 (35.9)
	Missing	8.9 (-)	6.6 (-)	7.4 (-)
Depression	Moderate	29.1 (38.1)	21.0 (28.5)	17.7 (25.8)
	Severe	15.0 (19.7)	20.2 (27.5)	20.4 (29.8)
	Missing	23.6 (-) §	26.6 (-)	31.6 (-)
High blood pressure	Yes	27.0	27.6	37.9
	Missing	0.9	0.5	0.8
Circulatory problems	Yes	39.3	40.5	44.0
	Missing	0.9	0.9	1.1
Stroke	Yes	20.6	23.2	25.5
	Missing	0.7	0.8	0.4
Diabetes	Yes	19.0	18.7	20.0
	Missing	0.9	0.9	-
Respiratory problems	Yes	25.3	25.5	26.1
	Missing	0.6	0.1	0.6
Joint & Arthritis	Yes	50.0	51.6	56.8
	Missing	0.9	0.5	0.6
Tumor or Cancer	Yes	17.4	17.8	18.1
	Missing	0.8	0.5	0.2
Bladder problems	Yes	31.1 (34.1)	29.1 (32.1)	25.7 (28.7)
	Missing	8.8 (-)	9.4 (-)	10.5 (-)
Stomach problems	Yes	26.1	24.8	27.6
	Missing	0.5	0.6	0.2
Sensory problems	Yes	24.4	23.5	28.4
	Missing	0.4	0.4	1.1
Functional limits (U)	Yes	56.7	75.3	78.9
	Missing	0.6	0.1	1.3
Functional limits (L)	Yes	75.0	62.1	67.6
	Missing	0.9	0.1	1.3
ADL-Mobility Disability	None	31.0	29.7	22.5
	1–2 activities	38.4	32.7	35.6
	+ 2 activities	30.6	37.5	41.5
	Missing	-	-	0.4
ADL-Personal Care Disability	None	58.4	63.2	57.7
	1–2 activities	22.2	16.1	16.0
	+ 2 activities	19.4	20.7	25.9
	Missing	-	-	0.4
IADL Disability	0–2 activities	35.6	32.7	31.2
	3–4 activities	26.6	29.0	21.5
	+4 activities	37.8	38.0	46.7
	Missing	-	-	0.6

* Entries in parentheses are the proportions generated by Mplus under the MAR hypothesis; proportions for other health indicators are not affected by missing values.
§ Of these 23.6% (n = 274) missing GDS scores, 14% (n = 163) are missing by design for severely cognitively impaired individuals. Real missing scores represent 8.7% only. At T1 and T2, 18.4% (n = 147) and 13.1% (n = 62) respectively are missing by design.
ADL = Activity of daily living; IADL = Instrumental activity of daily living; U = upper limbs; L = lower limbs The upper and lower limbs distinction was made to capture variability in functional ability.

Health State Profiles

LCA models estimated for each time point suggest that a 4-class solution provides the best overall fit and explanation of the observed health indicators frequencies. The classification identified at baseline was reproduced in the T1 and T2 LCA as well as in LTA. Accordingly, we describe the four health state profiles identified at baseline. We then present results for subsequent steps with reference to that basecase.

Table 3 presents model fit statistics for LCA models fitted at baseline, T1 and T2. At baseline, increasing the number of classes improved the classification up to the fifth class (i.e. the LMR-LRT are no longer significant). Information criterion statistics suggest the four-class model best fits the data. The quality of the classification for that model is high (entropy: 0.805), with no identification problem (condition number: 0.233) and no major violation of conditional independence. LCA performed at T1 and T2 also point to 4 health status profiles.

Table 3. Model fit statistics for latent class analysis at baseline, T1, T2

	T0				T1		T2	
	2 classes	3 classes	4 classes	5 classes	4 classes	5 classes	4 classes	5 classes
Sequential model comparisons	2 vs. 1 classes	3 vs. 2 classes	4 vs. 3 classes	5 vs. 4 classes	4 vs. 3 classes	5 vs. 4 classes	4 vs. 3 classes	5 vs. 4 classes
LMR LRT								
Log-likelihood value (c+1 classes)	13265.31	13595.67	12239.68	12093.26	8366.61	8119.41	4603.94	4449.05
-2 difference in log likelihood	1468.19	2711.98	292.85	107.55	494.40	108.374	309.774	77.03
p value	0.000	0.000	0.0006	0.307	0.000	0.286	0.000	0.198
Adjusted LMR LRT	1458.80	2375.46	290.97	106.86	490.03	107.64	265.99	76.43
p value	0.000	0.000	0.0006	0.309	0.000	0.289	0.000	0.198
Information criterion								
AIC	25148.43	24483.26	24192.52	24296.97	16244.82	16348.45	9072.11	9039.11
BIC	25365.96	24493.48	24207.69	24848.37	16258.87	16858.66	9426.66	9483.32
Adjusted BIC	25229.38	24487.13	24198.16	24502.15	16249.34	16512.53	9150.57	9137.40
Entropy	0.814	0.791	0.805	0.765	0.815	0.804	0.830	0.849
Condition number	0.0029	0.309	0.233	0.0012	0.0524	0.0014	0.0301	0.022

LMR-LRT = Lo-Mendell-Rubin Likelihood Ratio Test; AIC: Akaike Information Criterion; BIC: Bayesian Information Criterion.
Condition number = ratio of largest Eigen value to the smallest Eigen value for the Fisher information matrix. Values less than 10E^{-09} indicate problem with model identification.

Table 4 presents profile-specific health indicators probabilities for the four latent classes (λ; presented as %). These probabilities express how individuals within a profile differ from those in other profiles at each time point. For example, at T0, the first two profiles are characterized by high probabilities of cognitive problems

($\lambda = 0.687$ and $\lambda = 0.858$, respectively), whereas the other two are not ($\lambda = 0.078$ and $\lambda = 0.139$). Profiles are assigned a label to substantiate these differences.

Table 4. Health indicators distribution and conditional probabilities per health profile *

	Cog&Physic Impaired			Cognitively Impaired			Physically Impaired			Relatively Healthy		
	T0	T1	T2	T0	T1	T2	T0	T1	T2	T0	T1	T2
Cognition	68.7	58.8	74.2	85.8	71.1	89.2	7.8	2.5	12.1	13.9	6.6	2.4
Depression												
Moderate	43.5	25.9	28.1	24.7	43.8	18.3	45.3	30.3	26.0	32.2	24.1	26.4
Severe	35.7	45.6	46.8	8.3	12.8	12.9	25.1	38.9	38.2	9.9	9.8	19.3
Hypertension	24.7	25.7	33.0	13.8	17.3	17.5	32.9	34.9	45.6	27.8	26.1	41.1
Circulation	45.9	46.5	48.7	12.7	22.5	25.8	50.8	55.4	60.5	31.4	27.1	38.0
Stroke	37.1	36.6	39.4	11.5	23.4	23.1	21.1	20.4	17.4	11.1	14.4	21.8
Respiratory	26.7	33.2	27.9	9.0	10.9	5.5	35.7	30.4	38.9	19.0	19.1	21.3
Diabetes	20.9	21.4	19.1	15.7	9.3	6.9	20.2	18.0	19.3	18.2	22.1	25.6
Joint & Arthritis	46.4	52.6	47.9	19.2	24.3	33.6	71.9	72.7	79.5	40.1	38.8	58.4
Cancer	14.0	17.0	19.4	11.9	10.2	11.5	20.0	21.5	17.1	20.3	17.7	20.6
Bladder	24.3	24.9	19.3	11.8	1.5	6.4	43.6	50.3	42.6	36.8	28.8	33.2
Gastrointestinal	23.2	23.7	27.1	16.4	7.4	12.6	32.6	32.0	32.9	24.7	25.9	30.2
Sensory	48.4	44.7	56.4	23.2	24.7	21.7	18.7	16.1	15.1	14.0	11.9	15.6
Fx Limits (U)	84.4	95.3	92.9	12.8	22.9	8.0	78.6	86.9	94.2	28.6	20.1	50.3
Fx Limits (L)	98.4	99.1	100	47.3	62.8	58.7	94.3	96.6	92.6	47.4	33.4	59.1
ADL-Mobility												
None	0.0	0.0	1.6	32.3	17.8	25	11.4	20.5	0.0	77.6	74.3	56.7
1–2 activities	14.7	7.5	8.3	57.2	54.8	59.9	60.5	49.9	49.3	22.4	23.5	41.6
+2 activities	85.3	92.5	90.0	10.4	27.5	15.1	28.1	29.6	50.7	0.0	0.0	1.8
ADL-Personal care												
None	7.1	11.0	12.0	57.9	56.5	75.1	60.8	79.5	52.2	95.1	96.1	96.2
1–2 activities	21.7	20.4	10.8	31.4	28.4	18.9	33.7	17.7	38.0	4.9	3.9	3.8
+ 2 activities	71.2	68.5	77.2	10.7	15.1	6.0	5.5	2.8	9.8	0.0	0.0	0.0
IADL												
0–2 activities	0.0	0.0	0.0	4.7	0.0	18.3	27.8	36.3	11.7	84.8	74.7	79.9
3–4 activities	2.6	10.4	1.0	20.0	25.0	9.8	55.0	49.8	53.9	13.8	23.5	18.3
+4 activities	97.4	89.6	99.0	75.4	75.0	72.0	17.2	14.0	34.3	1.4	1.8	1.8

* Entries represent profile-specific probabilities (λ) × 100 of reporting problems for the index health indicator.
ADL refers to difficulty with activities of daily living; IADL refers to difficulty with instrumental activities of daily living; Fx limits U = functional limitations with upper limbs; L = lower limbs The upper and lower limbs distinction was made to capture variability in functional ability.

All selected health indicators significantly contribute to the classification (p < 0.001). However, the pattern of relationships along the cognitive and physical dimensions best describes the profiles' distinguishing features. Severe cognitive and physical impairments characterize the first health profile. Individuals have high probabilities of cognitive disorders, chronic conditions, stroke, sensory problems, and functional limitations. Their high probabilities of disability in IADL, personal care and mobility ADL capture the severity and combined consequences of these problems. This group is labeled "Cognitively & physically impaired" (Cog&Physic-Imp) and represents 23% of the sample at baseline. The second health profile is predominantly "COGNITIVELY IMPAIRED" (Cog-Imp), with minimal physical impairments. These individuals (11.4%) report relatively low probabilities for chronic conditions and functional limitations. The likelihood for them to present with ADL disability is comparatively low but high for severe IADL disability.

Individuals in the third health status profile have the highest probabilities for chronic conditions, but no cognitive problem. They are very likely to report depression, functional limitations and mobility disability, but unlikely to require help for personal care. We labeled this group "PHYSICALLY IMPAIRED" (Physic-Imp). Finally, we found a "RELATIVELY HEALTHY" (R-Healthy) profile. It comprises older people who report comparatively less chronic conditions (circulatory problems, respiratory diseases, arthritis, depression; $p < 0.01$) and who manifest low probabilities of disability, functional limitations and cognitive disorders. The later two profiles represent, respectively, 35.6% and 29.9% of the sample at baseline.

Comparison across time points of profile-specific health indicator probabilities (Table 4) shows that the profiles revealed at T1 and T2 correspond to constellations of health problems equivalent to those observed at baseline. When we compared T1 and T2 profiles with those obtained with T1 and T2 models constrained to have baseline conditional health indicator probabilities, the concordance in how individuals are grouped reaches 90% at T1, and 88% at T2. At T1, the concordance is 96% for the Cog&Physic-Imp, 92% for the Cog-Imp, 83% ($p < 0.01$) for the Physic-Imp and 86% ($p < 0.05$) for the R-Healthy. At T2, the concordance is 83% ($p < 0.05$) for the Cog&Physic-Imp (with 13% classified as Physic-Imp), 100% for the Cog-Imp, 69% ($p < 0.001$) for the Physic-Imp (with 27.4% classified as R-Healthy), and 95.6% for the R-Healthy. These differences in classification reflect the net progression of the sample as a whole towards a more compromised health state (as seen in tables 2 and 4). Although this results in an upward shift in the "severity" of health profiles at T1 and T2, each maintained its substantive meaning. Combined, these results confirm the stability of our classification, despite mortality and LTF.

The health indicator probabilities predicted by LCA are consistent with the proportion of people who have contributed each response patterns in the actual data. Furthermore, chi-square and contingency coefficients (Table 5) for health state profiles versus disability measures, comorbidity and cognitive problems confirm both the qualitative differences between health profiles and their coherence with key measures of health status. The correlation with the comorbidity measure is much lower.

The relationships between profile membership, covariates, and distal outcomes also support the validity of the classes. Results of LCA regression models indicate that Cog&Physic-Imp individuals are significantly older compared to those classified in the Physic-Imp and R-Healthy profiles. Relative to their younger peers (64–74 years), individuals in the age groups 75–84 and 85+ are 1.6 and 2.5 times more likely to be highly disabled as opposed to being "only" physically impaired (p < 0.01). These odds increase to 2 and 4, respectively, when compared to being relatively healthy (p < 0.001). Whereas women tend to be classified in the profiles with disability or be relatively healthy, men are significantly more likely to be classified in the Physic-Imp profile (OR range: 2.18–3.2; p < 0.01), where the probability of any type of disability is comparatively low despite high probability of chronic conditions. In both profiles characterized by cognitive problems, individuals are more likely not to live alone (OR range: 3.84–9.28; p < 0.001).

Table 5. Health indicators' prevalence by health status profile at T0 (%)

		Cog&Physic Imp	Cog-Imp	Physic-Imp	R-Healthy
ADL-Mobility n = 1164	No	0	35.1	9.5	79.3
	1–2 activities	14.1	56.7	63.3	20.7
	+ 2 activities	85.9	8.2	27.2	0
	χ^2 (p-value)		955.01 (p < 0.001)		
	Adjusted C*		0.822 (p < 0.001)		
ADL-Personal care n = 1164	No	6.3	59.0	60.9	95.7
	1–2 activities	23.0	31.3	33.7	4.2
	+ 2 activities	70.7	9.7	5.3	0
	χ^2 (p-value)		770.27 (p < 0.001)		
	Adjusted C		0.773 (p < 0.001)		
IADL n = 1164	0–2 activities	0	2.2	26.2	87.3
	3–4 activities	1.1	19.4	57.5	12.4
	+4 activities	98.9	78.4	16.3	0.3
	χ^2 (p-value)		1156.42 (p < 0.001)		
	Adjusted C		0.865 (p < 0.001)		
Cognitive problems n = 1159	None	30.4	10.7	92.9	86.1
	Moderate	6.5	13.2	3.9	8.8
	Severe	63.1	76.0	3.1	5
	χ^2 (p-value)		556.05 (p < 0.001)		
	Adjusted C		0.719 (p < 0.001)		
Comorbidity n = 1158	0	25.4	71.4	11.9	36.1
	1–2	48.9	24.1	45.3	42.5
	>3	25.7	4.5	42.8	21.4
	χ^2 (p-value)		208.70 (p < 0.001)		
	Adjusted C		0.479 (p < 0.0001)		

* C = Contingency coefficient with adjustment so it reaches a maximum of 1 (i.e. C/Sqr-root of k1-/k, where k is the number of rows or column, whichever is less)
ADL refers to difficulty with activities of daily living; IADL refers to difficulty with instrumental activities of daily living

Controlling for age and gender, we found that Cog&Physic-Imp individuals are significantly more likely to die within 22-months compared to Cog-Imp (OR 2.84; p < 0.002), Physic-Imp (OR 3.27; p < 0.001) and R-Healthy individuals (OR 4.75; p < 0.001). In turn, Cog-Imp individuals are more likely to die compared to the Physic-Imp (OR 1.15; p < 0.003) and the R-Healthy (OR 1.67; p < 0.003); and Physic-Imp individuals more likely to die than the R-Healthy

(OR 1.45; p < 0.001). Finally, we find that Cog&Physic-Imp individuals are less likely to use nursing home services within 22-months compared to Cog-Imp individuals (OR 0.72; p < 0.001). In turn, individuals in the later two profiles, i.e. characterized by cognitive impairments, are more likely to use nursing home services compared to Physic-Imp individuals (OR 2.87 and 4.37, respectively; p < 0.001) and R-Healthy individuals (OR 3.18 and 4.38, respectively; p < 0.001). The likelihood of using nursing home services for Physic-Imp relative to the R-healthy is not as marked (OR 1.11; p < 0.004). These results all point in the expected direction.

Latent Transition Analyses

Table 6 presents age-controlled latent transition probabilities from baseline to T1 and from T1 to T2. Among Cog&Physic-Imp individuals, 51.4% are predicted to remain in that state and 24.9% are predicted to die within a year. Around 5% are predicted to move towards less disabled health states: 3.0% towards a Cog-Imp state and 2.3% towards a Physic-Imp state. The main gender difference lies in the increased probability of death for men and LTF for women. Among Cog-Imp individuals, women are more likely to transition towards the more disabled state (24.1%) or improve (4%), whereas men are comparatively more likely to stay in their same state (47.8% vs. 24.3%) or die (11.2% vs. 9%). The competing risk for women to be LTF is again higher. Physic-Imp individuals are characterized by more stability (63.8%) with no striking gender difference. Overall, their probability of transitioning towards more disabled states is 4.3%, and somewhat higher for men; their transition probability towards the R-healthy state is 3.4%, and somewhat higher for women. Men are twice as likely to die. Although the R-Healthy also tend to be characterized by stability (66.2%), this applies most importantly to women. Over a year, men' transition probabilities towards more compromised health states is higher despite their higher competing risk for dying or being LTF.

The patterns of transitions between T1 to T2 are similar, with quantitative differences. Overall, transition probabilities towards more disabled states are higher, and improvements less likely. A noteworthy gender difference is the increased probability of R-Healthy women to transition to the Physic-Imp or Cog-Physic-Imp states, as opposed to stability. Also, Cog-Imp women have a 35% probability of becoming highly disabled whereas men in that state appear more stable. The probability of dying across health profiles is lower, except for Physic-Imp men. These differences must be considered in light of the increased competing risk of being LTF and the 10-month interval.

Table 6. Transition probabilities *

	Cog&Physic-Imp	Cog-Imp	12 months Physic-Imp	R-Healthy	Deceased	LTF
Baseline						
Cog&Physic-Imp	**0.514**	**0.030**	**0.023**	**0.000**	**0.249**	**0.184**
Female	0.550	0.013	0.033	0.000	0.184	0.220
Male	0.502	0.032	0.025	0.000	0.284	0.156
Cog-Imp	**0.181**	**0.438**	**0.001**	**0.023**	**0.101**	**0.256**
Female	0.241	0.243	0.001	0.040	0.090	0.385
Male	0.177	0.478	0.001	0.013	0.112	0.219
Physic-Imp	**0.043**	**0.000**	**0.638**	**0.034**	**0.064**	**0.221**
Female	0.036	0.000	0.692	0.033	0.040	0.199
Male	0.042	0.000	0.683	0.017	0.079	0.179
R-Healthy	**0.046**	**0.025**	**0.041**	**0.662**	**0.071**	**0.155**
Female	0.037	0.008	0.042	0.732	0.038	0.143
Male	0.061	0.036	0.059	0.551	0.107	0.185

	Cog&Physic-Imp	Cog-Imp	22-months Physic-Imp	R-Healthy	Deceased	LTF
12-months						
Cog&Physic-Imp	**0.412**	**0.013**	**0.000**	**0.000**	**0.153**	**0.422**
Female	0.477	0.006	0.000	0.000	0.104	0.413
Male	0.402	0.019	0.000	0.000	0.196	0.384
Cog-Imp	**0.256**	**0.329**	**0.000**	**0.018**	**0.037**	**0.361**
Female	0.353	0.171	0.000	0.028	0.029	0.420
Male	0.224	0.422	0.000	0.020	0.040	0.294
Physic-Imp	**0.062**	**0.003**	**0.479**	**0.022**	**0.076**	**0.358**
Female	0.065	0.001	0.521	0.025	0.055	0.333
Male	0.070	0.005	0.363	0.030	0.132	0.399
R-Healthy	**0.062**	**0.049**	**0.101**	**0.406**	**0.013**	**0.369**
Female	0.064	0.019	0.111	0.467	0.009	0.330
Male	0.057	0.065	0.064	0.471	0.017	0.326

*The overall transitions control for age and gender. Female and male transition probabilities represent marginal probabilities, also controlling for age. LTA ran under the assumption of MI.

Discussion

Our aim was to identify a meaningful latent classification that encompasses multiple dimensions of health and captures their synergistic effect on older people's health status. We identified four homogeneous health state profiles that are stable over time and sensitive to change.

The uncovered classification has face validity. It clearly distinguishes the physical and cognitive dimensions of health. And within each of these dimensions, a qualitative distinction along the disability dimension captures the consequences of diseases and impairments [1]. These findings generally agree with classifications obtained by other methods [39,41,40,42]. In elderly populations comparable to ours [39,40], published classifications revealed more nuanced groups (i.e. 5–7 profiles) but with meanings anchored in the same dimensions as those characterizing our profiles. In samples representative of community-living older people [41,42], an additional "Healthy" profile typically emerges. Given our target population's compromised health, we did not find nor expected a healthy profile. In our sample, four latent classes were sufficient to capture health status heterogeneity while maintaining interpretability and stability over time. There is no longitudinal evidence to determine whether the additional profiles of previously published classifications possess the later qualities.

The profiles' stability over and above observed changes in the sample's overall health substantiates the validity of our classification. The profiles are robust not only in being comparable across time points but also in holding despite high mortality and LTF. The differentiated and gender-specific patterns of transition probabilities demonstrate the profiles' sensitivity to change in health states. Although most individuals tended to remain in their health state or died, we found higher probabilities of unfavorable transitions for individuals in the more compromised health states, and lower probabilities of improvements. These observations concord with studies of change in disability [7,26,25], functional limitations [6,11,10] and frailty [60]. The consistent finding, across outcome measures, is a decreased probability of recovery and an increased probability of decline or death when more deficits are reported at baseline. For the Cog-Imp profile, our findings appear consistent with the course of disability progression as cognitive problems worsen: IADL are affected first, followed by basic activities of personal care [4,6]. Yet, a shortcoming of traditional functional measures is their inadequate ability to detect cognitive impairments, particularly when scaled with items influenced by physical ability [61]. Our results show that classification into homogenous health categories unmasked the differential relationship of physical and cognitive domains with progression in disability; gender specific analyses provide further insights.

Overall, the classification has good construct validity. We observed higher mortality and older individuals in the more vulnerable groups; an increased likelihood of nursing home use for cognitively impaired profiles; gender differences in transition probabilities, as well as high coherence between class membership and individual health indicators except comorbidity. The latter is consistent with previous work showing that assessment of disease alone is a weak marker of health status in older individuals, even when indicators of disease severity are considered [62].

In considering the generalizability of our findings, it is important to keep in mind that our reference population was selected to demonstrate the value of integrated services for older people with complex care needs [46,47]. The characteristics of our sample thus closely match those of the sub-population of community-living elderly targeted by such programs [63], not those of the general population; our classification reflects their compromised health status. Evidence shows that particular groups, namely frail elderly, may be more likely than others to benefit from better integration of care [46] but identifying them remains a challenge. Application of LCA may prove useful for doing so.

Three other issues deserve discussion. The first relates to missing depression and cognitive scores. Firstly, depression scores are mainly missing by design for the cognitively impaired. Assuming a MAR mechanism for depression did not significantly change how individuals are grouped yet we cannot exclude misclassification for individuals who also have a missing cognition score. Secondly, despite the known association between depressive symptoms and cognitive impairments [64], we could not assess the effect of our MAR assumption on subsequent cognitive decline.

The second issue pertains to attrition. LTF individuals did not differ on sociodemographic characteristics and most health indicators at baseline but differed on mobility disability. Combined with our inability to control for unobserved individuals effects (e.g. lifestyle, social support), this means that we cannot exclude a selection bias. Yet, there is no significant difference between profiles in the proportions of LTF individuals at T1 (χ^2:6.317; p = 0.097) or T2 (χ^2:2.544; p = 0.467). Moreover, the high concordance in how individuals are grouped confirmed the classification's stability overtime, which also points to its robustness to the competing risk of being LTF. For the transition analyses, we captured LTF individuals through an absorbing state. This provision does not inform us on the effect of change in health status on attrition—or vice versa. Nevertheless, it deals with the potential attrition biases introduced in transition studies when these individuals are excluded.

Conversely, a mortality bias is unlikely: we recorded death using administrative databases and captured these transitions through an absorbing state. Excluding deceased individuals would have yielded a healthier sample, overestimated stability and recovery, and underestimated progression relative to a more representative sample. Our modeling approach avoids this bias without having to modify the indicators, run separate analyses or use imputation techniques—all common shortcomings in geriatric studies [65,66].

Thirdly, we performed the transition analyses in two steps to avoid convergence problems due to the large number of missing data patterns. This approach is not as powerful as performing one LTA on 3 time points but it allowed us to account for the competing risks of death and being LTF in the same analyses. Moreover, because we constrained health profiles to have the same meaning across time points, this two-step strategy should yield valid transition patterns. To be sure, concurrent information on transition, death and LTF probabilities by health status provides critical information for the planning of longitudinal studies of aging.

Despite those limits, our work tackles a core challenge of gerontology research by making the multidimensional and dynamic nature of older people's health status computationally tractable. LCA capture multiple dimensions of health;

reveal the smallest number of health profiles that can explain away the associations among observed health dimensions; and makes no assumption about the distribution of health indicators or their relationships other than that of local independence [34]. On these methodological grounds, LCA supersedes classical statistical models by eliminating part of the endogeneity bias [67] introduced in multivariable modeling when indicators of diseases, cognition and disability enter in the model simultaneously. Dealing with this problem is even more pressing when measuring the dynamic of health status, which implies concomitant and interrelated changes in various factors over time [68]. LTA provides a useful empirical heuristic for studying this complex process because the measurement model is specifically developed for dynamic variables as an outgrowth of substantive theory [33,69].

Conclusion

In his seminal paper on the compression of morbidity, Fries argues that the means for affecting positive change in an aging population are to be found in the variability of the population, as well as in the average values [70]. Our study presents some means to identify and quantify inter-individual variability in health status. Notably, the important weight of the cognitive dimension in explaining this variability and transitions along the disability dimension underscores the importance of moving beyond "simple" functional measures if we are to comprehend the dynamic of elderly people health and social needs. The combination of chronic conditions, cognition and disability items for our LCA finds a parallel in the approaches used to develop the Frailty Index [71] and Clinical Frailty Scale [72]. Compared to our profiles, the former continuous measures of health status provide finer gradations likely to be pertinent to clinical practice and aging research. Conversely, our approach may be unwieldy for clinical use but finds its application at organization and policy levels where many issues call for classification of individuals into pragmatically meaningful groups. Econometric modeling has already demonstrated the sensitivity of such classifications to differences, and changes, in available patterns of health and social services in specific milieu [40,73]. Applications of LCA and LTA to larger, more representative samples are needed to confirm our findings and expand the methodological underpinnings of these approaches to study health status in older populations.

Abbreviations

ADL: Activity of Daily Living; AIC: Akaike Information Criteria; BIC: Bayesian Information Criteria; Cog-Imp: Cognitively Impaired; Cog&Physic-Imp:

Cognitively and Physically Impaired; GDS: Geriatric Depression Scale; GoM: Grade of Membership; IADL: Instrumental Activity of Daily Living; LCA: Latent Class Analysis; LMR-LRT: Lo-Mendell-Rubin Likelihood Ratio Test; LTF: Lost to Follow-up; LTA: Latent Transition Analysis; MAR: Missing at Random; Physic-Imp: Physically Impaired; R-Healthy: Relatively Healthy; SIPA: System of Integrated Care for Older People.

Competing Interests

The authors declare that they have no competing interests.

Authors' Contributions

LL conceived and designed the study, performed the statistical analyses, drafted and revised the manuscript. FB made substantial contributions to the conception and design of the study, acquisition of the original data and interpretation of the data. HB made substantial contributions to the acquisition of the original data. FB, HB and JA all critically revised the manuscript. All authors read and approved the final manuscript.

Acknowledgements

Special thanks to John Fletcher for his precious help with data preparation. This research was funded through a PhD dissertation grant supplied to the first author by the Quebec Network for Research on Aging.

References

1. Verbrugge LM, Jette AM: The disablement process. Soc Sci Med 1994, 38(1):1–14.

2. Fried LP, Ferrucci L, Darer J, Williamson JD, Anderson G: Untangling the concepts of disability, frailty, and comorbidity: implications for improved targeting and care. J Gerontol A Biol Sci Med Sci 2004, 59(3):255–263.

3. Hogan DB, MacKnight C, Bergman H, Steering Committee CIoFaA: Models, definitions, and criteria of frailty. Aging Clin Exp Res 2003, 15(3 Suppl):1–29.

4. Barberger-Gateau P, Alioum A, Peres K, Regnault A, Fabrigoule C, Nikulin M, Dartigues JF: The contribution of dementia to the disablement process and modifying factors. Dement Geriatr Cogn Disord 2004, 18(3–4):330–337.

5. Stuck AE, Walthert JM, Nikolaus T, Bula CJ, Hohmann C, Beck JC: Risk factors for functional status decline in community-living elderly people: a systematic literature review. [see comment]. Soc Sci Med 1999, 48(4):445–469.

6. Deeg DJ: Longitudinal characterization of course types of functional limitations. Disabil Rehabil 2005, 27(5):253–261.

7. Hardy SE, Dubin JA, Holford TR, Gill TM: Transitions between states of disability and independence among older persons. Am J Epidemiol 2005, 161(6):575–584.

8. Peres K, Verret C, Alioum A, Barberger-Gateau P: The disablement process: factors associated with progression of disability and recovery in French elderly people. Disabil Rehabil 2005, 27(5):263–276.

9. Mendes de Leon CF, Glass TA, Beckett LA, Seeman TE, Evans DA, Berkman LF: Social networks and disability transitions across eight intervals of yearly data in the New Haven EPESE. J Gerontol B Psychol Sci Soc Sci 1999, 54(3):S162–172.

10. Anderson RT, James MK, Miller ME, Worley AS, Longino CFJ: The timing of change: patterns in transitions in functional status among elderly persons. J Gerontol B Psychol Sci Soc Sci 1998, 53(1):S17–S27.

11. Béland F, Zunzunegui M-V: Predictors of functional status in older people living at home. Age Ageing 1999, 28(2):153–159.

12. Hébert R, Brayne C, Spiegelhalter D: Factors associated with functional decline and improvement in a very elderly community-dwelling population. Am J Epidemiol 1999, 150(5):501–510.

13. Comijs HC, Dik MG, Aartsen MJ, Deeg DJ, Jonker C: The impact of change in cognitive functioning and cognitive decline on disability, well-being, and the use of healthcare services in older persons. Results of Longitudinal Aging Study Amsterdam. Dement Geriatr Cogn Disord 2005, 19(5–6):316–323.

14. Leveille SG, Penninx BW, Melzer D, Izmirlian G, Guralnik JM: Sex differences in the prevalence of mobility disability in old age: the dynamics of incidence, recovery, and mortality. J Gerontol B Psychol Sci Soc Sci 2000, 55(1):S41–50.

15. Lamarca R, Ferrer M, Andersen PK, Liestol K, Keiding N, Alonso J: A changing relationship between disability and survival in the elderly population: differences by age. [erratum appears in J Clin Epidemiol. 2004 Mar;57(3):324]. J Clin Epidemiol 2003, 56(12):1192–1201.

16. Aguero-Torres H, von Strauss E, Viitanen M, Winblad B, Fratiglioni L: Institutionalization in the elderly: the role of chronic diseases and dementia.

Cross-sectional and longitudinal data from a population-based study. J Clin Epidemiol 2001, 54(8):795–801.

17. Wolff JL, Starfield B, Anderson G: Prevalence, expenditures, and complications of multiple chronic conditions in the elderly. Arch Intern Med 2002, 162(20):2269–2276.

18. Himelhoch S, Weller WE, Wu AW, Anderson GF, Cooper LA: Chronic medical illness, depression, and use of acute medical services among Medicare beneficiaries. Medical Care 2004, 42(6):512–521.

19. Fried TR, Bradley EH, Williams CS, Tinetti ME: Functional disability and health care expenditures for older persons. Arch Intern Med 2001, 161(21):2602–2607.

20. Borrayo EA, Salmon JR, Polivka L, Dunlop BD: Utilization across the continuum of long-term care services. Gerontologist 2002, 42(5):603–612.

21. Walsh EG, Wu B, Mitchell JB, Berkmann LF: Cognitive function and acute care utilization. J Gerontol B Psychol Sci Soc Sci 2003, 58(1):S38–S49.

22. Crimmins EM: Trends in the health of the elderly. Annu Rev Public Health 2004, 25:79–98.

23. Parker MG, Thorslund M: Health trends in the elderly population: getting better and getting worse. Gerontologist 2007, 47(2):150–158.

24. Ferrucci L, Guralnik JM, Simonsick E, Salive ME, Corti C, Langlois J: Progressive versus catastrophic disability: a longitudinal view of the disablement process. J Gerontol A Biol Sci Med Sci 1996, 51(3):M123–130.

25. Wolinsky FD, Armbrecht ES, Wyrwich KW: Rethinking functional limitation pathways. Gerontologist 2000, 40(2):137–146.

26. Romoren TI, Blekeseaune M: Trajectories of disability among the oldest old. Journal of Aging & Health 2003, 15(3):548–566.

27. Lunney JR, Lynn J, Foley DJ, Lipson S, Guralnik JM: Patterns of functional decline at the end of life. JAMA 2003, 289(18):2387–2392.

28. Lubke GH, Muthen B: Investigating population heterogeneity with factor mixture models. Psychol Methods 2005, 10(1):21–39.

29. Muthén B: Beyond SEM: General latent variable modeling. Behaviormetrica 2002, 29(1):81–117.

30. McLachlan GJ, Peel D: Finite Mixture Models. New York, Toronto: Wiley; 2000.

31. Muthén B, Muthén LK: Integrating person-centered and variable-centered analyses: growth mixture modeling with latent trajectory classes. Alcohol Clin Exp Res 2000, 24(6):882–891.

32. Madigson J, Vermunt JK: Latent class models for clustering: A comparison with K-means. Canadian Journal of Marketing Research 2002, 20:37–44.

33. Reboussin BA, Liang KY, Reboussin DM: Estimating equations for a latent transition model with multiple discrete indicators. Biometrics 1999, 55(3):839–845.

34. Hagenaars J, McCutcheon AL: Applied Latent Class Analysis. Cambridge: Cambridge University Press; 2002.

35. Bandeen-Roche K, Xue QL, Ferrucci L, Walston J, Guralnik JM, Chaves P, Zeger SL, Fried LP: Phenotype of frailty: characterization in the women's health and aging studies. J Gerontol A Biol Sci Med Sci 2006, 61(3):262–266.

36. Bandeen-Roche K, Miglioretti DL, Zeger SL, Rathouz PJ: Latent variable regression for multiple discrete outcome. J Am Stat Assoc 1997, 92(440):1375–1386.

37. Moran M, Walsh C, Lynch A, Coen RF, Coakley D, Lawlor BA: Syndromes of behavioral and psychological symptoms in mild Alzheimer's disease. Int J Geriatr Psychiatry 2004, 19(4):359–364.

38. Manton KG, Woodbury MA: Grade of Membership generalizations and aging research. Exp Aging Res 1991, 17(4):217–226.

39. McNamee P: A comparison of the grade of membership measure with alternative health indicators in explaining costs for older people. Health Econ 2004, 13(4):379–395.

40. Wieland D, Lamb V, Wang H, Sutton S, Eleazer GP, Egbert J: Participants in the Program of All-Inclusive Care for the Elderly (PACE) demonstration: developing disease-impairment-disability profiles. Gerontologist 2000, 40(2):218–227.

41. Portrait F, Lindeboom M, Deeg D: Health and mortality of the elderly: the grade of membership method, classification and determination. Health Econ 1999, 8(5):441–457.

42. Berkman L, Singer B, Manton K: Black/white differences in health status and mortality among the elderly. Demography 1989, 26(4):661–678.

43. Erosheva EA: Latent class representation of the grade of membership model. Seatle: University of Washington; 2006.

44. Melzer D, Izmirlian G, Leveille SG, Guralnik JM: Educational differences in the prevalence of mobility disability in old age: the dynamics of incidence,

mortality, and recovery. J Gerontol B Psychol Sci Soc Sci 2001, 56(5):S294–301.

45. Beckett LA, Brock DB, Lemke JH, Mendes de Leon CF, Guralnik JM, Fillenbaum GG, Branch LG, Wetle TT, Evans DA: Analysis of change in self-reported physical function among older persons in four population studies. Am J Epidemiol 1996, 143(8):766–778.

46. Béland F, Bergman H, Lebel P, Clarfield AM, Tousignant P, Contandriopoulos A-P, Dallaire L: A system of integrated care for older persons with disabilities in Canada: results from a randomized controlled trial. J Gerontol A Biol Sci Med Sci 2006, 61(4):367–373.

47. Bergman H, Béland F, Lebel P, Contandriopoulos AP, Tousignant P, Brunelle Y, Kaufman T, Leibovich E, Rodriguez R, Clarfield M: Care for Canada's frail elderly population: fragmentation or integration? CMAJ 1997, 157(8):1116–1121.

48. Hebert R, Carrier R, Bilodeau A: The Functional Autonomy Measurement System (SMAF): description and validation of an instrument for the measurement of handicaps. Age Ageing 1988, 17(5):293–302.

49. Pfeiffer E: A short portable mental status questionnaire for the assessment of organic brain deficit in elderly patients. J Am Geriatr Soc 1975, 23(10):433–441.

50. Yesavage JA, Brink TL, Rose TL, Lum O, Huang V, Adey M, Leirer VO: Development and validation of a geriatric depression screening scale: a preliminary report. J Psychiatr Res 1982, 17(1):37–49.

51. Nagi SZ: An epidemiology of disability among adults in the United States. Milbank Mem Fund Q Health Soc 1976, 54(4):439–467.

52. Mahoney FI, Barthel DW: Functional Evaluation: The Barthel Index. Md State Med J 1965, 14:61–65.

53. Fillenbaum GG, Smyer MA: The development, validity, and reliability of the OARS multidimensional functional assessment questionnaire. J Gerontol 1981, 36(4):428–434.

54. Muthén L, Muthén B: Mplus: statistical analysis with latent variables. Los Angeles: Muthén & Muthén; 1998.

55. Lazerfeld PF, Henry NW: Latent structure analysis. Boston: Houghton Mifflin; 1968.

56. Rubin RJA, Rubin DB: Statistical analysis with missing data. Second edition. Wiley Interscience; 2002.

57. Lo Y, Mendell NR, Rubin DB: Testing the number of components in a normal mixture. Biometrika 2001, 88:767–778.

58. Collins LM, Flaherty BP: Latent class models fongitudinal data. In Applied latent class analysis. Edited by: Hagenaars JA, McCutcheon AL. Cambridge: Cambridge University Press; 2002:287–303.

59. Nylund KL: Latent transition analysis: Modeling extensions and an application to peer victimization. PhD dissertation. Los Angeles: University of California; 2007.

60. Gill TM, Gahbauer EA, Allore HG, Han L: Transitions between frailty states among community-living older persons. Arch Intern Med 2006, 166(4):418–423.

61. Johnson RJ, Wolinsky FD: The structure of health status among older adults: disease, disability, functional limitation, and perceived health. J Health Soc Behav 1993, 34(2):105–121.

62. Hogan DB, Ebly EM, Fung TS: Disease, disability, and age in cognitively intact seniors: results from the Canadian Study of Health and Aging. J Gerontol A Biol Sci Med Sci 1999, 54(2):M77–82.

63. Johri M, Béland F, Bergman H: International experiments in integrated care for the elderly: a synthesis of the evidence. International Journal of Geriatric Psychiatry 2003, 18(3):222–235.

64. Ganguli M, Du Y, Dodge HH, Ratcliff GG, Chang CC: Depressive symptoms and cognitive decline in late life: a prospective epidemiological study. Arch Gen Psychiatry 2006, 63(2):153–160.

65. Brogan DJ, Haber M, Kutner NG: Functional decline among older adults: comparing a chronic disease cohort and controls when mortality rates are markedly different. J Clin Epidemiol 2000, 53(8):847–851.

66. Diehr P, Johnson LL, Patrick DL, Psaty B: Methods for incorporating death into health-related variables in longitudinal studies. J Clin Epidemiol 2005, 58(11):1115–1124.

67. Berg GD, Mansley EC: Endogeneity bias in the absence of unobserved heterogeneity. Ann Epidemiol 2004, 14(8):561–565.

68. Rockwood K: What would make a definition of frailty successful? Age Ageing 2005, 34(5):432–434.

69. Collins LM: The measurement of dynamic latent variables in longitudinal aging research: quantifying adult development. Gerodontology 1990, 9(4):13–20.

70. Fries JF: The compression of morbidity. Milbank Mem Fund Q Health Soc 1983, 61(3):397–419.

71. Mitnitski AB, Mogilner AJ, MacKnight C, Rockwood K: The mortality rate as a function of accumulated deficits in a frailty index. Mech Ageing Dev 2002, 123(11):1457–1460.

72. Rockwood K, Song X, MacKnight C, Bergman H, Hogan DB, McDowell I, Mitnitski A: A global clinical measure of fitness and frailty in elderly people. CMAJ 2005, 173(5):489–495.

73. Lafortune L, Béland F, Bergman H, Ankri J: Health state profiles and service utilization in community-living elderly. Medical Care 2009, in press.

CITATION

Effectiveness of a Mobile Smoking Cessation Service in Reaching Elderly Smokers and Predictors of Quitting

Abu Saleh M. Abdullah, Tai-Hing Lam, Steve K. K. Chan,
Gabriel M. Leung, Iris Chi, Winnie W. N. Ho
and Sophia S. C. Chan

ABSTRACT

Background

Different smoking cessation programs have been developed in the last decade but utilization by the elderly is low. We evaluated a pilot mobile smoking cessation service for the Chinese elderly in Hong Kong and identified predictors of quitting.

Methods

The Mobile Smoking Cessation Programme (MSCP) targeted elderly smokers (aged 60 or above) and provided service in a place that was convenient to the elderly. Trained counselors provided individual counseling and 4 week's free supply of nicotine replacement therapy (NRT). Follow up was arranged at 1 month by face-to-face and at 3 and 6 months by telephone plus urinary cotinine validation. A structured record sheet was used for data collection. The service was evaluated in terms of process, outcome and cost.

Results

102 governmental and non-governmental social service units and private residential homes for the elderly participated in the MSCP. We held 90 health talks with 3266 elderly (1140 smokers and 2126 non-smokers) attended. Of the 1140 smokers, 365 (32%) received intensive smoking cessation service. By intention-to-treat, the validated 7 day point prevalence quit rate was 20.3% (95% confidence interval: 16.2%–24.8%). Smoking less than 11 cigarettes per day and being adherent to NRT for 4 weeks or more were significant predictors of quitting. The average cost per contact was US$54 (smokers only); per smoker with counseling: US$168; per self-reported quitter: US$594; and per cotinine validated quitter: US$827.

Conclusion

This mobile smoking cessation program was acceptable to elderly Chinese smokers, with quit rate comparable to other comprehensive programs in the West. A mobile clinic is a promising model to reach the elderly and probably other hard to reach smokers.

Background

Cigarette smoking is the leading cause of premature mortality among older persons in Hong Kong [1,2] and elsewhere.[3] Many common diseases among older people are caused by tobacco use.[4,5] The World Bank has estimated that five hundred million people alive today will eventually be killed by tobacco.[6] Worldwide trends in mortality attributable to smoking will increase in both older men and women.[7]

The prevalence of cigarette smoking was 14% in Hong Kong people aged over 60 in the 1998 General Household Survey and there were a total of 129,600 older smokers at that time [8]. A higher prevalence of current smoking was reported in studies conducted among older people aged 60 and over by the Hong Kong Society for the Aged (19%) [9] and by the University of Hong Kong (19%) [10].

In Hong Kong, there has been a lack of smoking cessation services and there is no evidence whether such services could help older people to quit smoking. Nevertheless, about 15% of smokers aged 60 and older wanted to quit within the next 6 months [10] and evidence elsewhere shows that older smokers are more likely to be successful in quitting attempts than smokers aged 35–64.[11] This paper reports the acceptance and benefits of smoking cessation services among older smokers in Hong Kong.

Although a variety of smoking cessation programs have been developed in the last decade, utilization by the elderly is low.[12] A frequently cited reason is inconvenience in reaching the services, [13] because the service locations are not near to their living environment. On the other hand, many elderly people live alone or in elderly homes and traveling to a smoking cessation clinic far away is not practicable. A more accessible service should encourage more people to utilize the service and benefit from it. Mobile clinical service was useful in reaching the hard to reach population in other settings.[14] However, we found no such reports in the literature that targeted elderly smokers with a mobile smoking cessation service. We examined the effectiveness of a mobile smoking cessation service in reaching elderly Chinese smokers in Hong Kong and identified predictors of quitting. We aimed to answer four specific questions: (1) Would Chinese elderly smokers participate in a mobile smoking cessation program (MSCP)? (2) Is the program effective in promoting smoking cessation among elderly Chinese smokers? (3) What are the predictors of quitting among the Chinese elderly? (4) What are the costs of the program?

Methods

Mobile Smoking Cessation Programme (MSCP)

The Departments of Community Medicine and Nursing Studies and School of Public Health of the University of Hong Kong with funding from the Elderly Commission, Government of the Hong Kong Special Administration Region developed a Mobile Smoking Cessation Programme (MSCP) to reach elderly smokers (aged 60 or above). The MSCP started in November 2002 and continued till September 2004. The mobile team included a coordinator and 3 trained smoking cessation counselors. These counselors were registered nurses and had completed satisfactorily a smoking cessation counseling training program with assessment by written and practical examinations. The mobile team was supported by a project director specialized in smoking cessation. The MSCP included health talks, assessment of clients' smoking status and nicotine dependence level, provision of individually tailored behavioral counseling, prescription of nicotine replacement therapy, NRT (patch only), and arrangements for follow up (telephone and on-site). We recommended subjects to use NRT for 8 weeks and gave out free supply for the first 4 weeks. We followed social cognitive theory (SCT) to design the intervention of the program. SCT explains why a behavior occurs positing that there is a three-way reciprocal interaction between the environment, the individual and a behavior [15]. The SCT has been successfully applied in several clinical and community based studies of smoking cessation [16,17].

Target Population and Recruitment

The eligible subjects were current smokers who were attending 102 social service units or private residential homes (both Government and non-Government) throughout Hong Kong to receive health services or elderly care. All these service units were specialized in service provision for the elderly. We invited a social worker, if available, in each of these centers to act as our contact person. These social workers were trained by us on basic smoking cessation skills, the details of which were described elsewhere.[18,19] The respective social worker from each of these centers identified elderly smokers within their service areas and confirmed a date for the visit of the MSCP team. Elderly non-smokers and family members of elderly smokers who were interested to know about smoking and health issues were also encouraged to attend the health talks, but were not included in the analysis. Most of the private homes did not have social workers and some clients were referred by other staff members.

The health talks were organized in the premises of the social service centers. Each health talk continued for about an hour. The nurse counselors from the mobile team delivered pre-designed talks (about 30 minutes) and discussed on different aspects of smoking cessation. The content included the harms of tobacco use (both active and passive smoking), benefits of quitting smoking and tips for quitting. Informal discussion, experience sharing, and a question and answer session were conducted during the second half of the talk, including brief information about the MSCP.

After the health talks, all smokers were asked to enrol for an intensive smoking cessation service, including cognitive-behavoural stage matched counseling and use of NRT, which lasted for about half an hour, provision for free NRT supply for 4 weeks and follow up arrangements. Those who consented to participate were included in the program (Figure 1). Ethical approval for this study was obtained from the Ethics Committee of the Faculty of Medicine, the University of Hong Kong.

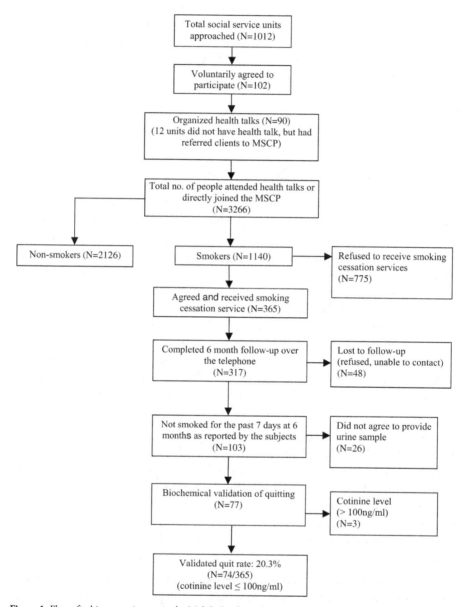

Figure 1. Flow of subject recruitment in the Mobile Smoking Cessation Programme (MSCP).

Data Collection

A structured questionnaire was used to collect data at baseline and at 1, 3 and 6 months. Details of the questionnaire were described elsewhere.[20,21] Briefly, the questionnaire included demographic information, smoking and quitting history, nicotine dependence level, and perceived motivation (self efficacy rating), confidence and difficulty of quitting smoking (perceived barriers). The client's satisfaction towards different aspects of MSCP was assessed at 3 month follow up, including counseling received in the center and/or over the telephone, and follow up arrangements. Counselors' satisfaction was assessed from the meeting notes, every 2 months. We asked the counselors questions about the process of service delivery, operational guidelines, work load, availability of support and job satisfaction. Satisfaction was rated on a four-point scale (very satisfactory, satisfactory, unsatisfactory and very unsatisfactory).

Follow Up Assessment

Follow up assessment and relapse prevention counseling was carried out at 1 month post-counseling by face-to-face and at 3 and 6 months by telephone for all who attended the MSCP. Both face-to-face and telephone follow up lasted for an average of 20 minutes. We also made a follow up call (lasting for 2–5 minutes) at 1 week to assess whether the elderly were having any problem with NRT use and encourage further use. The nurse counselors carried out follow up interviews. At 6 months, those who stopped smoking (not smoking for 7 days or more preceding the follow up interview) were invited to attend the nearest social service unit for biochemical validation (by measuring urinary cotinine level). For those who could not attend for validation, our research assistant visited the subjects to collect urine samples.

The Evaluation

The program was evaluated in terms of process, outcome and cost. The process evaluation comprised documentation of comments or suggestions from the nurse counselors and other members of the MSCP team, and satisfaction ratings of the subjects regarding counseling and follow up arrangements. The main outcome evaluation was based on the validated 7 day point prevalence quit rate at 6 months, and we also reported several other quitting outcomes as secondary outcomes. Other process outcomes included number of MSCP organized, number of elderly members attended the health talks, number of elderly smokers who had received intensive smoking cessation service (counseling and/or NRT) and contributions to smoking cessation research in Hong Kong and elsewhere. We calculated all the relevant costs (staff salary, stationery, NRT, travel for the mobile team, and urine cotinine tests) and divided the total costs by the total number of attendees/quitters and compared them with other relevant programs.

Data Analysis

Data were analyzed using SPSS for Windows, version 10.0. The baseline characteristics of clients were described. The prevalence of quitting in the MSCP was compared with those of studies elsewhere. The characteristics of quitters and non-quitters were compared by chi-square test. The variables which were significant in the bivariate analysis were tested by forward stepwise logistic regression modeling to identify predictors for quitting and to estimate adjusted odds ratios and 95% confidence intervals (CI).

Quitting smoking was defined as not smoking any cigarettes during the past 7 days at 6 month follow up as reported by the subjects and confirmed by urine cotinine validation (7 day point prevalence validated quit rate). All subjects who could not be contacted at 6 month follow up and those who failed the validation test (a urinary cotinine level > 100 ng/ml) [22] were considered as smokers (i.e. had no change from baseline) based on intention-to-treat analysis (a conservative approach). As secondary outcomes, we also measured 7 day point prevalence quit rate at 6 months without validation (defined as not smoking during the 7 days preceding the 6 month follow up), 24 hour point prevalence quit rate at 6 months without validation (defined as not smoking during the 24 hours preceding the 6 month follow up), continuous abstinence rate (abstinence from tobacco smoking continuously for the whole period prior to the interview at 6 months) [23] and reduction in smoking rate (reduction of the amount smoked by at least 50% at 6 month follow up).[24]

Results

Utilisation and Process Evaluation

During the study period, we contacted a total of 1012 governmental and non-governmental social service units and private residential homes for the elderly, and 102 participated in the MSCP. We organized 90 health talks (12 units did not require health talk but recruited smokers to receive our intensive counseling) with 1140 smokers attended. Of the 1140 smokers, 365 (32%) agreed to receive our smoking cessation service. The demographic, lifestyle and quitting characteristics of these subjects are shown in Table 1.

Table 1. Demographic, lifestyle, smoking and quitting related factors of 365 smokers to the mobile smoking cessation program (MSCP) participants (n = 365)

Characteristics	%
Demographics:	
Gender	
Male	71
Female	29
Occupational status	
Retired	86
Unemployed	6
Employed	1
Homemakers	7
Age	
60–69	21
70–79	41
80 or above	38
Educational attainment	
No formal education	39
Primary school	46
Secondary or above	15
Marital Status	
Single	15
Married	44
Divorced, separated and widowed	41
Tobacco use related:	
Daily cigarette consumption	
≤ 10	70
> 10	30
Age started smoking	
Under 20	62
20 or above	38
Nicotine Dependency level [a]	
Mild	65
Moderate	20
Severe	15
Number of other smokers in household	
Nil	71
1 or more	29
Smoking status of spouse	
No spouse/spouse not smoker	90
Spouse is smoker	10
Quitting History:	
Number of previous quitting attempt(s)	
Nil	41
1 attempt or more	59
Length of abstinence in the last quitting attempt	
Less than a day or not at all	5
> = 1 day	95
NRT related:	
Use of NRT for at least 1 day[b]	
No	23
Yes	77
Adherence to use NRT for 4 weeks or more [b, c]	
No	62
Yes	38
Other factors:	
Perceived importance on quitting (mean score = 73)[d]	
Less important (< mean)	48
More important (> = mean)	52
Perceived difficulties on quitting (mean score = 55)[e]	
Less difficult (< mean)	53
More difficult (> = mean)	47
Perceived confidence on quitting (mean score = 67)[f]	
Less confident (< mean)	42
More confident (> = mean)	58
Alcohol consumption[g]	
Regular user and occasional users	18
Never or rarely drink	82

Note: Total percentage may be more or less than 100 due to rounding of the figures.
[a] Nicotine dependency level was measured by Fagerstrom scale, then further divided into 3 levels: low (score 0–3), moderate (score 4–8) and severe (score 6–10).; n = 354, 11 missing.
[b] Included only those who were given NRT (n = 255).
[c] Subjects who reported using nicotine replacement therapy (NRT) for at least 4 weeks during the 3 month follow up was defined as adherent to NRT use.
[d] Total n = 352, 13 missing
[e] Total n = 356, 9 missing
[f] Total n = 355, 10 missing
[g] Total n = 345, 20 missing

Satisfaction

More than 90% of the respondents were satisfied with the counseling service and follow up arrangements. Eighty five percent of the subjects would probably or definitely recommend this program to other smokers. However, 80% of the subjects were asking for free full course of NRT, which was not possible as our funding could only support 4 week free supply.

Structured interview was also conducted with all the four counselors and the coordinator, which showed that all the staff members were satisfied with the counseling process, supervisory support and workload. The main barrier was the travel time and communication difficulties with the elderly. Repeated reminders were necessary as many elderly often forgot about their follow up arrangements. No complaints about the program and no other major difficulties were encountered.

Outcome Evaluation

Primary Outcome (Urine Cotinine Validated Quit Rate)

At 6 month follow up, 48/365 (13%) subjects could not be reached (left Hong Kong, telephone number changed or refused to talk) and 103 subjects reported that they did not smoke in the 7 days preceding the 6 month follow up (Table 2). All the 103 self-reported quitters were invited for biochemical validation with urine cotinine test (Nicalert test) and 77 gave urine samples. Three of these had a cotinine level of > 100 ng/ml and were considered as smokers. By intention-to-treat analysis, the validated (cotinine level = < 100 ng/ml) quit rate was 20.3% (74/365) (95% confidence interval, CI: 16.2%–24.8%).

Table 2. Quitting outcome of 365 smokers in the MSCP at 6 month follow up, by intention to treat

Quit rates	N	% (95% confidence interval)
Primary outcome:		
Biochemically (urine cotinine) validated quit rate	74	20.3 (16.2 – 24.8)
Secondary outcomes (self-reported)		
7 day point prevalence quit rate	103	28.2 (23.6–32.8)
24 hours point prevalence quit rate	110	30.1(25.4–34.8)
Continuous abstinence quit rate	91	24.9 (20.5–29.4)
Had not quit but had reduced smoking by at least 50% from the baseline level	94	25.8(21.3 – 30.2)

Secondary Outcomes

At 6 month follow up, by intention-to-treat analysis, of the 365 subjects, 28.2% did not smoke any cigarettes during the 7 days prior to the interview (7 day point prevalence), 30.1% did not smoke any cigarettes during the 24 hours prior to the interview (24 hour point prevalence), and 24.9% did not smoke any cigarettes during the six months prior to the interview (6 month continuous abstinence), and25.8% reported that they did not quit but had reduced daily smoking by at least 50% (reduction rate) (Table 2).

Factors Associated with Quitting at 6 Months

With the inclusion of those who did not return for follow-up as non-quitters, we carried out bi-variate analysis of all the eighteen variables in Table 1 to identify factors associated with quitting. Nine factors were significantly associated with quitting: smoking less than 11 cigarettes per day, having made one or more serious quitting attempts in the past, being moderately or mildly dependent on nicotine, quitting for at least a day in the last quitting attempt, using NRT for at least a day in the present quitting attempt, being adherent to NRT use for 4 weeks or more, perceiving more importance and confidence on quitting, and having an exhaled carbon monoxide level of below the mean (< 10 ppm) at the first visit. Stepwise logistic regression modeling on these nine factors showed that smoking less than 11 cigarettes per day and being adherent to NRT use for 4 weeks or more were the two significant independent predictors of quitting (Table 3).

Table 3. Final logistic regression (forward stepwise) model to predict successful quitting at 6 month follow up

Independent variables	OR (95% CI)	P value
Smoking less than 11 cigarettes per day	2.63 (1.37–5.06)	< 0.01
Adhered to NRT use for 4 weeks or more	3.57 (1.95–6.55)	< 0.001

Note: OR = adjusted odds ratio; CI = Confidence interval

Costs of the MSCP

The cost of the MSCP included mainly operation cost (staff salary and stationery), cost of equipment and NRT and travel cost for the mobile team. Some hidden costs such as time spent by the program director and other members of the project team (who were involved mainly in the planning, monitoring, evaluation and research), and client costs (travel time, travel cost, cost to other family members) were not included. The total expenditure for the operation of MSCP was US$ 61,162 (Table 4). The MSCP had 1140 smoker subjects for health talks and 365 smokers for smoking cessation service. At 6 month follow up, 103 smokers did not smoke any cigarettes during the past 7 days. 74/77 of the self-reported quitters who had cotinine validation were confirmed as quitters. Therefore, the cost per contact was US$53.65 (smokers only); cost per smoker with counseling was US$167.57; and cost per self-reported quitter was US$593.81 and per cotinine validated quitter, US$826.54.

Table 4. Costs for the Mobile Smoking Cessation Programme (MSCP)

Costs	HK$
Fixed capital costs	Not costed
Stationary and equipment	1,220
MSCP service costs	
Salary of part-time smoking cessation counsellors	328,569
Salary of other part-time supporting staff	18,671
NRT costs	
Costs for 4 week free supply (50% discounted price)	109,223
Biochemical validation costs	
Cost for urine cotinine test	7,294
General expenses	
Transportation	8,622
Photocopies	3,465
Hidden cost (Time spent by the programme director and other members of the project team)	Not costed
Total cost (in HK$)	**477,064**
Total cost (in US$)	**61,162** (US$47,159 excluding NRT costs)

Note: US$1 = HK$7.8
Cost per smoker contact = US$61162/1140 = US$53.65
Cost per subject with intensive counselling = US$61162/365 = US$167.57
Cost per successful self-reported quitter (intention to quit) = US$61162/103 = US$593.81 (US$457.82, excluding NRT costs)
Cost per validated quitter (intention to quit) = US$61162/74 = US$826.54 (US$637.18, excluding NRT costs)

Discussion

Our experience and findings from the mobile smoking cessation program (MSCP) suggest feasibility and acceptance of this outreach program among the Chinese elderly. Although the project was not designed as a controlled experiment, the results suggest that mobile smoking cessation programs can be effective for the hard to reach population such as the elderly, provided that the needs and difficulties of the targeted population are addressed. While a community-based study in the United States reported that smokers aged 65 and older were least likely to use a smoking cessation program, [8] our mobile service was reasonably accepted as reflected from clients' participation and enthusiasm. It was convenient for many elderly clients who could not travel to receive services far away. Few community smoking cessation projects used cotinine validation for evaluation of quitting outcomes. Our study has the strengths in the use of cotinine validation and the high percentage (75%) of acceptance of the validation. It is worth mentioning that our cessation service was not able reach all the elderly smokers in the 102 participating social service units. Although we trained social workers in each of these units to refer smokers to our program, we did not conduct any baseline survey of all the residence in the studies social service units nor did record data about what proportion of the smokers actually attended the program. However, based on our exploratory estimate, we assume that our program reached at least 70% of the smoker population in these social service units.

Our service (individual counseling and 4 week free supply of NRT) resulted in a self-reported 6-month point prevalence quit rate (by intention-to-treat) of 28%, which was comparable with the one year point prevalence self-reported quit rate (27%) among Chinese adult smokers who attended the Hong Kong Smoking Cessation Health Center, [25] and higher than the 6 month point prevalence self-reported quit rate (14.4%) in clinic based smoking cessation services in New Zealand [26] and the United States (22%).[27] Our quit rate was also comparable to the 7-day point prevalence self-reported quit rate (29%) among American elderly (aged 65 to 74 years) who also used nicotine patch for an average of 5 weeks.[28] While our quit rates seems better than the above studies abroad [26-28], few clarifications worth noting. Our subject included a higher proportion (70%) of those who were light smokers (smoked less than 10 cigarettes daily), however, based on our review of these studies [26,27] papers, a higher proportion of subjects in other studies [26-28] were moderate or heavy smokers. The average daily consumption of our subject was 10 cigarettes per day compared to the mean number of 25.4 cigarettes per day among American smokers [27] and a median of 20 (range 1–85) among smoker in New Zealand [26].

The cost per self-reported quitter (US$458, excluding cost for NRT) was 35% higher than that in the Hong Kong Smoking Cessation Health Center, which was the first such clinic in Hong Kong (US$339, excluding cost for NRT).[22] The higher cost was mainly due to the mobile nature of the service, longer duration of counseling needed for elderly clients, and the additional costs for cotinine validation.

We found that being a light smoker (smoking less than 11 cigarettes per day) and using NRT for four weeks or more were significant independent predictors of quitting. This suggests that heavy smokers might need to be targeted with more intensive programs.[29] Efforts to increase NRT adherence are also needed to improve quit rates.[20] Provision of free NRT supply for a longer duration (full course, 8 weeks) is suggested for those who want to use but cannot afford.

A major limitation of the study is the lack of a control group to compare the elderly who participated in our MSCP with those who did not. However, our validated quit rate of 20.3% is about two times the natural self-reported quit rate (10.0%) among the elderly aged 65 or above in the US general population.[30] Moreover, participation in the study was voluntary and this might have resulted in the recruitment of more motivated smokers from the general population. On the other hand, motivation of clients to a mobile service would be lower than that among those who travel to a clinic further away. Some of our smokers could have attended the service due to the pressure and/or encouragement from other family members or social workers from the social service units, and the prohibition of smoking inside most units. It is also possible that free counseling service and offer of 4-week's supply of NRT free of charge encouraged many smokers to attend the program.

This study has important public health implications. First, to the best of our knowledge, this is the first study to report promotion of smoking cessation program through mobile service targetting the elderly smokers, and our experiences and results should be important for other smoking cessation service providers. Second, the low cost of the program suggests that a mobile service could be promoted to attract more smokers in addition to the elderly. A timetable convenient to the target clients is needed and should be publicized through the health care facilities, elderly homes and other health centers. Health care and social service providers could be motivated to identify older smokers and, if no smoking cessation services are provided in their premises, could refer them to the appropriate mobile service scheduled nearer to their residents or clients. Mass media promotional activities would increase the coverage but the cost would be high and local publicity should be more affordable. The setting up of a smoking cessation service can provide a golden opportunity for publicity. For example, we held two exhibitions, which attracted about 800 people to visit our booths and collect self-help materials. Integration of the smoking cessation service with other existing mobile service (if any) can reach more clients with shared costs. Finally, it would be useful to test the effectiveness and cost effectiveness of mobile smoking cessation service for other vulnerable population groups (such as pregnant women and young people).

Conclusion

We conclude that a mobile smoking cessation program is a feasible approach to reach elderly smokers. The quit rate is comparable to other comprehensive programs in the West. We identified several predictors of quitting smoking through the mobile program which could guide the future service provision. A mobile clinic is a promising model to reach the elderly and probably other hard to reach smokers.

Competing Interests

The authors declare that they have no competing interests.

Authors' Contributions

ASMA and THL originated the study, developed study protocol and supervised all aspects of its implementation. ASMA performed data analysis and drafted the manuscript. SKC and WNH assisted in data collection and in the analysis. THL provided in-depth comments in the earlier draft of the manuscript. GML, IC and

SCC contributed to the interpretation of data and revising the final draft of the manuscript. All authors reviewed drafts of the manuscript and approved the version to be published.

Acknowledgements

We thank Pfizer Consumer Health (Hong Kong) for providing NRT supplies with a 50% discount rate. However, Pfizer played no role in the results/data analysis/preparation or approval of the manuscript. We also thank the social service units for the elderly for their logistic support. The study was funded by the Elderly Commission Community Partnership Scheme of the Government of the Hong Kong SAR (Ref no. 01-071).

References

1. Lam TH, Ho SY, Hedley AJ, Mak KH, Peto R: Mortality and smoking in Hong Kong: case-control study of all adult deaths in 1998. Br Med J 2001, 323:361–2.

2. Lam TH, Li ZB, Ho SY, Chan WM, Ho KS, Tham MK, Cowling BJ, Schooling CM, Leung GM: Smoking, quitting and mortality in an elderly cohort of 56000 Hong Kong Chinese. Tob Control 2007, 16:182–9.

3. Janssen F, Kunst AE: The Netherlands Epidemiology and Demography Compression of Morbidity research group. Cohort patterns in mortality trends among the elderly in seven European countries, 1950–99. Int J Epidemiol 2005, 34(5):1149–59.

4. LaCroix AZ, Lanag J, Scherr P, Wallace RB, Cornoni-Huntley J, Berkman L, Curb JD, Evans D, Hennekens CH: Smoking and mortality among older men and women in three communities. N Engl J Med 1991, 324(23):1619–1625.

5. Panginini-Hill A, Hsu G: Smoking and mortality among residents of a California retirement community. Am J Public Health 1994, 84:992–5.

6. World Bank: Curbing the epidemic: Government and the Economic of Tobacco Control. Washington DC: The World Bank; 1999.

7. Peto R, Lopez AD, Borchan J, Thun M, Heath C Jr, Doll R: Mortality from smoking worldwide. Br Med J 1996, 52:12–21.

8. Census and Statistics Department, Hong Kong Government: Special Topics Report No. 20. General Household Survey 1998. Hong Kong: Government printer; 1998.

9. Leung EM, Lo MB: Social and health status of elderly people in Hong Kong. In The Health of the Elderly in Hong Kong. Edited by: Lam SK. Hong Kong: Hong Kong University Press; 1997:43–61.

10. Lam TH, Chan B, Ho SY, Chan SK: Report on Healthy Living Survey 1999. [http://www.info.gov.hk/gia/general/200004/27/0427111.htm], Department of Community Medicine, The University of Hong Kong (funded by the Hong Kong Department of Health); 1999.

11. Hatziandreu RJ, Pierce JP, Letkopoulou M: Quitting smoking in the United States in 1986. J Natl Cancer Inst 1990, 82:1402–6.

12. Brown DW, Croft JB, Schenck AP, Malarcher AM, Giles WH, Simpson RJ Jr: Inpatient smoking-cessation counseling and all-cause mortality among the elderly. Am J Prev Med 2004, 26:112–8.

13. Abdullah AS, Simon JL: Health promotion in older adults: evidence based smoking cessation programs for use in primary care settings. Geriatrics 2006, 61:30–4.

14. Bandura A: Social Learning Theory. Englewood Cliffs, NJ: Prentice Hall; 1997.

15. Miller NH, Smith PK, BeBusk RF, Sobel DS, Taylor DB: Smoking cessation in hospitalized patients. Results of a randomized trial. Arch Intern Med 1997, 157:409–15.

16. Zheng P, Guo F, Chen Y, Fu Y, Fu H: A randomized controlled trial of group intervention based on social cognitive theory for smoking cessation in China. J of Epidemiology 2007, 17(5):147–155.

17. Jin AJ, Martin D, Maberley D, Dawson KG, Seccombe DW, Beattie J: Evaluation of a mobile diabetes care telemedicine clinic serving Aboriginal communities in Northern British Columbia, Canada. Int J Circumpolar Health 2004, 63:124–8.

18. Johnston JM, Chan SSC, Chan SK, Chan SK, Woo PP, Chi I, Lam TH: Training nurses and social workers in smoking cessation counseling: a population needs assessment in Hong Kong. Prev Med 2005, 40:389–406.

19. Leung GM, Chan SSC, Johnston JM, Chan SK: Effectiveness of an elderly smoking cessation counseling training program for social workers: A longitudinal study. Chest 2007, 131:1157–65.

20. Lam TH, Abdullah AS, Chan SSC, Hedley AJ: Adherence versus quitting smoking: predictors among Chinese smokers. Psychopharmacology 2005, 177:400–8.

21. Abdullah AS, Lam TH, Chan SSC, Hedley AJ: Which smokers use the smoking cessation Quitline in Hong Kong and how effective is the Quitline? Tob Control 2004, 13:415–21.

22. Gariti P, Rosenthal DI, Lindell K, Hansen-Flaschen J, Shrager J, Lipkin C, Alterman AI, Kaiser LR: Validating a dipstick method for detecting recent smoking. Cancer Epidemiol Biomarkers Prev 2002, 11:1123–5.

23. Platt S, Tannahill A, Watson J, Fraser E: Effectiveness of antismoking telephone helpline: follow up survey. Br Med J 1997, 314:1371–5.

24. Hurt RD, Croghan GA, Wolter TD, Croghan IT, Offord KP, Williams GM, Djordjevic MV, Richie JP Jr, Jeffrey AM: Does smoking reduction result in reduction of biomarkers associated with harm? A pilot study using a nicotine inhaler. Nicotine Tob Res 2000, 2:327–36.

25. Abdullah AS, Hedley AJ, Chan SSC, Lam H: Establishment and evaluation of a smoking cessation clinic in Hong Kong: a model for the future service provider. J Public Health Med 2004, 26:239–44.

26. Town GI, Fraser P, Graham S, McSweeney W, Brockway K, Kirk R: Establishment of a smoking cessation program in primary and secondary care in Canterbury. N Z Med J 2000, 113:117–9.

27. Croghan IT, Offord KP, Evans RW, Schmidt S, Gomez-Dahl LC, Schroeder DR, Patten CA, Hurt RD: Cost-effectiveness of treating nicotine dependence: the Mayo Clinic experience. Mayo Clin Proc 1997, 72:917–24.

28. Orleans CT, Resch N, Noll E, Keintz MK, Rimer BK, Brown TV, Snedden TM: Use of transdermal nicotine in a state-level prescription plan for the elderly. A first look at 'real-world' patch users. JAMA 1994, 271:601–7.

29. Gourlay SG, Benowitz NL: The benefits of stopping smoking and the role of nicotine replacement therapy in older patients. Drugs Aging 1996, 9(1):8–23.

30. Salive ME, Cornoni-Huntley J, LaCroix AZ, Ostfeld AM, Wallace RB, Hennekens CH: Predictors of smoking cessation and relapse in older adults. Am J Public Health 1992, 82:1268–71.

CITATION

Factors Associated with Self-Rated Health in Older People Living in Institutions

Javier Damián, Roberto Pastor-Barriuso
and Emiliana Valderrama-Gama

ABSTRACT

Background

Although self-rated health has been extensively studied in community older people, its determinants have seldom been investigated in institutional settings. We carried out a cross-sectional study to describe the physical, mental, and social factors associated with self-rated health in nursing homes and other geriatric facilities.

Methods

A representative sample of 800 subjects 65 years of age and older living in 19 public and 30 private institutions of Madrid was randomly selected through stratified cluster sampling. Residents, caregivers, physicians, and nurses were interviewed by trained geriatricians using standardized instruments to assess self-rated health, chronic illnesses, functional capacity, cognitive status, depressive symptoms, vision and hearing problems, and social support.

Results

Of the 669 interviewed residents (response rate 84%), 55% rated their health as good or very good. There was no association with sex or age. Residents in private facilities and those who completed primary education had significantly better health perception. The adjusted odds ratio (95% confidence interval) for worse health perception was 1.18 (1.07–1.28) for each additional chronic condition, 2.37 (1.38–4.06) when comparing residents with moderate dependency to those functionally independent, and 10.45 (5.84–18.68) when comparing residents with moderate/severe depressive symptoms to those without symptoms. Visual problems were also associated with worse health perception. Similar results were obtained in subgroup analyses, except for inconsistencies in cognitively impaired individuals.

Conclusion

Chronic conditions, functional status, depressive symptoms and socioeconomic factors were the main determinants of perceived health among Spanish institutionalized elderly persons. Doubts remain about the proper assessment of subjective health in residents with altered cognition.

Background

Self-rated health is a complex variable that captures multiple dimensions of the relation between physical health and other personal and social characteristics. It is very consistent its capacity to predict mortality [1] and functional loss [2,3], independently of objective health, psychosocial, and demographic variables. It has also been strongly associated with successful aging [4] and evidence of biologic roots has been recently shown [5,6]. Self-rated health is very easy to obtain through a single-item question and, consequently, it is often included in health surveys and as an outcome in many studies, resulting in a large body of research. However, few studies have focused on the determinants of self-rated health in institutionalized older people. One early study found a positive association between subjective and objective health in community older persons, but not in those living in institutions [7]. Another study confirmed the association of self-rated health with mortality in institutionalized Chinese elderly, but did not offer relevant information on its determinants [8].

The singular physical, psychosocial, and environmental characteristics of institutionalized older people justify the study of determinants of subjective health in these populations. The present study aims at identifying the principal determinants of self-rated health among a representative sample of older people living in institutions in Madrid, Spain. We also aimed at exploring potential effect modifications by sex, type of facility, and cognitive status, since differences in socioeconomic status and delivery of care among residents of public and private facilities, as well as the ability of cognitively impaired subjects to make self evaluations, might condition the effect of other variables on subjective health.

Methods

Population and Selection of Participants

Between June 1998 and June 1999, we conducted a survey based on a probabilistic sample of residents 65 years of age and older of public and private nursing homes in the city of Madrid and a surrounding area of 35 km. Study participants were selected through stratified cluster sampling. Sampling was stratified by the funding of the institution: one stratum included 22 public and 25 concerted (privately owned but publicly funded) nursing homes, and the other stratum included 139 private institutions. In the first sampling stage, we sampled 19 public/concerted and 30 private institutions with probability proportional to its size. In the second stage, we randomly sampled 10 men and 10 women within each selected public/ concerted facility (6 large facilities comprised two clusters each, thus selecting 20 men and 20 women), yielding a sample of 500 residents from public/concerted nursing homes. Similarly, we sampled 5 men and 5 women from each selected private nursing home, totalling 300 residents in this stratum. As a result of this selection procedure, residents of public institutions and men were over-sampled in order to improve precision in these subgroups.

The institutional review board of the "Carlos III" Health Institute approved the study. Informed consent was obtained from all subjects (or next of kin) studied.

Data Collection

Trained geriatricians or residents in geriatrics used structured questionnaires to collect information by interviewing the residents, their main caregivers, and the physician (or nurse).

Socio-Demographic Variables

Each resident's sex, age, and educational level (less than primary, primary -8 years-, and secondary -12 years- or more) were obtained by interview.

Self-Rated Health

Residents were asked about their health through the question: "In general terms, how would you describe your health: very good, good, fair, poor, or very poor?" No health-related questions that could influence the response were made before asking people to rate their health.

Chronic Conditions

Physicians (or nurses in 8% of cases) were asked whether the residents suffered from any of the following conditions: arthritis (including severe osteoarthritis), obstructive pulmonary disease (emphysema, asthma, or other chronic obstructive pulmonary disorder), diabetes, hypertension, anaemia, ischemic heart disease, congestive heart failure, peripheral arterial disease, arrhythmias, stroke, depression, Parkinson's disease, Alzheimer's disease, other dementias, epilepsy, and cancer.

Functional Status

We used the Barthel index as modified by Shah et al [9]. Subjects (55%) or their main caregivers (if assigned, 45%) were asked as to the degree of dependency in performing the following basic activities of daily living: eating, going to the toilet, personal hygiene, bathing/showering, dressing/undressing, transferring, walking, use of stairs, and urinary/faecal continence. Residents were classified into the following categories: independent (100 points), mild dependency (91–99 points), moderate dependency (21–90 points), and severe dependency (0–20 points).

Cognitive Status

Residents were subject to the Pfeiffer's Short Portable Mental Status Questionnaire [10] (SPMSQ, range 0–10 errors) with some modifications to adapt to the institutional setting. We also used the Minimum Data Set Cognition Scale (MDS-COGS) [11,12], which obtains an assessment from the main caregiver of the resident's cognitive status based on selected Minimum Data Set questions (0–10 point scale from intact to very severe impairment). With both scales we created a two-category variable: impaired cognition comprised persons with more than 3 education-adjusted errors in SPMSQ or more than 2 points in the MDS-COGS, while the rest formed the normal cognition group.

Depressive Symptoms

Residents were asked to respond to a 10-item version of the Geriatric Depression Scale [13], with 0–3, 4–7, and 8–10 point ranges indicating normal, mild, and moderate/severe depressive symptoms, respectively [14].

Vision and Hearing

These conditions were assessed by means of the two corresponding Minimum Data Set four-category questions [15], and then dichotomized as good/mild versus moderate/severe impairment. Residents or caregivers were asked to assess the residents' ability to see in adequate light and with glasses, if used; as well as their hearing status even with hearing appliance, if used.

Social Interaction

These aspects were appraised with three questions, further dichotomized. One inquired on the degree of relationships and participation in activities in the institution (four categories collapsed into frequent/normal versus rare/nothing). Another asked for the frequency of outside contacts with family or friends (five categories collapsed into daily/weekly/monthly versus lower than monthly/nothing). The third one questioned on whether the resident had a family member in the facility.

Analysis

We used ordered logistic regression models to obtain the odds of reporting worse health perception (five category response variable) as a function of multiple explanatory variables. The main model included sex, age (65–84 or ≥ 85 years), education (less than primary, primary, or secondary or more), type of facility (public/concerted or private), number of chronic conditions, and functional status (independent, mild, moderate, or severe dependency) as explanatory variables. We assessed the association of cognitive status, depressive symptoms, vision and hearing problems, and social interactions (internal, external contacts, and relatives in facility), adjusted for the above main model covariates. Finally, to evaluate potential effect modifications, additional models were fitted separately for each sex, type of facility, and cognition category group. Interaction tests were performed by contrasting product terms in models adjusted for main covariates.

Due to the complex sampling and the different selection probabilities of study participants, all analyses were weighted to the underlying population distribution and accounted for the effect of stratification and clustering on point and interval estimates. Analyses were run using Stata 8.1 statistical software [16].

Results

We obtained information on self-rated health in 669 of the 800 residents in the initial sample (overall response rate 84%). In our population (weighted and design-corrected estimates), 75% of residents were women, and the mean (SD) age was 83.4 (7.3) years. Forty-three percent had no formal education and 40% completed the primary level. Health was declared as very good, good, fair, poor, and very poor by 15, 40, 30, 10, and 5% of residents, respectively. The mean (SD) number of chronic conditions was 2.9 (2.0), and 51% were independent or mild dependent in basic activities of daily living. In addition, 41% had cognitive impairment, 31% showed depressive symptoms, 12.8 and 12.9% had moderate/severe vision and hearing impairments, respectively, and for 43% internal contacts were scarce or none.

Table 1 displays results from the main model. Self-rated health was similar by sex and age group. Residents in private facilities and those who completed primary education had better health perception (adjusted odds ratios 0.57 and 0.64, respectively). The adjusted odds for worse health perception increased 18% for each additional chronic condition. Residents with any degree of dependency had between 1.81 and 2.37 times higher odds of worse perception than those functionally independent. Results further adjusted for cognitive status and depressive symptoms remained virtually unchanged (not shown).

We also assessed the independent associations of mental and social variables with self-rated health (Table 2). Cognitive status showed no association. The odds ratios for worse health perception were 4.04 and 10.45 for residents with mild and moderate/severe depressive symptoms compared to those without depressive symptoms, respectively. We also found an association with vision problems (odds ratio 2.05), but not with hearing impairment. In the social support realm, only internal contacts showed an independent association (odds ratio 1.42).

Table 1. Factors associated with worse self-rated health in institutionalized older persons.

Factor	OR* (95% CI)	
	Unadjusted	Adjusted†
Sex		
Men	I	I
Women	1.00	1.02 (0.73 – 1.44)
Age (years)		
65 – 84	I	I
≥85	1.00	0.87 (0.63 – 1.20)
Education		
None	I	I
Primary	0.62	0.64 (0.43 – 0.95)
Secondary or more	0.51	0.66 (0.40 – 1.10)
Facility		
Public/concerted	I	I
Private	0.55	0.57 (0.37 – 0.87)
Chronic conditions		
I unit increase	1.21	1.18 (1.07 – 1.28)
Functional dependency		
Independent	I	I
Mild	1.74	1.94 (1.11 – 3.37)
Moderate	2.00	2.37 (1.38 – 4.06)
Severe	1.74	1.81 (0.96 – 3.41)

* Odds ratio (95% confidence interval) of worse health perception for each category compared to the corresponding reference group (men, 65 – 84 years, no education, public facility, and functionally independent), except for chronic conditions, showing the odds ratio for an increase of one chronic condition.
† Adjusted for the remaining variables in the table.

Table 2. Mental and social factors associated with worse self-rated health in institutionalized older persons.

Factor	Subjects	OR* (95% CI)
Cognitive status	484	
Normal		I
Impaired		1.01 (0.60 – 1.72)
Depressive symptoms	588	
Normal		I
Mild		4.04 (2.50 – 6.60)
Moderate/severe		10.45 (5.84 – 18.68)
Vision	556	
Good/mild		I
Moderate/severe		2.05 (1.36 – 3.08)
Audition	542	
Good/mild		I
Moderate/severe		1.21 (0.74 – 1.98)
Internal contacts	619	
Frequent/normal		I
Rare/nothing		1.42 (1.03 – 1.97)
External contacts	617	
Daily/weekly/monthly		I
Lower than monthly/nothing		1.08 (0.82 – 1.42)
Relatives in facility	622	
Yes		I
No		0.73 (0.48 – 1.12)

* Odds ratio (95% confidence interval) of worse health perception for each category compared to the corresponding reference group (normal cognition, no depressive symptoms, good/mild vision and audition, and present of internal, external contacts, and relatives in facility), adjusted for sex, age, education, type of facility, number of chronic conditions, and functional dependency.

Table 3 shows results stratified by sex, type of facility, and cognitive function. The pattern of determinants was similar between men and women, except for education and functional dependency. The better health perception in those with primary education was mainly attributed to women (p value for interaction 0.005), whereas the worse perception in functionally dependent residents was stronger for men (p value for interaction 0.082). We found no significant departures from the main effects by type of facility, although a qualitative modification of the sex effect was noted: in public facilities women rated their health worse than men, and the opposite came about in private nursing homes (p value for interaction 0.179). We finally explored the profile in each subgroup of cognitive status. In those with normal function effects were similar to those previously reported, but somewhat erratic patterns were observed in cognitively impaired residents. In particular, the effect of functional dependency changed its direction, with functionally dependent subjects showing better health perception than those independent (p value for interaction 0.315). Nevertheless, the reduced number of cognitively impaired individuals resulted in imprecise interval estimates in this subgroup and low statistical power of interaction tests.

Table 3. Factors associated with worse self-rated health by sex, type of facility, and cognition in institutionalized older persons.

Factor	OR* (95% CI)					
	Sex		Facility		Cognitive function	
	Men (n = 298)	Women (n = 332)	Public/concerted (n = 425)	Private (n = 205)	Normal (n = 308)	Impaired (n = 176)
Sex						
Men			I	I	I	I
Women			1.24 (0.87 – 1.77)	0.72 (0.38 – 1.36)	1.17 (0.74 – 1.86)	0.96 (0.53 – 1.76)
Age (years)						
65 – 84	I	I	I	I	I	I
≥85	0.70 (0.42 – 1.15)	0.92 (0.62 – 1.35)	0.89 (0.59 – 1.33)	0.75 (0.43 – 1.30)	0.59 (0.37 – 0.93)	1.13 (0.61 – 2.10)
Education						
None	I	I	I	I	I	I
Primary	1.17 (0.64 – 2.13)	0.53 (0.34 – 0.82)	0.58 (0.35 – 0.97)	0.71 (0.37 – 1.37)	0.82 (0.47 – 1.42)	0.52 (0.24 – 1.13)
Secondary	0.51 (0.22 – 1.17)	0.75 (0.38 – 1.51)	0.76 (0.40 – 1.44)	0.63 (0.29 – 1.34)	0.86 (0.48 – 1.53)	0.36 (0.12 – 1.11)
Facility						
Public/concerted	I	I			I	I
Private	0.68 (0.33 – 1.40)	0.51 (0.30 – 0.87)			0.54 (0.32 – 0.90)	0.58 (0.25 – 1.32)
Chronic conditions						
1 unit increase	1.25 (1.07 – 1.46)	1.16 (1.06 – 1.27)	1.23 (1.10 – 1.37)	1.09 (0.93 – 1.28)	1.20 (1.04 – 1.39)	1.12 (0.95 – 1.31)
Functional dependency						
Independent	I	I	I	I	I	I
Mild	3.09 (1.46 – 6.47)	1.61 (0.81 – 3.22)	2.29 (1.14 – 4.62)	1.37 (0.36 – 1.25)	1.75 (0.87 – 3.51)	0.49 (0.09 – 2.68)
Moderate	2.59 (1.31 – 5.09)	2.29 (1.14 – 4.62)	2.10 (1.19 – 3.73)	2.55 (1.71 – 9.15)	2.21 (1.17 – 4.18)	0.56 (0.10 – 3.12)
Severe	2.79 (1.16 – 6.70)	1.55 (0.74 – 3.22)	1.88 (0.76 – 4.58)	1.74 (0.49 – 6.25)	1.51 (0.60 – 3.81)	0.57 (0.10 – 3.24)

* Odds ratio (95% confidence interval) of worse health perception adjusted for sex, age, education, type of facility, number of chronic conditions, and functional dependency.

Discussion

To our knowledge, this is the first study that examines factors associated with self-rated health in older people living in institutions. Apart from a strong effect

of depressive symptoms, the main determinants of self-rated health in our institutionalized population were the number of illnesses and functional dependency.
We also found an independent association with socioeconomic variables, such as
type of facility and education, particularly among women.

In general terms, and compared to other institutionalized populations [17],
this population can be described as being in relatively good level of health, functioning, and cognition. The facilities in our area are assisted living, skilled nursing,
and mixed. Overall, 55% residents had no assigned caregiver, thus indicating a
low need of care, assimilated to residential care.

A comparison of self-rated health in elders participating in our study with
community-dwelling elders in a study conducted with similar methods in the
same geographic area showed higher health perception ratings among institutionalized participants (55% versus 49% with very good and good self-rated health)
[18]. This unexpected finding may be partly explained by general health status
and functioning of these populations. In our facilities there is a large portion of
residents with low needs for care, while there are also persons in the community
that should be better in institutions, but remain in their homes. In addition, self-
rated health, as a subjective matter, implies some comparative assessments [19]
making context a relevant issue. When rating their health, individuals may make
some "adjustment" by comparing their situation with that of others around. This
phenomenon may lead to overly positive ratings in institutionalized populations,
thus limiting its value. On the other hand, we think that some people uncon-
sciously incorporate certain degree of complaint to their answers. We hypothesize
that this group is proportionately higher in the community.

We did not find overall associations of self-rated health with demographic
variables, but some subgroup results are worth mentioning. In the community
[18], we previously reported a clear better health perception in the oldest group ($\geq$
85 years) as compared with younger individuals (65–74 years), a behavior consistent with other reports [20]. In the present study we found a similar effect only in
the subgroup of residents with normal cognition. Regarding sex, we appreciated
some differences between men and women concerning the effects of education
and type of facility, with a better health perception more clearly associated with
higher educational levels and private facilities among women than men. Though
these effect modifications were weak, it seems that sex and socioeconomic status
may interact in determining health perception.

Cognitive function is an important variable that could play a role in institutionalized populations. Although some degree of misclassification is likely, we
found no relevant effect of cognitive status on health perception. In addition, the
pattern of determinants remained very similar after adjusting for cognitive status
and also after excluding cognitively impaired residents. On the other hand, we

found clear inconsistencies in those cognitively impaired, but random errors derived from the small sample size of this subgroup impede a proper appraisal. Some work suggests that cognitively impaired individuals can provide equivalent assessment of their health status [21,22], but in-depth studies on the role of cognitive function as determinant of self-rated health could add valuable knowledge.

Depression was highly prevalent and showed a very strong, independent effect on self-rated health. Similar results were observed in our population of community-dwelling older people [18], as well as in other communities [23]. Accordingly, this great dependence on depressive symptoms must always be considered when interpreting health ratings.

The strengths of this study include the use of a probabilistic sample, the high response rate, and the analysis of a wide panel of relevant variables. However, some limitations are worth mentioning. First, the SPMSQ and MDS-COGS cognition scales used in this study, as well as the Geriatric Depression Scale, have not been validated for Spanish population, but the translation of most questions is straightforward and no important misclassification biases are expected when applied to our setting. Second, an important part of the distinctive value of studies conducted in nursing homes resides on cognitively impaired residents and our work achieved low power in this subgroup.

Conclusion

In summary, we found that some consistent determinants of health perception in the community, such as chronic conditions and functional dependency, are also relevant in institutions, but new features emerge in this setting, particularly the very strong effect of depressive symptoms and the important role of type of facility and educational level. It remains important to elucidate issues in cognitively impaired persons through an adequately powered study. Finally, it should be pointed out that our population, with a favourable health and function profile, can differ from other nursing home populations, making highly constructive the research on health perceptions in various institutional settings to provide sensible contrasts.

Competing Interests

The author(s) declare that they have no competing interests.

Authors' Contributions

All authors contributed to concept and design, analysis and interpretation of data, preparation of manuscript, and have given final approval for publication. EV and JD also contributed to the fieldwork and acquisition of data.

Acknowledgements

This study was supported by grant 96/0201 from the Spanish "Fondo de Investigación Sanitaria."

References

1. Idler EL, Benyamini Y: Self-rated health and mortality: a review of twenty-seven community studies. J Health Soc Behav 1997, 38:21–37.

2. Idler EL, Kasl SV: Self-ratings of health: do they also predict change in functional ability? J Gerontol B Psychol Sci Soc Sci 1995, 50:S344–S353.

3. Mor V, Wilcox V, Rakowski W, Hiris J: Functional transitions among the elderly: patterns, predictors, and related hospital use. Am J Public Health 1994, 84:1274–1280.

4. Roos NP, Havens B: Predictors of successful aging: a twelve-year study of Manitoba elderly. Am J Public Health 1991, 81:63–68.

5. Jylha M, Volpato S, Guralnik JM: Self-rated health showed a graded association with frequently used biomarkers in a large population sample. J Clin Epidemiol 2006, 59:465–471.

6. Lekander M, Elofsson S, Neve IM, Hansson LO, Unden AL: Self-rated health is related to levels of circulating cytokines. Psychosom Med 2004, 66:559–563.

7. Fillenbaum GG: Social context and self-assessments of health among the elderly. J Health Soc Behav 1979, 20:45–51.

8. Leung KK, Tang LY, Lue BH: Self-rated health and mortality in chinese institutional elderly persons. J Clin Epidemiol 1997, 50:1107–1116.

9. Shah S, Vanclay F, Cooper B: Improving the sensitivity of the Barthel index for stroke rehabilitation. J Clin Epidemiol 1989, 42:703–709.

10. Pfeiffer E: A short portable mental status questionnaire for the assessment of organic brain deficit in elderly patients. J Am Geriatr Soc 1975, 23:433–441.

11. Hartmaier SL, Sloane PD, Guess HA, Koch GG: The MDS Cognition Scale: a valid instrument for identifying and staging nursing home residents with dementia using the Minimum Data Set. J Am Geriatr Soc 1994, 42:1173–1179.

12. Gruber-Baldini AL, Zimmerman SI, Mortimore E, Magaziner J: The validity of the minimum data set in measuring the cognitive impairment of persons admitted to nursing homes. J Am Geriatr Soc 2000, 48:1601–1606.

13. D'Ath P, Katona P, Mullan E, Evans S, Katona C: Screening, detection and management of depression in elderly primary care attenders. I: The acceptability and performance of the 15 item Geriatric Depression Scale (GDS15) and the development of short versions. Fam Pract 1994, 11:260–266.

14. Shah AK, Phongsathorn V, Bielawska C, Katona C: Screening for depression among geriatric inpatients with short versions of the Geriatric Depresion Scale. Int J Geriatr Psychiatry 1996, 11:915–918.

15. Morris JN, Hawes C, Fries BE, Philips CD, Mor V, Katz S, Murphy K, Drugovich ML, Friedlob AS: Designing the national resident assessment instrument for nursing homes. Gerontologist 1990, 30:293–307.

16. StataCorp: Stata Statistical Software: Release 8. College Station, TX: StataCorp LP; 2003.

17. Ribbe MW, Ljunggren G, Steel K, Topinkova E, Hawes C, Ikegami N, Henrard JC, Jonnson PV: Nursing homes in 10 nations: a comparison between countries and settings. Age Ageing 1997, 26 Suppl 2:3–12.

18. Damian J, Ruigomez A, Pastor V, Martin-Moreno JM: Determinants of self assessed health among Spanish older people living at home. J Epidemiol Community Health 1999, 53:412–416.

19. Fayers PM, Sprangers MA: Understanding self-rated health. Lancet 2002, 359:187–188.

20. Idler EL: Age differences in self-assessments of health: age changes, cohort differences, or survivorship? J Gerontol 1993, 48:S289–S300.

21. Walker JD, Maxwell CJ, Hogan DB, Ebly EM: Does self-rated health predict survival in older persons with cognitive impairment? J Am Geriatr Soc 2004, 52:1895–1900.

22. Scocco P, Fantoni G, Caon F: Role of depressive and cognitive status in self-reported evaluation of quality of life in older people: comparing proxy and physician perspectives. Age Ageing 2006, 35:166–171.

23. Mulsant BH, Ganguli M, Seaberg EC: The relationship between self-rated health and depressive symptoms in an epidemiological sample of community-dwelling older adults. J Am Geriatr Soc 1997, 45:954–958.

CITATION

Discomfort and Agitation in Older Adults with Dementia

Isabelle Chantale Pelletier and Philippe Landreville

ABSTRACT

Background

A majority of patients with dementia present behavioral and psychological symptoms, such as agitation, which may increase their suffering, be difficult to manage by caregivers, and precipitate institutionalization. Although internal factors, such as discomfort, may be associated with agitation in patients with dementia, little research has examined this question. The goal of this study is to document the relationship between discomfort and agitation (including agitation subtypes) in older adults suffering from dementia.

Methods

This correlational study used a cross-sectional design. Registered nurses (RNs) provided data on forty-nine residents from three long-term facilities. Discomfort, agitation, level of disability in performing activities of daily living (ADL), and severity of dementia were measured by RNs who were well acquainted with the residents, using the Discomfort Scale for patients with Dementia of the Alzheimer Type, the Cohen-Mansfield Agitation Inventory, the ADL subscale of the Functional Autonomy Measurement System, and the Functional Assessment Staging, respectively. RNs were given two weeks to complete and return all scales (i.e., the Cohen-Mansfield Agitation Inventory was completed at the end of the two weeks and all other scales were answered during this period). Other descriptive variables were obtained from the residents' medical file or care plan.

Results

Hierarchical multiple regression analyses controlling for residents' characteristics (sex, severity of dementia, and disability) show that discomfort explains a significant share of the variance in overall agitation (28%, p < 0.001), non aggressive physical behavior (18%, p < 0.01) and verbally agitated behavior (30%, p < 0.001). No significant relationship is observed between discomfort and aggressive behavior but the power to detect this specific relationship was low.

Conclusion

Our findings provide further evidence of the association between discomfort and agitation in persons with dementia and reveal that this association is particularly strong for verbally agitated behavior and non aggressive physical behavior.

Background

Dementia is not an inevitable consequence of ageing but the risk of dementia increases sharply with advancing age and prevalence is expected to increase dramatically over the coming decades [1]. Dementia features an alteration of memory and at least one other cognitive disorder such as aphasia, agnosia, apraxia or a disturbance in executive functioning [2]. Various etiologies are related to dementia, such as strokes, head trauma, Parkinson's disease, and substance abuse. Yet Alzheimer's disease is considered the most widespread form of all senile dementias, representing more than half of all cases [3]. The onset of dementia of the Alzheimer's type (DAT) is gradual and involves continuing cognitive decline [2].

In addition to cognitive symptoms, persons with dementia often present behavioral and psychological symptoms which may increase their suffering, be difficult to manage by caregivers, and precipitate institutionalization [4]. Behavioral symptoms of dementia include wandering, screaming, and hitting, while psychological symptoms include hallucinations, delusions, and depression. Between 50 and 90% of dementia patients present with behavioral or psychological symptoms [5]. The term "agitation" is often used in reference to behavioral symptoms associated with dementia [6]. Agitation was originally defined as any inappropriate verbal, vocal or motor activity which, according to an outside observer, does not result directly from the needs or the confusion of the agitated person [7]. Behavior which constitutes agitation can be broadly classified as aggressive vs non-aggressive and physical vs verbal [8]. A factor analysis of a measure of agitation used with nursing home residents produced three factors which make it possible to distinguish various forms of agitation [9]: aggressive behavior (AB, e.g., hitting), non-aggressive physical behavior (NAPB, e.g., pacing), and verbally agitated behavior (VAB, e.g., complaining).

The specific determinants of agitation remain unclear [6]. Predisposing factors may include gender, personality, poor health, functional impairment of activities of daily living, as well as cognitive and neurological deterioration [10,11]. Other factors may precipitate the occurrence of agitation and include the characteristics of the physical and social environment (e.g., too much noise, not enough social interaction) as well as physical needs such as hunger, thirst, and discomfort. Some of these variables, such as sex, the severity of cognitive impairment and the level of dependence in performing activities of daily living, are well documented in the literature in terms of their relationship with different types of agitation. For example, males are more likely to be aggressive than females, NAPB is more likely to be manifested by persons who are more cognitively impaired, and VAB tends to be exhibited by persons who are more functionally impaired [8]. Other variables, such as hunger and light intensity, may be associated with agitation but this association remains hypothetical [12-14].

Several studies have examined environmental or contextual determinants of agitation [15,16]. Considerably less attention has been devoted to internal states which may also trigger difficult behaviors. Discomfort, which is defined as a negative emotional or physical state subject to variation in response to internal or environmental conditions [17], may act as an internal factor which precipitates the occurrence of agitation [10,12]. Persons suffering from dementia may behave in ways that are disruptive to those around them but which, for these patients, serve to communicate the discomfort they feel. For example, a patient suffering from moderate dementia and who gradually becomes aphasic may revert to shouting, emit odd noises, become unruly or hit those around him to let them know he feels pain during dressing and bathing. Various studies have shown the importance of discomfort in persons suffering from dementia [18-20] and a majority of patients must deal with painful acute or chronic ailments such as cancer, depression, cardiovascular disease and musculo-skeletal disorders [21].

We searched the literature (PsycINFO, CINAHL, Ageline, and Medline databases) and the reference list of selected papers for empirical studies of the relationship between discomfort and agitation in dementia. Two such studies were found. The first study, by Buffum et al. [20], has shown that more discomfort is associated with more agitation. However, the authors considered agitation as a single construct without considering its subtypes (i.e., AB, NAPB and VAB). Given that the importance of the relationship between different factors varies according to these subtypes, [8,11,12], discomfort may be more strongly related to certain subtypes of agitation than others. In their review of the literature, Cohen-Mansfield and Deutsch [8] indicated that all subtypes of agitation may be associated with discomfort. However, the available evidence strongly suggests this association in the case of VAB whereas the reasons for both AB (e.g., cognitive impairment, personality style, unsuccessful communication with caregivers) and NPAB (e.g., need for exercise or stimulation, performing previous roles) appear more diversified. In the second study, Young [22] used the Cohen-Mansfield Agitation Inventory (CMAI) [9] and reported low but significant positive correlations between discomfort and both overall agitation and AB. Unfortunately, these results are difficult to interpret because it is unclear whether the original 29-items version or a shorter 14-items version of the CMAI was used and what scale was used for rating each item. Moreover, the exact items used in calculating scores for each agitation subtype are not reported. Two additional scores of overall agitation, obtained from the Functional Abilities Checklist [23] and the Minimum Data Set, were also correlated with discomfort.

The goal of this study is to document the relationship between discomfort and agitation (including agitation subtypes) in older adults suffering from dementia. The following hypotheses were tested: (a) the frequency of overall agitation is related positively to the degree of discomfort, and (b) the degree of discomfort is related positively to frequency of VAB. To ensure more accurate results, sex, the severity of dementia and disability in performing activities of daily living, three variables whose relationship with agitation is well documented in the literature, were measured for control purposes.

Methods

Participants

This correlational study used a cross-sectional design. Thirteen registered nurses (RNs) from three long-term care facilities provided data on forty-nine residents. The RNs were all women, had a mean of 15 years of education and 22 years of experience working with older adults.

Residents were selected by the RNs according to the following criteria: (a) being at least 65 years old, (b) having a diagnosis of dementia documented by the medical file (all had a diagnosis of DAT), and (c) having been living in the same facility for at least three months. Residents suffering from delirium or any form of psychosis were excluded. Fifty-five residents were initially selected but data could not be collected for six of them because a respondent was not available during the data collection period. The final sample therefore included 49 residents, which is near the sample size estimate of 50 indicated by a power analysis for Pearson correlation coefficient (r) conducted prior to the study. This sample size estimate was obtained for a power of 0.80, an alpha level of 0.05, and an effect size of 0.40 determined from the correlation obtained between discomfort and overall agitation by Buffum et al. [20]. An effect size of 0.40 is halfway between medium (0.30) and large (0.50) effect size conventions for r proposed by Cohen [24]

Measures

Descriptive Variables

The following information was drawn from the resident's medical file and care plan: date of birth, sex, date of admittance to the facility, medical conditions, as well as diagnosis and type of dementia. Analgesics taken on a daily basis were also recorded for control reasons because these can influence the experience of discomfort. RNs, who filled out questionnaires for the residents, were also asked to provide descriptive information (gender, years of education, and years of experience working with older adults).

Severity of Dementia

The level of cognitive impairment was measured using the Functional Assessment Staging (FAST) developed by Reisberg and colleagues [25,26]. The scale comprises seven stages measuring function loss associated with cognitive deterioration. Each resident was therefore given a score between 1 and 7 where 1 refers to a normal cognitive function and 7 to severe dementia with very severe cognitive impairment. Intraclass correlation coefficients indicate good test-retest reliability (0.86) and interrater agreement (0.87) [27]. Previous studies have indicated strong concurrent validity between FAST and various psychometric and mental status assessments [26,27]. For example, Sclan and Reisberg [27], found that increasing functional disability assessed by FAST is significantly related to decreasing cognitive functioning (r = -0.79).

Disability

The level of disability in performing activities of daily living (ADL) was measured using a subscale of the Functional Autonomy Measurement System (SMAF) [28,29]. This ADL subscale helps measure the level of impairment for five activities (feeding, washing, dressing, personal hygiene and using the bathroom), as well as urinary and fecal incontinence. Each item is evaluated using a four-point Likert-type scale where 0 indicates that the person is autonomous while -3 denotes dependency. Hébert and colleagues have found that the SMAF has acceptable interrater reliability (mean Cohen's weighted kappa of 0.75) and that the scale can be reliably used by both nurses and social workers in either community or institutional settings [28,29]. Results obtained by Desrosiers et al. show an intraclass correlation coefficient of 0.95 for test-retest and 0.96 for interrater reliability [30]. Concurrent validity is supported by a strong correlation between SMAF scores and the amount of nursing time required for care (r = 0.88) [28].

Agitation

Agitation was evaluated using the French version of the Cohen-Mansfield Agitation Inventory (CMAI) [31,32]. The frequency of all 29 items of the CMAI is evaluated for the previous two weeks using a Likert-type scale and scores given for each item varies from 1 ("never") to 7 ("several times per hour"). For each resident, an overall agitation (OA) score as well as a score for each one of the three factors of the CMAI (AB, NAPB, and VAB) were calculated by summing across items. These factors have been identified for the French version of the CMAI [31] and the corresponding items are very similar to those of the factors of the original version [9]: AB (6 items: hitting, grabbing, kicking, scratching, pushing, and spitting), NAPB (6 items: pacing, trying to get to a different place, handling things inappropriately, hoarding, hiding, general restlessness), and VAB (6 items: complaining, repetitive sentences or questions, negativism, constant requests for attention, cursing or verbal aggression, screaming). Deslauriers et al. [30] have demonstrated the interrater reliability (r = 0.72), test-retest reliability (r = 0.72), internal consistency (Cronbach's alpha from 0.75 to 0.77), concurrent validity (r = 0.74) and construct validity of the French version of the CMAI completed by nurses.

Discomfort

The Discomfort Scale for patients with Dementia of the Alzheimer Type (DS-DAT) [17] is composed of nine behavioral indicators of discomfort determined following interviews with caregivers working with persons suffering from dementia. Thus, the authors of the scale gave particular attention to content validity by identifying the kinds of behavior most frequently associated with a sign of discomfort in this population. The nine indicators are (a) noisy breathing, (b) negative vocalization, (c) content facial expression, (d) sad facial expression, (e) frightened facial expression, (f) frown, (g) relaxed body language, (h), tense body language, and (i) fidgeting. Each item comes with a list of observable forms of behavior which helps evaluators observe and record the signs of discomfort as objectively as possible. This tool makes it possible to evaluate the frequency (from 0 to ≥3), intensity (high or low) and duration (long or short) of the nine indicators associated with discomfort, as perceived by the observer, in the course of an observation period usually lasting five minutes. The level of discomfort is then derived from the value attributed to these three components. Each of the nine items is evaluated independently on a scale from 0 ("no observed discomfort") to 3 ("high level of observed discomfort"). Content validity was insured by initially generating twenty-six items from interviews with nursing staff from inpatient Alzheimer units and then retaining eighteen items rated by independent nurses holding advanced nursing degrees as relevant with an operational definition of discomfort (the number of items was later reduced to nine). The internal consistency of the DS-DAT has been evaluated in two studies [17,22] with results showing Cronbach's alpha varying from 0.74 to 0.89. The interrater reliability (r varying between 0.61 and 0.98) and test-retest reliability (r = 0.6) have also been verified [17,22,33]. Nurses acted as raters in studies reporting the psychometric properties of the DS-DAT [22,33].

Procedure

RNs working on the residents' units were asked to collaborate in this study. A group meeting was called with the principal investigator (ICP) to explain the study to the RNs, obtain their consent, and provide them with information so that they would be able to identify residents matching the study criteria based on each resident's file. The RNs were also given a presentation to demonstrate how to complete each of the scales used in the study. Because the administration and scoring of the DS-DAT is complex, a large portion of the meeting was devoted to demonstrating how to use this tool. Each item of the scale was explained in detail, case examples were presented, and staff members had the opportunity to practice scoring. These instructions included Hurley et al.'s [17] recommendations regarding the timing of the assessment.

The DS-DAT, the CMAI, the ADL subscale of the SMAF and the FAST were completed by the RN most familiar with each resident. They were given two weeks to complete and return all scales (i.e., the CMAI was completed at the end of the two weeks and all other scales were answered during this period). The principal investigator could be contacted during this period to answer questions or provide additional information regarding the procedure.

Staff members were the participants in this study. Indeed, the main variables reflect their perception of selected residents. The residents themselves did not participate directly in the data collection. Staff members did provide the researchers with factual information obtained from the residents' file to which they had authorized access as part of their work. Prior to the study, the researchers were granted permission to obtain this information by the Professional Services Director of each participating facility. Residents' rights were protected by concealing their identity (a code was used instead of their name on each document related to the study). Université Laval's Research Ethics Committee approved this project.

Results

Data analysis was performed using SPSS software, version 11.0 for Windows, on an Intel Pentium 4 computer with Microsoft Windows XP, Home Edition, Version 2002, as the operating system.

Residents characteristics are shown in Table 1. It should be noted that the sample was mostly comprised of women and that the majority of residents had a medical condition. In addition, nearly 60% of residents took an analgesic on a daily basis. For two of the 28 residents taking analgesics, the prescription was to take this medication if needed (prn) while the remaining residents took regularly scheduled analgesics at least once a day. This information suggests that an important proportion of the residents experienced chronic pain. Most residents were rated in the three last stages of the FAST which indicates that the cognitive functioning of most residents was highly impaired. The mean score on the SMAF (-15.5) was near the lowest score possible (-21), indicating that these residents were very dependent on the nursing staff.

Table 2 shows the Pearson correlation coefficients (r) between each score of agitation, discomfort, and the remaining descriptive variables. Sex was coded as follows: men = 1 and women = 2. Positive and significant correlation coefficients are observed between discomfort and each of the following agitation scores: OA, NAPB and VAB. No other correlation coefficient was significant. The power to detect significant correlation coefficients with our data (n = 49) was examined using GPower software, version 3.0. for Windows [34]. Post hoc power analyses were conducted for t tests with alpha level fixed at 0.05. Power analyses for OA and VAB were for one-tailed tests because of our hypotheses regarding these variables, while analyses for two-tailed tests were done for NAPB and AB. Power was high for tests using OA (0.99; effect size = 0.55), VAB (0.99; effect size = 0.53), and PNAB (0.95; effect size = 0.47) but the test using AB had low power (0.37; effect size = 0.23).

Table 1. Residents characteristics

Characteristics	Percentage	m	Min, Max	SD
Sex				
Male	10.2			
Female	89.8			
Age (years)		82.7	67, 98	7.8
Duration of institutionalization (months)		38.7	3, 198	37.9
Number of medical conditions		5.2	1, 13	2.6
Daily administration of an analgesic	57.2			
Stage of dementia				
1	0.0			
2	0.0			
3	4.1			
4	2.0			
5	16.3			
6	53.1			
7	24.5			
Disability in performing ADL		-15.5	-1, -21	5.8
Agitation				
OA		41.5	29, 90	14.2
AB		7.2	6, 18	2.5
NAPB		10.0	6, 29	5.5
VAB		10.7	6, 32	6.1
Discomfort		4.2	0, 14	3.5

Note. OA = Overall agitation; AB = Aggressive behavior; NAPB = Non aggressive physical behavior; VAB = Verbally agitated behavior.

Table 2. Correlation coefficients between agitation scores, discomfort and the other descriptive variables

	Sex	Severity of dementia	Disability in performing ADL	Level of discomfort
OA	-0.25	-0.00	0.04	0.56***
AB	-0.08	.014	-0.16	0.23
NAPB	-0.28	-0.10	0.06	0.47**
VAB	-0.24	-0.13	0.21	0.54***

Note. OA = Overall agitation; AB = Aggressive behavior; NAPB = Non aggressive physical behavior; VAB = Verbally agitated behavior.
**$p < 0.01$
***$p < 0.001$

Hierarchical multiple regression analyses were next performed to specify the relationship between discomfort and each agitation score (except AB) while statistically controlling for sex, severity of dementia and disability in performing ADL. For each analysis, these three last variables were introduced in a first block of the regression (Step 1), followed by the level of discomfort at the next step (Step 2). Results are shown in Table 3 which presents variance explained (R2) at Step 1, incremental variance (ΔR2) as a result of entering discomfort at Step 2, and total variance. The other statistics presented are the unstandardized (B) regression coefficients, the standard error of regression coefficients (SE B), and the standardized beta (β) regression coefficients at Step 2. The estimated β coefficients indicate the relative contribution of each independent variable in the prediction of the dependent variable (agitation). As shown, discomfort contributed significantly to the prediction of the variance of OA, NAPB and VAB, beyond the residents' other characteristics that were statistically controlled in each regression equation. The variance of the different agitation scores specifically attribuable to discomfort (ΔR^2) varied between 18% and 30%. Further, estimated standardized beta coefficients show that only discomfort carries significant weight in the regression equation for OA, NAPB and VAB.

Table 3. Summary of hierarchical multiple regression analysis for variables predicting agitation, non aggressive physical behavior, and verbally agitated behavior

Variable	B	SE B	β
Overall agitation			
Sex	-4.17	6.01	-0.09
Severity of dementia	.94	3.09	0.06
Disability (ADL)	.40	0.50	0.17
Level of discomfort	2.20	0.52	0.55***
R^2 for Step 1 = 0.06, $F(3,45)$ = 0.98 (p = 0.648)			
ΔR^2 for Step 2 = 0.28, $F(1,44)$ = 18.21***			
R^2 Total = 0.34, $F(4,44)$ = 5.57***			
Non aggressive physical behavior			
Sex	-3.00	2.46	-0.17
Severity of dementia	-1.14	1.27	-0.19
Disability (ADL)	-0.04	0.21	-0.05
Level of discomfort	0.70	0.21	0.45**
R^2 for Step 1 = 0.09, $F(3,45)$ = 1.55 (p = 0.214)			
ΔR^2 for Step 2 = 0.18, $F(1,44)$ = 10.89**			
R^2 Total = 0.28, $F(4,44)$ = 4.14**			
Verbally agitated behavior			
Sex	-0.98	2.51	-0.05
Severity of dementia	0.57	1.29	0.09
Disability (ADL)	0.38	0.21	0.36
Level of discomfort	0.99	0.22	0.57**
R^2 for Step 1 = 0.09, $F(3,45)$ = 1.45 (p = 0.241)			
ΔR^2 for Step 2 = 0.30, $F(1,44)$ = 21.04***			
R^2 Total = 0.38, $F(4,44)$ = 6.83***			

Note. Sex, severity of dementia and disability were introduced in a first block of the regression (Step 1), followed by the level of discomfort at the next step (Step 2). Statistics shown for each variable are those at Step 2.
**p < 0.01
***p < 0.001

Discussion

The purpose of this study was to document the relationship between discomfort and the various types of agitation in older adults suffering from dementia. As hypothesized, our results show a positive and significant relationship between the degree of discomfort and the frequency of overall agitation. This confirms results obtained previously by Buffum and colleagues [20] and Young [22]. Our results further demonstrate that the relationship with discomfort varies according to the type of agitation. Also as hypothesized, we found that the degree of discomfort is associated both positively and significantly with VAB. Furthermore, our results show a positive and significant relationship with NAPB. No significant relationship was observed between discomfort and aggressive behavior but it should be noted that the power to detect this specific relationship was low and that the correlation coefficient obtained from our data (0.23) is similar to the one reported by Young (0.25).

Various authors contend that discomfort acts as an internal factor which precipitates the occurrence of agitation [10,12]. VAB possibly acts as a means of communicating the patient's discomfort [8,35]. Matteau and colleagues [35] have found that patients that display VAB also present more language difficulties. Through behaviors such as complaining and screaming, patients may attempt to attract their caregivers' attention in the hope that they will provide them some relief. The relationship between discomfort and NAPB was unexpected and is less clear. It is possible that several types of disruptive behavior related to this type of agitation, such as wandering and pacing, result from the discomfort experienced by the patient. For example, a patient who feels sad and depressed because he or she is homesick may attempt to leave his residential facility to reduce this discomfort [8].

Algase and colleagues [12] have suggested that behavioral problems associated with dementia result from unmet needs which the patient expresses using his remaining abilities. Given our results, the need for comfort appears as one such need. From a practical point of view, the occurrence of agitation deserves particular attention on the part of caregivers since it may communicate discomfort. Identifying and treating the cause of this discomfort may help reduce agitation. Kovach and colleagues [36] have recently reported findings which support this assertion. In their study, nursing home residents with dementia were treated using the Serial Trial Intervention which selects an appropriate treatment based on physical and affective needs assessments. Compared to a control group, the treated group had less discomfort and more frequently had behavioral symptoms return to baseline. Barton and colleagues [37] have suggested a similar hierarchical approach to the management of inappropriate vocalization.

The mean score on the DS-DAT in our sample (4.2) is lower than mean scores found in some studies [17,20] but similar to that reported by other researchers [22,33]. For example, a mean score of 4.64 was reported by Miller et al. [33] when their participants were assessed in situations of unlikely discomfort (the equivalent of what Hurley et al. [17] refer to as baseline discomfort). Also, the range of scores on the DS-DAT reported in Table 1 is broad (min to max: 0 to 14), suggesting that the DS-DAT was sensitive to various levels of discomfort in our sample.

The scores on the CMAI in our sample also deserve some comments. Looking at other's work, we found that CMAI scores are reported for clinical samples in which participants are selected because they present agitation [38,39]. These scores are higher than for the participants in our study who were not selected on the basis of agitation. For example, mean CMAI total scores vary between 65 and 78 across different samples in which participants present at least mild behavioral symptoms [39] compared to 41 in our sample. However, this mean score is similar to what we found previously in a separate but similar sample (mean scores of 42.6 and 40.7 at two separate times) [32]. Although there is no official cutoff for agitation on the CMAI, the range of scores (min to max: 29 to 90) is quite broad and suggests that some participants presented some agitation while others did not.

Limitations to the generalization of our findings are the relatively small sample size and the fact that the data were compiled only during the day shift. Some authors have indicated that the different types of agitation occur at different periods of the day [40,41]. It is unclear whether the same factors are equally important contributors to agitation at different periods of the day and this should be investigated. Another limitation is that it is uncertain whether the FAST is valid and reliable when used by nurses. However, it should be noted that the FAST is widely utilized by healthcare professionals of various backgrounds and requires minimal training because of its' face validity [42]. Other limitations have to do with the DS-DAT and have been pointed out in detail elsewhere [43]. This measure of discomfort provides no specific information as to the nature and origin of this experience. Discomfort is a broad concept referring to a negative emotional or physical state. Various conditions, including pain, distress, depression, loneliness, lack of stimulation, and lack of sleep can contribute to discomfort. Identifying which of these sources of discomfort play a role in agitation is essential for selecting appropriate treatment. Future studies, therefore, should identify the determinants of discomfort that are related to agitation in dementia patients and compare findings across different periods of the day. Finally, we did not check reliability because of restrictions in the availability of the RNs. The DS-DAT is a complex measure and we did provide raters with detailed instructions. Moreover, this limitation does not seem to pose a threat to our conclusions since the results confirm our hypotheses and are consistent with those of other researchers.

Dementia creates a paradoxical context in which patients are more vulnerable to various sources of discomfort while, at the same time, being less able to modify these by themselves or to communicate their discomfort directly to their caregivers. Further understanding of the internal determinants of behavioral symptoms in dementia will ultimately lead to finer assessment and more effective treatment of patients.

Conclusion

Our findings provide further evidence of the association between discomfort and agitation in persons with dementia and reveal that this association is particularly strong for VAB and NAPB.

Competing Interests

The author(s) declare that they have no competing interests.

Authors' Contributions

ICP conceived and carried out the study and drafted the manuscript. PL contributed to the conception of the study and helped to draft the manuscript. Both authors read and approved the final manuscript.

Acknowledgements

The authors would also like to express special thanks to Philippe Voyer and René Verreault for their comments on a previous version of this manuscript, as well as to the administrators and nursing staff members of the participating long-term care facilities for their generous involvement in this study.

The authors are grateful to the Center de recherche du Center Hospitalier Affilié Universitaire de Québec (through the Unité de recherche en gériatrie de l'Université Laval) for funding of the translation of the manuscript.

References

1. Canadian study of health and aging: study methods and prevalence of dementia CMAJ 1994, 150(6):899–913.

2. American Psychiatric Association: Diagnostic and statistical manual of mental disorders (4th edition, revised). Washington, DC , Author; 2000.

3. Ritchie K: Epidemiology of the dementias and Alzheimer's Disease. In Evidence-based dementia practice. Edited by: Qizilbash N, Schneider LS, Chui HTP, Brodaty H, Kaye J, Erkinjuntti T. Oxford , Blackwell; 2002:238–240.

4. Finkel S Costa E Silva, J., Cohen, G., Miller, S., & Sartorius, N: Behavioral and psychological signs and symptoms of dementia: A consensus statement on current knowledge and implications for research and treatment. Int J Geriatr Psychiatry 1997, 12:1060–1061.

5. Rousseau F: Principes généraux d'évaluation et de traitement des symptômes comportementaux et psychologiques de la démence. In Symptômes comportementaux et psychologiques de la démence. Edited by: Landreville P, Rousseau F, Vézina J, Voyer P. Acton Vale, QC , Edisem; 2005:61–108.

6. Cohen-Mansfield J: Agitation in the Elderly: Definitional and theoretical conceptualizations. In Agitation in patients with dementia: A practical guide to diagnosis and management. Edited by: Hay DP, Klein DT, Grossberg GT, Kennedy JS. Washington, DC , American Psychiatric Publishing; 2003:1–21.

7. Cohen-Mansfield J, Billig N: Agitated behaviors in the elderly, I: A conceptual review. J Am Geriatr Soc 1986, 34:711–721.

8. Cohen-Mansfield J, Deutsch LH: Agitation: Subtypes and their mechanisms . Semin Clin Neuropsychiatry 1996, 1:325–339.

9. Cohen-Mansfield J, Marx MS, Rosenthal AS: A description of agitation in a nursing home. J Gerontol 1989, 44:M77–M84.

10. Kolanowski AM: An overview of the need-driven dementia-compromised behavior model . J Gerontol Nurs 1999, 25:7–9.

11. Cohen-Mansfield J, Marx MS, Rosenthal AS: Dementia and agitation in nursing home residents: How are they related? Psychol Aging 1990, 5:3–8.

12. Algase DL, Beck C, Kolanowski A, Whall A, Berent S, Richards K, Beattie E: Need-driven dementia compromised behavior: An alternative view of disruptive behavior. Am J Alzheimers Dis Other Demen 1996, 11:10–19.

13. Cohen-Mansfield J, Marx MS, Werner P: Agitation in elderly persons: An integrative report of findings in a nursing home. Int Psychogeriatr 1992, 4:221–240.

14. Kiely DK, Morris JN, Algase DL: Resident characteristics associated with wandering in nursing homes. Int J Geriatr Psychiatry 2000, 15:1013–1020.

15. Aubert J, Vézina J, Landreville P, Brochu C, Primeau G, Imbeault S, Laplante C: Éléments contextuels associés à l'émission de comportements d'agitation verbale présentés par des personnes âgées institutionnalisées atteintes de démence. Eur Rev Appl Psychol 2007, 57:157–165.

16. Burgio LD, Butler FR, Roth DL, Hardin JM, Hsu CC, Ung K: Agitation in nursing home residents: The role of gender and social context. Int Psychogeriatr 2000, 12:495–511.

17. Hurley AC, Volicer BJ, Hanrahan PA, Houde S, Volicer L: Assessment of discomfort in advanced Alzheimer patients. Res Nurs Health 1992, 15:369–377.

18. Feldt KS, Warne MA, Ryden MB: Examining pain in aggressive cognitively impaired older adults . J Gerontol Nurs 1998, 24:14–22.

19. Galloway S, Turner L: Pain assessment in older adults who are cognitively impaired. J Gerontol Nurs 1999, 25:34–39.

20. Buffum MD, Miaskowski C, Sands L, Brod M: A pilot study of the relationship between discomfort and agitation in patients with dementia. Geriatr Nur 2001, 22:80–85.

21. Geda YE, Rummans TA: Pain: Cause of agitation in elderly individuals with dementia. Am J Psychiatry 1999, 156:1662–1663.

22. Young DM: Pain in institutionalized elders with chronic dementia. Volume Ph.D. Iowa City , The University of Iowa; 2001.

23. Swanson E, Maas M, Buckwalter K: Alzheimer's residents cognitive and functional measurements: Special and traditional care unit comparison. Clin Nurs Res 1994, 3:27–41.

24. Cohen J: Statistical power analysis for the behavioral sciences (2nd ed.). Hillsdale, NJ, Erlbaum; 1988.

25. Reisberg B, Ferris SH, Fransson E: An ordinal functional assessment tool for Alzheimer's-type dementia. Hosp Community Psychiatry 1985, 36:593–595.

26. Reisberg B: Functional assessment staging (FAST). Psychopharmacol Bull 1988, 24:653–659.

27. Sclan SG, Reisberg B: Functional assessement staging (FAST) in Alzheimer's disease: Reliability, validity, and ordinality. Int Psychogeriatr 1992, 4:55–69.

28. Hébert R, Carrier R, Bilodeau A: The functional autonomy measurement system (SMAF): Description and validation of an instrument for the measurement of handicaps. Age Ageing 1988, 17:293–302.

29. Hébert R, Carrier R, Bilodeau A: Le Système de mesure de l'autonomie fonctionnelle (SMAF). Revue Geriatr 1988, 13:161–167.

30. Desrosiers J, Bravo G, Hébert R, Dubuc N: Reliability of the revised functional autonomy measurement system (SMAF) for epidemiological research. Age Ageing 1995, 24:402–406.

31. Landreville P, Casault L, Julien E, Dicaire L, Verreault R, Lévesque L: Structure factorielle de l'Inventaire d'agitation de Cohen-Mansfield. Eur Rev Appl Psychol 2007, 57:167–174.

32. Deslauriers S, Landreville P, Dicaire L, Verreault R: Validité et fidélité de l'inventaire d'agitation de Cohen-Mansfield. Can J Aging 2001, 20:373–384.

33. Miller J, Neelon V, Dalton J, Ng'andu N, Bailey D, Layman E, Hosfeld A: The assessment of discomfort in elderly confused patients: A preliminary study. J Neurosci Nurs 1996, 28:175–182.

34. Faul F, Erdfelder E, Lang AG, Buchner A: G*Power 3: A flexible statistical power analysis for the social, behavioral, and biomedical sciences . Behav Res Meth Ins C 2007, 39:175–191.

35. Matteau E, Landreville P, Laplante L, Laplante C: Disruptive vocalizations: A means to communicate in dementia? Am J Alzheimers Dis Other Demen 2003, 18:147–153.

36. Kovach CR, Logan BR, Noonan PE, Schlidt AM, Smerz J, Simpson M, Wells T: Effects of the Serial Trial Intervention on discomfort and behavior of nursing home residents with dementia. Am J Alzheimers Dis Other Demen 2006, 21:147–155.

37. Barton S, Findlay D, Blake RA: The management of inappropriate vocalisation in dementia: a hierarchical approach. Int J Geriatr Psychiatry 2005, 20:1180–1186.

38. Kunik ME, Edwards M, Molinari VA, Hale DD, Orengo CA: Outcomes of decreased length of hospital stay among geriatric patients with dementia. Psychiatr Serv 2001, 52:376–378.

39. Rabinowitz J, Davidson M, De Deyn PP, Katz I, Brodaty H, Cohen-Mansfield J: Factor analysis of the Cohen-Mansfield Agitation Inventory in three large samples of nursing home patients with dementia and behavioral disturbance. Am J Geriatr Psychiatry 2005, 13:991–998.

40. Cohen-Mansfield J, Marx MS, Werner P, Freedman L: Temporal patterns of agitated nursing home residents. Int Psychogeriatr 1992, 4:197–206.

41. Burgio LD, Scilley K, Hardin JM, Hsu C: Temporal patterns of disruptive vocalization in elderly nursing home residents. Int J Geriatr Psychiatry 2001, 16:378–386.

42. Auer S, Reisberg B: The GDS/FAST staging system. Int Psychogeriatr 1997, 9, Suppl.1:167–171.

43. Herr K, Bjoro K, Decker S: Tools for assessment of pain in nonverbal older adults with dementia: A state-of-the-science review. J Pain Symptom Manage 2006, 31:170–192.

CITATION

Originally published under the Creative Commons Attribution License. Pelletier IC, Landreville P. "Discomfort and agitation in older adults with dementia," in BMC Geriatrics 2007, 7:27. © 2007 Pelletier and Landreville; licensee BioMed Central Ltd. doi:10.1186/1471-2318-7-27.

Factors Influencing Elderly Women's Mammography Screening Decisions: Implications for Counseling

Mara A. Schonberg, Ellen P. McCarthy, Meghan York,
Roger B. Davis and Edward R. Marcantonio

ABSTRACT

Background

Although guidelines recommend that clinicians consider life expectancy before screening older women for breast cancer, many older women with limited life expectancies are screened. We aimed to identify factors important to mammography screening decisions among women aged 80 and older compared to women aged 65–79.

Methods

Telephone surveys of 107 women aged 80+ and 93 women aged 65–79 randomly selected from one academic primary care practice who were able to communicate in English (60% response rate). The survey addressed the following factors in regards to older women's mammography screening decisions: perceived importance of a history of breast disease, family history of breast cancer, doctor's recommendations, habit, reassurance, previous experience, mailed reminder cards, family/friend's recommendations or experience with breast cancer, age, health, and media. The survey also assessed older women's preferred role in decision making around mammography screening.

Results

Of the 200 women, 65.5% were non-Hispanic white and 82.8% were in good to excellent health. Most (81.3%) had undergone mammography in the past 2 years. Regardless of age, older women ranked doctor's recommendations as the most important factor influencing their decision to get screened. Habit and reassurance were the next two highly ranked factors influencing older women to get screened. Among women who did not get screened, women aged 80 and older ranked age and doctor's counseling as the most influential factors and women aged 65–79 ranked a previous negative experience with mammography as the most important factor. There were no significant differences in preferred role in decision-making around mammography screening by age, however, most women in both age groups preferred to make the final decision on their own (46.6% of women aged 80+ and 50.5% of women aged 65–79).

Conclusion

While a doctor's recommendation is the most important factor influencing elderly women's mammography screening decisions, habit and reassurance also strongly influence decision-making. Interventions aimed at improving clinician counseling about mammography, which include discussions around habit and reassurance, may result in better decision-making.

Background

There is great heterogeneity in health among older women leading to substantial differences in life expectancy [1]. For instance, women aged 80–84 in the top quartile of health have 13 years of life expectancy while women aged 80–84 in the lowest quartile of health have only 4.6 years of life expectancy [1]. Meanwhile, experts generally agree that a woman needs 5 to even 10 years of life expectancy to potentially benefit from mammography screening [1,2]. Benefits of mammography screening among older women include possibly prolonging life or preventing morbidity associated with advanced breast cancer [3,4]. However, potential risks include complications and anxiety related to finding and treating breast cancers that would never have become clinically significant in an older woman's lifespan [1]. Therefore, guidelines recommend that clinicians consider older women's life expectancy and comorbidities before recommending mammography screening [5,6]. Increasingly more women aged 80 and older are undergoing mammography screening and evidence suggests it is not being targeted to the oldest women in the best health and most likely to benefit [7,8].

To better understand elderly women's mammography screening decisions, we interviewed women aged 80 and older and physicians who cared for these women using qualitative methods [9]. In that study, we developed a conceptual framework of the factors that influence mammography screening decisions of women aged 80 and older, including: 1) Patient factors (e.g., risk perception, habit, history of breast disease, etc.); 2) System factors (e.g., access, mailed reminders); 3) Social Influences (e.g., daughter's encouragement, family/friends' experience with breast cancer); and 4) Physician influences. We also found that physicians feel uncomfortable discussing stopping screening with women aged 80 and older. Since qualitative data cannot be used to determine the prevalence of attitudes or beliefs in a population and since qualitative methods do not allow for statistical comparisons between groups, we designed a telelphone survey to determine which factors identified in our qualitative study were most important to older women's mammography screening decisions. We were most interested in factors that influence elderly women to choose screening since these factors may need to be addressed before elderly women can feel comfortable stopping screening. We compared responses for women aged 80 and older with women aged 65–79 to see if certain factors need to be specifically addressed among the oldest women. In addition, we examined whether older women preferred that their physician make the decision whether or not they should get sreened or whether they preferred to make the decision on their own or share the decision with their physician.

Methods

Study Sample

We telephone surveyed English speaking women aged 65 or older that received their primary care at a hospital based general internal medicine practice in Boston to learn about their mammography screening decisions. The practice consists of approximately 50 faculty internists, over 100 internal medicine residents, and

10 nurse practitioners that provide care to approximately 34,000 patients. We excluded patients who had a history of dementia, significant hearing loss, or were terminally ill as determined by chart review and/or by patients' primary care physicians. We also excluded patients whose physicians thought that answering survey questions would be too psychologically disturbing (e.g., patient had just lost a spouse or was mentally ill). Our initial electronic search identified 716 women aged 80 and older and 1,962 women aged 65–79 who had at least one primary care billing record from a clinic visit in the past year. Since we anticipated greater exclusion criteria among women aged 80 and older, we randomly identified 400 women aged 80 and older and 275 women aged 65–79 from these lists to reach a targeted sample of 200 women (100 per age group). We obtained consent from each patient's primary care physician and we sent women deemed eligible a letter informing them of the study with an opt-out card. The 30-minute survey was administered by one of two study investigators (MS, MY) or by one research assistant. The institutional review board at the Beth Israel Deaconess Medical Center approved our study.

Data Collection

Our survey first asked women whether or not they had received a screening mammogram in the past two years. We then asked women who had undergone mammography screening in the past two years how important each of the following factors (identified in our qualitative study) were in their screening decision, including: history of breast disease (if applicable), family history of breast cancer (if applicable), a doctor's recommendation, habit (meaning that the woman always got a mammogram every year or so), reassurance (meaning that a normal mammogram would reassure a woman about her health), a mailed reminder card, a family member's recommendation, a friend's recommendation, a friend's experience with breast cancer, age, health, and the media [9]. We asked women who had not undergone mammography screening in the past two years how important their previous experience with mammography, doctor's counseling, habit, health, and the media were in their decision not to get a mammogram. Women rated the influence of each factor on their mammography screening decision on a 4-point scale (essential, very important, somewhat important or not at all important to their decision) [10]. After evaluating each measure individually, we then asked women to rank from 1 to 3, in order of importance, the factors that most influenced their decision to get a mammogram or not to get a mammogram. We also asked women who did not undergo mammography screening recently how strongly they agreed with the statement "I am not concerned about breast cancer" on a 5-point Likert scale. Women could rank lack of concern about breast cancer as one of their reasons for choosing not to be screened.

In addition, we asked women their preferred role in decision-making around mammography screening using a scale created by Degner et al [11]. Responses were categorized into 3 groups: the patient prefers to make her own decision about mammography screening, the patient prefers her doctor makes the decision, or the patient prefers that she share the decision with her doctor. We additionally obtained data on patient's race/ethnicity, education, income, marital status, functional dependency, and perceived health status [12]. We pre-tested the survey on 10 women who were identified using the methods above and we amended the survey based on these interviews.

To compare respondents with non-respondents we obtained data on patients' insurance, race/ethnicity, illness burden, and receipt of mammography in the past 2 years, from patients' medical records. We reviewed one year of patient clinic notes and their current problem list to collect data on patients' illness burden to calculate a Charlson Comorbidity Index (CCI) [13].

Statistical Analyses

Chi-square statistics were used to compare race/ethnicity, education, insurance coverage, income, perceived health status, CCI, and functional dependency between women aged 80 and older and women aged 65–79. For analyses involving participants we only used data collected from telephone interviews for consistency. Chi-square statistics were also used to compare respondent characteristics with non-respondents. We additionally used chi-square statistics to compare the relative importance of different factors (habit, reassurance, history of breast disease, reminder cards, doctor's recommendations, family history, health, friend's experience with breast cancer, age, family member recommendation, friend recommendation, and the media) on older women's decisions to undergo mammography screening and to compare the relative importance of different factors (habit, doctor counseling, age, previous experience with mammography, health, media, and no concern about breast cancer) on older women's decisions not to be screened. To construct a list of factors influencing older women's screening decision by age and order of importance, we weighted whether a woman ranked a factor as 1, 2, or 3 and then averaged the score provided to each factor across all women within the age group (first choices were scored as 3 and the third choices were scored as 1). We also used the two sample t-test and/or the Wilcoxon Rank Sum Test to compare whether the score calculated for each factor differed significantly by age. Finally, we compared older women's preferred role in decision-making around mammography screening with younger women and we compared receipt of mammography screening by women's preferred role in decision-making. All statistical analyses were performed using SAS statistical software, version 9.1 (SAS Institute, Cary, NC).

Results

Of the 166 women aged 80 and older and 170 women aged 65–79 ultimately eligible for our study, 102 women aged 80 and older and 98 women aged 65–79 agreed to be interviewed, resulting in a combined response rate of 60%. Non-respondents (n = 136) were similar to respondents with regard to age, race/ethnicity, insurance, CCI, number of clinic visits in the past year, and receipt of mammography. Five women initially identified in the 65–79 age group had turned 80 years by the time they were interviewed and their responses were included with women aged 80 and older. Of the 339 women excluded from the study, 115 did not speak English, 82 had dementia, 77 had left the practice, 24 were deceased, 17 were hearing impaired, 15 were terminally ill, and 9 had physicians who thought the survey would be too psychologically disturbing.

Of the 200 women who participated in the survey, 53.5% were aged 80 and older (mean age 85.3 years) and 46.5% were aged 65–79 (mean age 71.5 years). The majority were non-Hispanic white (65.5%) and were in good to excellent health (82.8%). Most (81.3%) had undergone mammography screening in the previous 2 years (88.2% of women aged 65–79 and 75.2% of women aged 80 and older, p = 0.02). Women aged 80 and older were significantly more likely than women aged 65–79 to report a household income less than $20,000 per year (55.7% vs. 28.8%) and to have a functional dependency (47.7% vs. 14.0%) (Table 1).

Table 1. Characteristics of respondents by age group.

		Age Group		p value
		65–79 (n = 93) %	80+ (n = 107) %	
Race:	Non-Hispanic White	60.2	66.4	0.35
	Non-Hispanic Black	30.1	29.0	
	Other	9.7	4.7	
Education:	< High School	16.1	14.2	0.39
	High School Graduate	28.0	35.9	
	Some College	28.0	18.9	
	College Graduate or beyond	28.0	31.1	
Income: (n = 168)	<$20K	28.8	55.7	<0.01
	$20K–<$35K	21.2	15.9	
	$35K or more	50.0	28.4	
Insurance:	Private	78.5	85.1	0.23
	Other	21.5	14.9	
Marital Status:	Married	43.0	17.8	<.01
	Widowed	22.6	60.8	
	Other	34.4	21.5	
Clinic Visits in past year:	1	12.9	11.2	0.74
	2–4	49.5	45.8	
	5+	37.6	43.0	
CCI:	0	49.5	43.9	0.15
	1	29.0	22.4	
	2+	21.5	33.6	
Function:	None	86.0	52.3	<0.01
	IADL dependency only[a]	12.9	37.4	
	ADL dependency[a]	1.1	10.3	
Perceived Health: (n = 192)	Excellent	31.2	18.2	0.08
	Very Good	26.9	30.3	
	Good	29.0	30.3	
	Fair	8.6	19.2	
	Poor	4.3	2.0	
Screening Mammogram in the past 2 years		88.2	75.2	0.02

[a] ADL = Activity of Daily Living, IADL = Instrumental Activity of Daily Living

Table 2 illustrates the proportion of women in each age group that considered various factors essential/very important to their mammography screening decision. Although there were no statistical differences by age, among women who were recently screened (n = 162) the majority considered a history of breast disease, a doctor's recommendation, receipt of a reminder card, reassurance, and habit as essential/very important factors in their decision (Table 2). Fewer women considered family history of breast cancer, family member or friend's recommendation, friend's experience with breast cancer, health, age, or the media as essential/very important to their decision. Only small numbers of older women chose not to be screened and there were no statistical differences in factors influencing this decision by age (Table 3). However, the majority of women aged 80 and older considered age as essential/very important to their decision not to be screened. In addition, 80.8% of women aged 65 and older who chose not to be screened were not concerned about breast cancer.

Table 2. Proportion of women who considered each factor essential/very important in decision to get a mammogram.

	65–79 (n = 82)	80+ (n = 80)	p Value
Habit	76.2	87.0	0.08
Reassurance	81.0	73.0	0.23
History of Breast Disease (n = 20 aged 65–79 and n = 28 aged 80+)	65.0	85.7	0.09
Reminder Card (n = 62 aged 65–79 and n = 38 aged 80+)	72.6	73.7	0.90
MD Recommendations	60.2	66.2	0.43
Family History	42.3	38.1	0.77
Health	36.6	47.1	0.19
Friend's Experience with Breast Cancer	30.8	25.0	0.57
Age	28.2	29.0	0.92
Family Member Recommendation	15.9	13.9	0.73
Friend Recommendation	16.9	6.3	0.06
Media	15.5	16.0	0.93

Table 3. Proportion of women who considered each factor essential/very important in decision NOT to get a mammogram.

	65–79 (n = 11) % (proportion n's)	80+ (n = 26) % (proportion n's)	P Value
Habit	50.0% (3/6)	40.0% (8/20)	0.66
MD Counseling	28.6% (2/7)	50.0% (11/22)	0.32
Age	50.0% (3/6)	56.5% (13/23)	0.77
Previous Experience with Mammography	28.6% (2/7)	18.8% (3/16)	0.60
Health	50.0% (3/6)	19.1% (4/21)	0.13
Media	0% (0/6)	8.3% (2/24)	0.46
Not Concerned about Breast Cancer	83.3% (5/6)	80.0% (16/20)	0.86

Figure 1 demonstrates how older women ranked the importance of different factors on their decision to undergo mammography screening. Both women aged 65–79 and women aged 80 and older ranked their doctor's recommendation as the most important factor influencing their decision. Habit and reassurance were the next two factors influencing older women's decision to get screened. There were no significant differences in scores given to factors that influence older women's mammography screening decisions by age. Women aged 80 and older who did not get screened with mammography in the past two years ranked age and then doctor's counseling as the most important factors influencing this decision. Women aged 65–79 ranked a previous negative experience with mammography as the most important factor influencing their decision not to get screened and then a lack of concern about breast cancer. Women aged 80 and older were significantly more likely than younger women to score age highly as a factor influencing their decision not to get screened (p = 0.02).

65-79 (n=82)	80+ (n=80)
1. Doctor's Recommendation	1. Doctor's Recommendation
2. Reassurance	2. Habit
3. Habit	3. Reassurance
4. Mailed Reminder Card	4. Personal history of breast disease
4. Personal history of breast disease	5. Age
6. Family history of breast cancer	6. Mailed reminder card
7. Friend's experience with breast cancer	6. Family history of breast cancer
8. Age	8. Media
9. Friend's recommendation	9. Friend's recommendation
9. Media	9. Friend's experience with breast cancer
11. Family member's recommendation	11. Family member's recommendation
12. Health	12. Health
12. Costs	12. Costs

Figure 1. How older women ranked factors influencing their decision to undergo mammography screening in the past two years.* * To create these rankings we asked women to rate the 3 factors most influential to their decision to undergo mammography in the past 2 years. We then weighted whether a woman ranked a factor as 1, 2, or 3 and averaged the score given to each factor across all women. We listed the factors by highest average score to lowest average score.

Table 4 demonstrates receipt of mammography screening among older women by their preferred role in decision-making around screening and by their age. There were no statistically significant differences in women's preferred role in decision-making by age. However, most women in both age groups preferred to make the final decision on their own about whether or not to undergo mammography screening. Women aged 80 and older who preferred to share their mammography screening decision with their doctor were less likely to be screened (59.1%) than women aged 65–79 who preferred to share their decision with their doctor (96.2%). In addition, in post hoc analyses using the Fischer Exact test, we found that women aged 80 and older who preferred to share their decision and perceived themselves to be in good to excellent health were significantly more likely to be screened (76.5%, n = 13/17) than those who preferred to share their decision and perceived themselves to be in fair or poor health (0% n = 0/4) (p = 0.01). Receipt of screening did not differ by perceived health among women aged 80 and older who did not prefer to share their decision making around screening with their physician.

Table 4. Receipt of mammography screening in the past 2 years by preferred role in decision-making around screening and by age.

| | Women aged 65–79 (n = 93) | | Women aged 80 and Older (n = 103) | | |
	Overall	(Proportion who reported being screened)	Overall	(Proportion who reported being screened)	P value*
Preferred Role in Decision Making:*					
Doctor makes the final decision	21.5% (n = 20)	(19/20 or 95.0%)	32.0% (n = 33)	(27/32† or 84.8%)	0.25
Patient makes the final decision	50.5% (n = 47)	(38/47 or 80.9%)	46.6% (n = 48)	(36/48 or 75.0%)	0.49
Shared decision	28.0% (n = 26)	(25/26 or 96.2%)	21.4% (n = 22)	(13/22 or 59.1%)	<0.01

* The P value represents the Chi-square test for reported receipt of mammography screening by age stratified by preferred role in decision-making. There is no significant difference by age in older women's preferred roles in decision making (p = 0.22).
†One woman was uncertain whether or not she had a mammogram in the past 2 years

Discussion

In this study we identified factors, such as a doctor's recommendation, habit, and the need for reassurance, that are very important to older women when deciding to undergo mammography screening. Understanding the influence of these factors in older women's decision-making is important to improve counseling around mammography screening. Ideally, communication and decision-making would improve such that most women aged 80 and older in poor health and unlikely to benefit would choose not to undergo screening while most women aged 80 and older in good health would choose to undergo screening. In this study we found that a doctor's recommendation is the most highly ranked factor influencing older women's decision to get screened. Moreover, screening is common among women who prefer that they alone or their doctor makes the final decision whether or not they get screened regardless of health. In contrast, among women aged 80 and older who preferred to share decision-making around screening with their physicians, patient health plays a significant role in decision-making. This may indicate that when physicians have the opportunity to discuss the risks and benefits of mammography screening, and patients have an opportunity to discuss their values and perspectives, older women feel comfortable stopping screening. Interventions aimed at improving clinician counseling about mammography screening with older women, which include discussions around habit and reassurance, may result in more optimal decision-making.

Although several studies offer clinicians advice on how to encourage patients to undergo mammography screening [14-16] and two studies provide clinicians data to help determine which of their older patients may benefit from screening [1,17], we are unaware of studies that offer clinicians advice on how to discuss stopping screening with their patients. As it is important for clinicians to learn how to recommend screening, it is also important for clinicians to learn how to discuss stopping screening, especially since older patients report that they want to have these discussions with their clinicians [18]. When discussing stopping screening with elderly women in poor health, clinicians may want to acknowledge they understand how hard it must be for a patient to stop going for mammography when they have been doing so for years. Clinicians may also want to explain that just as there is a time to start screening, there is also a time to stop screening. Clinicians should discuss the risks of screening elderly women in poor health (e.g., unknown benefit of screening and complications from work-up and treatment of breast cancer). In addition, since a need for reassurance about one's health is important to patients, clinicians should be sure to offer patients reassurance by saying they will focus on preventive health measures (e.g., screening for geriatric health issues) that are more likely to benefit these women. This may help prevent patients from feeling like their doctor is giving up on them.

Although it is difficult to draw conclusions about factors that influence older women to choose not to be screened due to the low numbers of these women in our study, women aged 80 and older who chose not to be screened ranked age and then doctor's counseling as the most important factors influencing their decision. In fact, women aged 80 and older were significantly more likely than younger women to score age highly in their decision not to get screened. Resnick et al. also found that age and a lack of a doctor's recommendation were the most common reasons older adults living in a retirement community gave for not undergoing screening [19]. Physicians may need to explain to older women in good health who prefer not to be screened, that health rather than age should influence their decision. This is especially important since observational studies have found a mortality benefit of regular mammography screening for women aged 80–84 in good health [3,20].

We additionally found that shared decision-making around mammography screening between women aged 80 and older and their clinicians may result in better targeting of screening to older women in good health. Shared decision making occurs when clinicians involve patients as active partners in clarifying and choosing acceptable medical options [21]. Experts recommend that clinicians engage in shared decision making with patients when there is insufficient evidence about the risk-benefit ratio of a test; such as in the case of mammography screening among women aged 80 and older [21]. Interventions or tools, such as decision aids, aimed at improving shared decision-making around mammography screening between clinicians and their elderly patients may result in more optimal use of screening [22].

There are several limitations to this study. We interviewed English speaking women at one academic primary care practice in Boston and our results may not be generalizable to other women, especially since mammography screening was more common among elderly women in this study compared to national studies. However, the factors influencing older women's decisions to choose screening should be similar to other older women well connected to primary care and for whom counseling about screening is most likely to occur. Not all women eligible for the study chose to participate, however, non-respondents were similar to respondents with respect to age, race/ethnicity, illness burden, and previous mammography screening. In addition, our response rate is similar to other telephone surveys assessing older adults screening decisions [23,24]. Since few women in our practice chose not to get screened with mammography, we have less data on why women choose not to undergo mammography screening than on why women choose to undergo screening. Finally, Degner's decision-making preference scale may oversimplify women's thoughts about their preferred role in decision-making around mammography screening [25]. Despite these limitations, this study provides important information on factors that influence elderly women's mammography screening decisions that may guide future interventions.

Conclusion

In summary, we found that several factors are highly important to older women's mammography screening decisions, including their doctor's recommendation, habit, and the need for reassurance; yet, personal health was not factored into these decisions. Physicians may want to address the influential roles of habit and reassurance and the unappreciated role of health to improve mammography screening discussions. Optimal decision-making may occur when elderly women have the opportunity to participate in shared decision-making around mammography screening with their clinicians.

Competing Interests

The author(s) declare that they have no competing interests.

Authors' Contributions

MS designed the study, participated in data collection, analyzed and interpreted the data, and prepared the manuscript. EP helped to design the study, analyze and interpret the data, and prepare the manuscript. MY collected the data and helped to prepare the manuscript. RD helped to design the study, analyze the data, and prepare the manuscript. EM helped to design the study, analyze and interpret the data, and prepare the manuscript.

Acknowledgements

This work was funded by an Older Americans Independence Center Grant Pilot Project and a Harvard/Hartford Foundation Junior Faculty Development Grant, a National Research Service Award from the National Cancer Institute (1 F32 CA110424-01), and a 2007 Society of General Internal Medicine T. Franklin Williams Scholars Award in Geriatrics. The funding bodies played no role in study design; in the collection, analysis, and interpretation of data; in the writing of the manuscript; and in the decision to submit the manuscript for publication.

References

1. Walter LC, Covinsky KE: Cancer screening in elderly patients: a framework for individualized decision making. JAMA 2001, 285:2750–2756.

2. Fletcher SW, Elmore JG: Clinical practice. Mammographic screening for breast cancer. New Eng J Med 2003, 348:1672–1680.

3. McCarthy EP, Burns RB, Freund KM, Ash AS, Shwartz M, Marwill SL, Moskowitz MA: Mammography use, breast cancer stage at diagnosis, and survival among older women. J Am Geriatr Soc 2000, 48:1226–1233.

4. Taplin SH, Ichikawa L, Yood MU, Manos MM, Geiger AM, Weinmann S, Gilbert J, Mouchawar J, Leyden WA, Altaras R, Beverly RK, Casso D, Westbrook EO, Bischoff K, Zapka JG, Barlow WE: Reason for late-stage breast cancer: absence of screening or detection, or breakdown in follow-up? J Natl Cancer Inst 2004, 96:1518–1527.

5. American Geriatrics Society Clinical Practice Committee: Breast cancer screening in older women. J Am Geriatr Soc 2000, 48:842–844.

6. Screening for breast cancer: recommendations and rationale Ann Intern Med 2002, 137(5 Part 1):344–346.

7. Walter LC, Lindquist K, Covinsky KE: Relationship between health status and use of screening mammography and Papanicolaou smears among women older than 70 years of age. Ann Intern Med 2004, 140:681–688.

8. Schonberg MA, McCarthy EP, Davis RB, Phillips RS, Hamel MB: Breast cancer screening in women aged 80 and older: results from a national survey. J Am Geriatr Soc 2004, 52:1688–1695.

9. Schonberg MA, Ramanan RA, McCarthy EP, Marcantonio ER: Decision-Making and Counseling around Mammography Screening for Women aged 80 or Older. J Gen Int Med 2006, 21:979–985.

10. Laine C, Davidoff F, Lewis CE, Nelson EC, Nelson E, Kessler RC, Delbanco TL: Important elements of outpatient care: a comparison of patients' and physicians' opinions. Ann Intern Med 1996, 125:640–645.

11. Degner LF, Kristjanson LJ, Bowman D, Sloan JA, Carriere KC, O'Neil J, Bilodeau B, Watson P, Mueller B: Information needs and decisional preferences in women with breast cancer. JAMA 1997, 277:1485–1492.

12. National Health Interview Survey [http://www.cdc.gov/nchs], 2000.

13. Charlson ME, Pompei P, Ales KL, MacKenzie CR: A new method of classifying prognostic comorbidity in longitudinal studies: development and validation. J Chronic Dis 1987, 40:373–383.

14. Grady KE, Lemkau JP, McVay JM, Reisine ST: The importance of physician encouragement in breast cancer screening of older women. Prev Med 1992, 21:766–780.

15. Hawley ST, Earp JA, O'Malley M, Ricketts TC: The role of physician recommendation in women's mammography use: is it a 2-stage process? Med Care 2000, 38:392–403.

16. Fox SA, Murata PJ, Stein JA: The impact of physician compliance on screening mammography for older women. Arch Intern Med 1991, 151:50–56.

17. Lee SJ, Lindquist K, Segal MR, Covinsky KE: Development and validation of a prognostic index for 4-year mortality in older adults. JAMA 2006, 295:801–808.

18. Lewis CL, Kistler CE, Amick HR, Watson LC, Bynum DL, Walter LC, Pignone MP: Older adult' attitudes about continuing cancer screening later in life:

a pilot study interviewing residents of two continuing care communities. BMC Geriatrics 2006, 6:10.

19.   Resnick B: Health promotion practices of the older adult. Public Health Nursing 2000, 17:160–168.

20.   McPherson CP, Swenson KK, Lee MW: The effects of mammographic detection and comorbidity on the survival of older women with breast cancer. J Am Geriatr Soc 2002, 50:1061–1068.

21.   Sheridan SL, Harris RP, Woolf SH, the Share Decision-Making Workgroup for the US Preventive Services Task Force: Shared decision making about screening and chemoprevention; a suggested approach from the US Preventive Services Task Force. Am J Prev Med 2004, 26:56–66.

22.   Volk RJ, Spann SJ, Cass AR, Hawley ST: Patient education for informed decision making about prostate cancer screening: a randomized controlled trial with 1-year follow-up. Ann Fam Med 2003, 1:22–28.

23.   Messina CR, Lane DS, Grimson R: Colorectal cancer screening attitudes and practices preferences for decision making. Am J Prev Med 2005, 28:439–446.

24.   Weinberg DS, Turner BJ, Wang H, Myers RE, Miller S: A survey of women regarding factors affecting colorectal cancer screening compliance. Prev Med 2004, 38:669–675.

25.   Davey HM, Lim J, Butow PN, Barratt AL, Redman S: Women's preferences for and views on decision-making for diagnostic tests. Soc Sci Med 2004, 58:1699–1707.

CITATION

Do Social Networks Affect the Use of Residential Aged Care Among Older Australians?

Lynne C. Giles, Gary F. V. Glonek, Mary A. Luszcz
and Gary R. Andrews

ABSTRACT

Background

Older people's social networks with family and friends can affect residential aged care use. It remains unclear if there are differences in the effects of specific (with children, other relatives, friends and confidants) and total social networks upon use of low-level residential care and nursing homes.

Methods

Data were drawn from the Australian Longitudinal Study of Ageing. Six waves of data from 1477 people aged ≥ 70 collected over nine years of follow-up were used. Multinomial logistic regressions of the effects of specific and total social networks on residential care use were carried out. Propensity scores were used in the analyses to adjust for differences in participant's health, demographic and lifestyle characteristics with respect to social networks.

Results

Higher scores for confidant networks were protective against nursing home use (odds ratio [OR] upper versus lower tertile of confidant networks = 0.50; 95%CI 0.33–0.75). Similarly, a significant effect of upper versus lower total network tertile on nursing home use was observed (OR = 0.62; 95%CI 0.43–0.90). Evidence of an effect of children networks on nursing home use was equivocal. Nursing home use was not predicted by other relatives or friends social networks. Use of lower-level residential care was unrelated to social networks of any type. Social networks of any type did not have a significant effect upon low-level residential care use.

Discussion

Better confidant and total social networks predict nursing home use in a large cohort of older Australians. Policy needs to reflect the importance of these particular relationships in considering where older people want to live in the later years of life.

Background

At any point in time in Australia, around one in ten older people have left their home to receive either respite or permanent care in a residential care facility [1]. The Australian aged care system is a tiered system that comprises both community and residential aged care places. Residential aged care may be provided as either 'high-level' or 'low-level' care, depending on clients' needs. In the Australian aged care system, high-level care is equivalent to nursing home care in other countries, and reflects high levels of medical and personal care needs. Low-level residential care (also referred to as 'hostel care' in the Australian system) provides help and housing to older people who do not need continual, high level access to nursing care but have physical, medical, psychological or social care needs that cannot be met through living in the community [2].

Both high-level and low-level residential aged care services are predominantly funded and regulated by the Australian Government [1]. Currently a total of 88 residential aged care places per 1,000 people aged 70 years or more is provided in the Australian aged care system [1]. Religious and charitable organizations deliver the majority of residential aged care services in Australia, although publicly listed companies and small community-based organizations also deliver residential aged care services to a significant number of older people.

Unlike some other countries with specific taxation levies or social insurance programs, the Australian Government funded services are financed from general taxation revenue and user contributions [1]. From both individual and societal perspectives, there are high personal and financial costs associated with admission to residential care [3]. For the Australian Government, the costs of supplying aged care services are forecast to increase from $7.8 billion in 2002–2003 to $106.8 billion in 2042–2043 [4].

A substantial body of U.S. dominated research has identified factors including increasing age, female gender, lack of a marital partner, greater income, better education, lack of home ownership, more comorbid conditions, poorer self-rated health, prior nursing home use, more physical disability, and poorer cognitive status as significant predictors of residential care use, as reviewed recently [5,6]. The findings have been drawn from both cross-sectional and longitudinal studies with follow-up time that varied between one and twenty years, with the median length of follow-up equal to three years.

Social networks with family and friends may be particularly important in providing care to older people, and may thereby delay or prevent admissions to residential care [7-12]. However, few studies have distinguished between social networks with family and those with friends and separately examined their effects on use of residential care. Among those studies that have made this distinction, findings are conflicting. For example, Wolinksy et al. [13] reported non-kin networks were protective but kin networks were not, whereas Freedman [9] demonstrated networks with daughters and siblings, but not sons, were protective against nursing home use. The meta-analysis by Gaugler et al. [5] demonstrated that a greater number of children was protective against nursing home admission, although it is worth noting that only three studies were available for pooling in their analysis of the effects of children, reflecting the paucity of evidence in this area.

It is also likely that social networks are themselves related to some of the factors that have been demonstrated as predictors of residential care use. Therefore it is difficult to make a clear interpretation of the effects of specific and total social networks on residential care based on the existing literature.

Surprisingly little is known about the factors that predict residential care use in Australia. Two recent Australian studies have examined some health, function and lifestyle risk factors for entry to nursing homes. McCallum et al. [13] showed increasing age, incontinence, impaired respiratory flow, more disability, depression, male gender and lower alcohol intake were associated with nursing home use over a 14 year period. Wang et al. [14] found older age, poorer self-rated health, walking disability, current smoking, and lower alcohol consumption were risk factors for admission to nursing homes. However, both studies were set in narrowly defined geographic regions in the same Australian state and were focussed on cardiovascular and ophthalmologic factors respectively. Thus the generalisability of the findings to the wider Australian population remains unclear. Furthermore, social networks and low-level residential care were not considered in these studies. These important gaps in knowledge are addressed in the research reported here.

The primary aim of the present study was to consider the effects of specific types of networks (i.e. those with children, other relatives, friends, confidants, and total social networks) upon use of both low-level residential care and nursing homes in a large sample of older Australians, adjusting for the effects of a wide range of health, function and lifestyle variables. A secondary aim was to examine the effects of putative risk factors on use of nursing homes to add to the knowledge gained from the two previous Australian studies.

Methods

Sample

This study uses data from the Australian Longitudinal Study of Ageing (ALSA), a large epidemiological study which aims to increase our understanding of how social, biomedical, behavioral, economic and environmental factors are associated with age-related changes in the health and social well-being of older persons. The study has been described in detail elsewhere [15,16]. In brief, ALSA began in 1992 and is continuing with survivors of the original cohort. The primary sample for ALSA was randomly selected from the South Australian Electoral Roll, and was stratified by Local Government Area (LGA), gender, and five year age groups from 70–74 years through to 85 years and over. Older males aged 85 years or more were deliberately over-sampled to provide sufficient numbers of males for longitudinal follow-up. Persons identified through the Electoral Roll were defined as eligible for the study if they were resident in the Adelaide Statistical Division and were aged 70 years or more on 31 December 1992. Both community-dwelling and people living in residential care were eligible to take part in ALSA, although the majority of participants (91%) were living in the community at baseline interview. A total of 1477 eligible people took part in wave 1 (56% response rate).

Ethical approval for the study was granted by the relevant institutional ethics committee, and each study participant provided written informed consent.

Data Collection and Measures

Eight waves of data have been collected from participants between 1992 and 2005, with fieldwork for a ninth wave due to commence in late 2007. In the present article, data from the first six waves were analyzed. Waves 1 to 4 were annual interviews that began in 1992, and consenting participants were re-interviewed in 1993, 1994 and 1995. Wave 5 occurred in 1998, and wave 6 was conducted in 2000–2001. Waves 1, 3 and 6 involved detailed personal interviews that covered demographic, medical, psychological, social and economic areas of participants' lives. As well, clinical assessments of participants were carried out in these waves. The clinical examination included anthropometric, neuropsychological, physical performance, balance, and gait measures. Both the interview and clinical assessment were carried out in the participant's usual place of residence. Waves 2, 4 and 5 each consisted of a brief telephone interview that concentrated mainly on health and lifestyle.

Residential Care Use

At each interview, participants were classified as living in the community, low-level residential care, or nursing home. For waves 2 through 6, participants were classified as missing if they refused an interview or were untraceable. Ongoing searches of the database of official death certificates identified the participants who had died, and this approach has been validated previously by the authors [17].

Variables that summarized any use of low-level care or nursing homes over the nine-year study period were also created. For low-level care use, participants were classified as never using, using low-level care, already in nursing home at wave 1, or missing. Participants who died without known use of residential care were classified as never using. An analogous variable was created to reflect nursing home use. Participants' status was classified as missing if use of the relevant residential care could not be ascertained from the available data.

Social Networks

Adapting the approach of Glass and colleagues [18], confirmatory factor analyses of the wave 1 data were used to develop measures of social networks with children, other relatives, friends, confidants and total social networks. The derivation of the social network variables has been reported previously [19]. Briefly, the children network combined information on the number and proximity of children, and frequency of personal and phone contact with children. The relatives network was composed of the number of relatives, apart from spouse and children, the participant felt close to, and the frequency of personal and phone contact with such relatives. Similarly, the friends network captured the number of close friends, and frequency of personal and phone contact with friends. The confidant network reflected the existence of confidants and whether the confidant was a spouse. A total social network score was calculated as the sum of the children, relatives, friends, and confidant scores. Social network variables were then categorized according to their tertiles, and the tertile classification for each social network was used in further analyses.

Propensity Scores

A range of personal, health, and lifestyle variables were considered important covariates (see Table 1). Geographic area (with 24 levels that designate locality) was also included as a covariate, but is excluded from Table 1 for space considerations. There were many covariates and their distributions were unbalanced among the social network categories, in that participants in different social network categories (i.e. low, medium or high) tended to have different demographic, health and lifestyle characteristics.

In randomized controlled trials, group assignment is, by definition, randomly allocated and so the differences in observed covariates between treatment groups should be minimized. However, in observational studies such as ALSA, there is no manipulation of 'treatment' assignment, and so there is the potential for large differences between observed covariates in the different treatment groups. Ignoring these differences could potentially lead to biased estimates of treatment effects. Traditional methods of adjusting for observed covariates in analyses, such as matching or stratification, may be difficult to use if there are a large number of covariates. Regression adjustment can also be problematic. Missing values in one or more covariates for an individual will result in all data for that individual being dropped from a regression analysis unless estimation of the missing covariate values is carried out. Another potential problem in regression adjustment is that finding and fitting an appropriate functional form for each covariate may be difficult.

Propensity scores have been proposed as an alternative method to adjust for a set of covariates [20,21]. Most applications of propensity scores to date have involved simple cross-sectional studies with binary treatments. More recent work [21] that extends the derivation of propensity scores to treatments with multiple categories is applicable in the present study. In our study the 'treatment,' social network tertile, has three categories corresponding to the low, mid, or high tertile of the relevant social network score for each of the specific and total social networks.

We turn now to a more formal definition of propensity scores, and first consider their derivation for a binary treatment. Let Z_i be an indicator variable of assignment to a treatment for individual i, such that Math.

Table 1. Summary of baseline covariates and association with any nursing home use over study period

Variable	Classification	n (%)	Odds ratio (95%CI) (n = 909)[a]
Age group	70–74	379 (25.7%)	Referent
	75–79	352 (23.8%)	4.2 (1.7 – 10.3)
	80–84	341 (23.1%)	3.7 (1.5 – 9.0)
	85+	405 (27.4%)	4.1 (1.7 – 9.8)
Gender	Male	928 (62.8%)	Referent
	Female	549 (32.2%)	0.7 (0.4 – 1.2)
Education	Left school >14 yrs	633 (42.9%)	Referent
	Left school ≤14 yrs	830 (56.2%)	1.0 (0.6 – 1.5)
	Missing	14 (0.9%)	
Marital status	Married/de facto	771 (52.2%)	Referent
	Widowed	586 (40.0%)	1.2 (0.5 – 2.8)
	Single	120 (8.1%)	1.4 (0.8 – 2.4)
Household income	>$AUD12,000	779 (52.7%)	Referent
	≤$AUD12,000	590 (39.9%)	2.0 (1.2 – 3.2)
	Missing	108 (7.3%)	
Home ownership	Owns home	1038 (71.0%)	Referent
	Renting	242 (16.4%)	0.8 (0.5 – 1.5)
	Other	50 (3.4%)	0.4 (0.1 – 1.8)
	In residential care	137 (9.3%)	4.8 (2.5 – 9.0)
Number of chronic conditions[b]	0	264 (17.9%)	Referent
	1	494 (33.4%)	1.3 (0.7 – 2.4)
	2	421 (28.5%)	0.9 (0.5 – 1.8)
	3+	298 (20.2%)	0.6 (0.3 – 1.2)
Self-rated health	Excellent/very good	563 (38.1%)	Referent
	Good	440 (29.8%)	1.2 (0.7 – 2.0)
	Fair/poor	469 (31.8%)	1.4 (0.8 – 2.5)
	Missing	5 (0.3%)	
Hearing difficulty	No	726 (49.2%)	Referent
	Yes	746 (50.5%)	1.5 (1.0 – 2.3)
	Missing	5 (0.3%)	
Difficulty with (corrected) vision [43]	No	1035 (70.1%)	Referent
	Yes	375 (25.4%)	1.3 (0.8 – 2.1)
	Missing	67 (4.5%)	
Mobility disability [44,45]	No disability	949 (64.3%)	Referent
	Disability	506 (34.3%)	1.5 (0.9 – 2.4)
	Missing	22 (1.4%)	
Depressive symptoms CES-D [46]	<17/60	1181 (80.0%)	Referent
	≥17/60	219 (14.8%)	1.2 (0.7 – 2.1)
	Missing	77 (5.2%)	
Cognitive function [40, 47]	>16/21	1221 (82.7%)	Referent
	≤16/21	219 (14.8%)	1.6 (0.9 – 2.7)
	Missing	37 (2.5%)	
Alcohol consumption (AUDIT) [48]	<8/10	1401 (94.9%)	Referent
	≥8/10	65 (4.4%)	1.0 (0.3 – 3.0)
	Missing	11 (0.7%)	
Exercise status [16, 49]	Exerciser	794 (53.8%)	Referent
	Sedentary	663 (44.9%)	1.0 (0.7 – 1.6)
	Missing	20 (1.4%)	
Smoking status	Never	661 (44.8%)	Referent
	Former	667 (45.2%)	0.6 (0.4 – 1.0)
	Current	123 (8.3%)	0.5 (0.2 – 1.2)
	Missing	16 (1.1%)	

a: analysis based on data from 909 participants with complete information on both nursing home use (n = 1078) and risk factors (n = 1243)
b: self-reported ever suffering from arthritis, cancer, chronic bronchitis or emphysema, diabetes, fractured hip, heart attack, heart condition, hypertension, osteoporosis, stroke

The propensity score $p(x_i)$ is defined as the conditional probability of assignment to treatment versus control given a vector of observed covariates xi. More formally $p(x_i) = \Pr(Z_i = 1 | X_i = x_i)$ under the assumption that, given the Xi, the Zi are independent—that is,

$$\Pr\left(Z_1 = z_1, ..., Z_n = z_n \mid X_1 = x_1, ... X_n = x_n\right) = \prod_{i=1}^{n} p\left(x_i\right)_i^{z} \left\{1 - p\left(x_i\right)\right\}^{1-z_i}$$

In other words, $p(x_i)$ is a measure of the probability that an individual would have been treated based on only the individual's covariate information [22]. Propensity scores 'balance' the observed covariates, in that the conditional distribution of X_i given $p(x_i)$ is the same for individuals, irrespective of whether they receive treatment or control. In other words, Z_i and X_i are conditionally independent given $p(x_i)$. The success of the propensity scores in balancing the covariates can be checked through simple comparisons of the treatment and control groups that adjust for the propensity scores in the analyses [22].

Applications of propensity score adjustment with more than two treatment categories have not been widely reported. For three or more levels of treatment, Joffe and Rosenbaum [21] showed that if the distribution of treatment doses given X_i is accurately described by McCullagh's proportional odds model, then stratifying on $b(x_i) = x_i'\beta$ where β is a $p \times 1$ vector of parameters, will balance X_i across several dose groups. More generally, it is possible that the distribution of doses Z given a large number of covariates may depend on the covariates through only a small number of linear functions of X, say XG for some matrix G. Then controlling for the several variables in XG will 'tend to balance the ... variables in X' [21]. For example, if a multinomial logistic regression model was adequate to describe $Pr(Z_i = z | X_i = x_i)$ for some $z = 0, 1, ..., c$, then XG would be an $n \times c$ matrix in which each of the c columns defined a propensity score for level c of the treatment dose. Because of the linear dependence of the cth propensity score on the first c-1 propensity scores, analyses would adjust for the first c-1 propensity scores.

To obtain propensity scores in the present analysis, an ordinal logistic regression of each of the social network tertiles on the covariates was initially fit for participants with complete covariate data (n = 1243). A pragmatic approach was adopted so that if, for a given participant, an observation was missing for at least one of the covariates, a propensity score was estimated using the subset of covariates with complete data for that participant. In this way, a propensity score was estimated for every participant, not only those participants with complete data for all covariates, and a propensity score was estimated for each pattern of missing covariate observations. Thus for every participant, the propensity scores were estimated using the maximum covariate information available for that participant.

Ordinal logistic models were appropriate for the distribution of the children, relatives, friends and total social network variables given the observed covariates, but not for confidants. For the confidant network variable, a multinomial logistic regression model was used. The resulting conditional probabilities of being in each of the three confidant network categories defined the three propensity scores for the confidant social network [23,24]. Because of the linear dependence of the third propensity score on the other two, only the first two propensity scores were included in subsequent analyses of the effect of the confidant social network upon use of residential care.

Rosenbaum and Rubin [20], based on work by Cochran [25], stated that five strata based on the propensity score would remove over 90% of the bias in each of the covariates. Thus when an ordinal logistic model was used, participants were sub-classified into quintiles based on the propensity scores. When the multinomial logistic model was used, participants were sub-classified into nine strata based on the joint distribution of their first two propensity scores. The propensity score strata for each participant was included in all analyses.

The balance of the covariates in each of the propensity score strata in the present study was examined by chi-square tests of association of each covariate with each of the categorized social network variables [23]. A total of 25 out of 484 comparisons for balance status of the covariates (5%) were statistically significant at $P < 0.05$. This indicated that the propensity score method produced balance in the observed covariates similar to that which would be expected by randomization of these covariates across the social network tertiles. On this basis, it was determined that the propensity scores provided an adequate adjustment.

Statistical Analysis

Several analyses of place of residence were conducted. First, the effects of the putative risk factors on any nursing home use over the study period were examined to enable comparison with previous Australian studies [13,14]. An unordered, multinomial, multiple logistic regression model that included the factors shown in Table 1 and geographic area was fit.

Second, separate logistic regression models of i) any low-level care use and ii) nursing home use across the study period on each social network variable were fit, adjusting for propensity score strata. Sensitivity analyses, in which missing values were imputed as never used (most conservative) or all used (most extreme), were conducted to compare the effects of different assumptions regarding missing values with the analyses that used only available data.

Finally, the place of residence across the six study waves was longitudinally analyzed, with response categories of community, low-level care, nursing home, or dead possible at each wave. A separate multinomial logistic regression model of place of residence at waves 2–6 on each social network was fit, adjusting for propensity score strata, study wave, and place of residence at the previous wave. The Huber-White robust variance estimator was used to account for the repeated observations ($n \leq 5$ observations corresponding to waves 2–6) from each participant [26,27].

Results

Table 1 summarizes the baseline characteristics of the 1477 participants. The average age at selection was 79.8 years (SD = 6.9), and close to two-thirds of the sample were male. More than half of the participants had left school before the age of 15 years, and approximately half of the sample was married/partnered. Participants most commonly had one morbid condition, and 15% of participants showed some signs of cognitive deficits. More than half of the participants were former or current smokers, and almost half of the participants were sedentary. Also shown in Table 1 are odds ratios that describe the association of any nursing home use with each of the risk factors. These results are described later in this section.

Across the entire study period, a total of 778 participants (53%) did not use low-level care, or died without use, and a further 136 participants (9%) were either in a nursing home at wave 1 or moved directly to a nursing home from the community. Low-level care was known to have been used over the study period by 189 participants (13%). Information on use of low-level care could not be ascertained for 374 participants (25%) because they were alive but not interviewed at one or more waves, and thus their use of residential care at the missing wave(s) could not be determined.

Over the nine years of the study, 883 participants (60%) never used a nursing home or died without use, while 195 participants (13%) used a nursing home. Nursing home use could not be ascertained for the remaining 399 participants (27%), for the same reason as those with missing low-level care information.

The place of residence at each wave is shown in Table 2. The percentage of the surviving cohort living in the community decreased over the nine years from 91% at wave 1 to 82% at wave 6. Between 6% and 8% of participants lived in low-level care at each of the waves. The proportion of participants who were resident in nursing homes increased over time, from 3% at wave 1 to 12% at wave 6.

A total of 909 participants had complete data concerning any nursing home use across the nine-year study period and the putative risk factors. As summarized in Table 1, age group, lower household income, lack of home ownership and hearing difficulty were significant risk factors for nursing home use over the study period.

The effects of social networks on use of low-level care and nursing home use were then explicitly considered. As shown in Table 3, better social networks with children, confidants and total social networks appeared protective against any nursing home use across the study period, after adjusting for propensity score strata. However, only the upper tertile of children networks in comparison to the lower tertile had a significant effect on any nursing home use. Moreover, there was no evidence of a gradient of the effect of children networks on any nursing home use.

Table 2. Place of residence at each wave

Year	Wave	Community	Low-level care	Nursing Home	Missing	Dead
1992	1 (n)	1,340	92	45	0	0
	(% all)	91	6	3	0	0
1993	2 (n)	1,126	83	51	137	80
	(% all)	76	6	4	9	5
	(% alive)	89	7	4		
1994	3 (n)	1,030	80	61	113	193
	(% all)	70	5	4	8	13
	(% alive)	88	7	5		
1995	4 (n)	900	74	62	150	291
	(% all)	61	5	4	10	20
	(% alive)	87	7	6		
1998	5 (n)	646	64	51	210	506
	(% all)	44	4	4	14	34
	(% alive)	85	8	7		
2000	6 (n)	412	28	60	215	762
	(% all)	28	2	4	14	52
	(% alive)	82	6	12		

Shown are number and per cent of all participants (% all) and surviving participants (% alive).

Table 3. Summary of effect of social networks upon any nursing home use and any low-level care use

	Low-level care[a]		Nursing Home[b]	
	OR	95% CI	OR	95% CI
Any use over study period				
Children				
Mid tertile	1.44	0.97 – 2.15	1.02	0.69 – 1.50
Upper tertile	0.97	0.64 – 1.46	0.60	0.40 – 0.90
Overall χ^2_2 [c]		4.81		8.60*
Relatives				
Mid tertile	0.98	0.66 – 1.44	0.81	0.56 – 1.18
Upper tertile	1.00	0.65 – 1.53	0.76	0.50 – 1.17
Overall χ^2_2		0.02		1.81
Friends				
Mid tertile	1.05	0.71 – 1.55	0.78	0.54 – 1.14
Upper tertile	1.29	0.85 – 1.95	0.70	0.46 – 1.06
Overall χ^2_2		1.56		3.13
Confidants				
Mid tertile	1.53	1.04 – 2.24	0.67	0.46 – 0.97
Upper tertile	1.10	0.67 – 1.79	0.49	0.31 – 0.77
Overall χ^2_2		5.16		10.75*
Total				
Mid tertile	1.49	1.01 – 2.21	0.55	0.37 – 0.81
Upper tertile	1.25	0.80 – 1.95	0.54	0.35 – 0.83
Overall χ^2_2		4.05		12.21*

Lower tertile is referent category in all analyses
a: Complete data available for 1103 cases.
b: Complete data available for 1078 cases.
c. *: χ^2 on 2 df; values > 5.99 significant at $P < 0.05$

There was no significant effect of the specific or total social network variables upon low-level care use. The findings were robust to assumptions regarding the use of residential care by participants with missing data, as the sensitivity analyses did not differ substantively from the main results.

Table 4 summarizes the longitudinal analysis of the effect of social networks on place of residence, adjusted for propensity score strata, study wave, and residence at previous wave. As these results show, specific and total social networks did not have a significant effect upon low-level care use. The significant effect observed for the friends network was due to the protective effect of better friend networks upon survival. Higher scores for confidant networks appeared protective against nursing home use (odds ratio [OR] upper versus lower tertile of confidant networks = 0.50; 95%CI 0.33–0.75). Similarly, a significant effect of upper versus lower tertile for the total social network was observed (OR = 0.62; 95%CI 0.43–0.90).

Table 4. Summary of effects of social networks upon place of residence across study period

	Low-level care OR	95% CI	Nursing home OR	95% CI	Dead OR	95% CI
Children						
Mid tertile	1.03	0.74 – 1.44	1.24	0.89 – 1.72	1.11	0.87 – 1.42
Upper tertile	0.68	0.48 – 0.96	0.85	0.60 – 1.21	1.02	0.79 – 1.31
Overall χ_6^2 = 8.6[a]						
Relatives						
Mid tertile	0.93	0.67 – 1.29	0.72	0.53 – 1.00	0.95	0.75 – 1.20
Upper tertile	0.92	0.64 – 1.33	0.74	0.51 – 1.07	1.07	0.82 – 1.38
Overall χ_6^2 = 13.5						
Friends						
Mid tertile	1.28	0.91 – 1.81	0.99	0.70 – 1.38	0.92	0.73 – 1.17
Upper tertile	1.22	0.83 – 1.78	0.74	0.52 – 1.07	0.72	0.56 – 0.93
Overall χ_6^2 = 26.4						
Confidants						
Mid tertile	0.95	0.69 – 1.32	0.70	0.51 – 0.97	0.84	0.67 – 1.06
Upper tertile	0.86	0.57 – 1.31	0.50	0.33 – 0.75	0.73	0.56 – 0.94
Overall χ_6^2 = 31.0						
Total						
Mid tertile	0.77	0.55 – 1.08	0.57	0.40 – 0.81	0.76	0.60 – 0.97
Upper tertile	0.94	0.67 – 1.34	0.62	0.43 – 0.90	0.84	0.65 – 1.08
Overall χ_6^2 = 30.7						

Lower tertile is referent category in all analyses. Community dwelling is referent response category.

a: χ_6^2 test of effect of social network variable; values > 12.59 significant at P < 0.05.

Discussion

The effects of specific and total social networks on residential care use were examined over a nine year period, using propensity score methods to adjust for a broad range of covariates. Longitudinal analyses showed better confidant networks and better total social networks were associated with reduced odds of nursing home admission over the course of the study. There was weaker evidence of a significant effect of better children networks on reduced odds of nursing home use, and there was no evidence of an effect of children networks in the longitudinal analyses. There was no significant effect of social networks with other relatives or friends on nursing home use. Furthermore, the results suggested specific and total social networks had little effect on use of low-level residential care over the period of the study.

Increasing age, lower income, and hearing difficulty were shown to be significant risk factors for nursing home use across the course of the study, adding to previous Australian research in this area. The finding regarding hearing difficulty adds more evidence to the need for adequate assessment of sensory impairments at the time of assessment for nursing home placement [14] and ongoing monitoring of auditory acuity. In contrast to visual acuity, hearing difficulties may go unnoticed, as the behavioral consequences may not be immediately obvious in the context of competing demands on staff time and attention. The effect of income on risk of nursing home admission is equivocal in the international literature, with some authors reporting reduced income to increase risk [28,29], while others have shown higher income is a risk factor for nursing home admission [30-33]. The results for income reported here possibly reflect that older Australians with a lower income may not be able to purchase support services to assist them to continue to live in the community, and so are more likely to move to residential care. Furthermore, higher income may be a disincentive to nursing home use in Australia. Substantial entry costs or ongoing costs in addition to the Australian Age Pension can be levied by individual facilities according to means-tested criteria. Issues of equity and access to residential care must remain high on the agenda for Australian aged care policy makers.

Confidant networks were significantly protective against nursing home use in this study, suggesting a close, emotionally supportive relationship with another person is beneficial in preventing or delaying nursing home use. The importance of a confidant to mental and physical health is well known [34-36] but the translation of that effect to a reduction in risk of nursing home use has not been shown previously. Further research is clearly warranted to examine the repeatability of this finding in other settings and countries.

Social networks with relatives and friends had no significant effect on use of residential care. Children networks appeared to have some protective effect against any nursing home use over the study period, but this finding did not extend to the results from the longitudinal analyses. Those with fewer non-kin social supports may have smaller networks of human resources to draw upon for maintenance of community living status [12,37]. Other research has shown significant protection against nursing home use arising from having daughters and siblings [8]. Our research suggests that the core network of confidants, and to a lesser extent children, is more important than other specific networks in delaying or preventing use of nursing homes in Australia. The striking impact of absence of confidants may reflect the consequences of reduced emotional support that permitted continued residence in the community, which would be consistent with Carstensen's socioemotional selectivity theory [38]. Social networks of any of the types considered here had minimal effect upon use of low-level care.

Several limitations to the study must be acknowledged. ALSA non-respondents may have been more socially isolated than participants, although non-response bias has been demonstrated as minimal in other analyses of ALSA data [15,39,40]. The analyses were based on self-reported data and adjusted for covariates that were measured at wave 1. Social networks may have changed over time, but the social networks considered in the present study were based on only wave 1 data. However, total network size has been demonstrated as relatively stable over a long follow-up period in a study of older Dutch people [41]. Furthermore, disentangling the effects of time-varying social networks may be difficult as changes in social networks may be a consequence of changes in place of residence. A final limitation is that date of entry to residential care was not available, and thus residential care use between study waves was not reflected in the data.

Arguably these limitations are balanced by ALSA's strengths, which include the rich baseline data that enabled propensity score adjustment, the broad sample, and the Australian setting, which expands the generalisability of the role of social networks in the use of residential care. The follow-up time in the present study is also notably longer than that of many other international studies in this area. Our results add not only to the general body of knowledge concerning risk factors for residential care use, but also extend the literature to encompass the specific role played by social networks in this important transition. In future research, we will track place of death for study decedents which will reduce the proportion of missing data concerning the use of residential care over time.

ALSA took place against a background of reforms in Australian aged care [42] that may have had an impact on the use of residential care services independent of the risk factors considered here. One of the most significant reforms saw the assessment for entry to low-level and high-level residential aged care merged into one system in 1997. It is important to note that the policy changes did not affect an individual's eligibility for residential aged care, but streamlined the administrative processes concerning assessment criteria. An individual's eligibility for residential aged care is ascertained by Aged Care Assessment Teams (ACAT) against standardized criteria that include functional status, health and living arrangements. The persistent effects of social networks on use of nursing homes over a long period of follow-up and over and above the effects of a range of other variables suggest that an individual's social milieu needs to be reflected more strongly in eligibility criteria, particularly for high-level residential care. The results of the present study also highlight the importance of recognizing that social networks go beyond a simple ascertainment of marital status or number of children. It may be possible to incorporate the findings from the present study in better screening assessments by ACATs for residential care eligibility. Policymakers may need to reconsider whether social relationships have been given adequate weight in the current assessment and entry process.

The effects of social networks on residential care use have not been examined previously in an Australian context. We have shown that social networks with children and total social networks, especially those with confidants, predict nursing home use over nine years in a large cohort of older Australians. Policy needs to reflect the importance of these particular relationships, and incorporate these along with the expectations of future cohorts of older people about where they want to live in later life.

Competing Interests

The author(s) declare that they have no competing interests.

Authors' Contributions

LG conceived of the present study, participated in the design and conduct of the statistical analysis plan and had primary responsibility in drafting the manuscript. GG participated in the design of the study and the statistical analysis plan and participated in drafting of the manuscript. ML participated in the design of the study and in drafting of the manuscript. GA conceived of and directed the Australian Longitudinal Study of Ageing, and participated in the design of the present study.

LG, GG and ML read and approved the final manuscript. GA commented on early drafts of the manuscript but was unable to approve the final version due to his death in May 2006.

Acknowledgements

We wish to thank the participants in the Australian Longitudinal Study of Ageing, who have given their time over many years, and without whom the present study would not have been possible. This study was supported in part by grants from the South Australian Health Commission, the Australian Rotary Health Research Fund, the US National Institute on Aging (Grant No. AG 08523-02), and the National Health and Medical Research Council Health Services Research Program. Sabine Schreiber of the Center for Ageing Studies, Flinders University, and staff in the Epidemiology Branch of the Department of Health in South Australia are also thanked for their assistance with tracing participants and identifying deaths. We also wish to acknowledge the helpful comments and suggestions made by Professor Maria Crotty on an earlier draft of this manuscript.

References

1. Australian Government Department of Health and Ageing: Aged Care in Australia. Canberra, Commonwealth of Australia; 2006.

2. South Australian Network for Research on Ageing: Fact sheet 22: Residential aged care. 1999.

3. Gaugler JE, Zarit SH, Pearlin LI: Caregiving and institutionalization: Perceptions of family conflict and socioemotional support. International Journal of Aging and Human Development 1999, 49:1–25.

4. Hogan WP: Review of pricing arrangements in Residential Aged Care. Canberra , Department of Health and Ageing; 2004:1–2.

5. Gaugler JE, Duval S, Anderson KA, Kane RL: Predicting nursing home admission in the US: a meta-analysis. BMC Geriatrics 2007, 7:13.

6. Miller EA, Weissert WG: Predicting elderly people's risk for nursing home placement, hospitalization, functional impairment and mortality: a synthesis. Medical Care Research and Review 2000, 57(3):259–297.

7. Coward RT, Netzer JK, Mullens RA: Residential differences in the incidence of nursing home admissions across a six year period. Journal of Gerontology: Social Sciences 1996, 51(5):S258–S267.

8. Freedman V: Family structure and the risk of nursing home admission. Journal of Gerontology: Social Sciences 1996, 51B:S61–S69.

9. Steinbach U: Social networks, institutionalization, and mortality among elderly people in the United States. Journal of Gerontology: Social Sciences 1992, 47(4):S183–S190.

10. Wan T, Weissert WG: Social support networks, patient status and institutionalization. Research on Aging 1981, 3:240–256.

11. Wilmoth JM: Unbalanced social exchanges and living arrangement transitions among older adults. Gerontologist 2000, 40(1):64–74.

12. Wolinsky FD, Callahan CM, Fitzgerald JF, Johnson RJ: The risk of nursing home placement and subsequent death among older adults. Journal of Gerontology: Social Sciences 1992, 47:S173–S182.

13. McCallum J, Simons LA, Simons J, Friedlander Y: Patterns and predictors of nursing home placement over 14 years: Dubbo study of elderly Australians. Australasian Journal on Ageing 2005, 24(3):169–173.

14. Wang JJ, Mitchell P, Smith W, Cumming RG, Leeder SR: Incidence of nursing home placement in a defined community. Medical Journal of Australia 2001, 174(6):271–275.

15. Andrews GR, Clark MS, Luszcz MA: Successful ageing in the Australian Longitudinal Study of Ageing: Applying the MacArthur model cross-nationally. Journal of Social Issues 2002, 58(4):749–765.

16. Finucane P, Giles LC, Withers RT, Silagy CA, Sedgwick A, Hamdorf PA, Halbert JA, Cobiac L, Clark MS, Andrews GR: Exercise profile and mortality in an elderly population. Australian and New Zealand Journal of Public Health 1997, 21:155–158.

17. Giles LC, Glonek GFV, Luszcz MA, Andrews GR: Effect of social networks on 10-year survival in very old Australians: the Australian Longitudinal Study of Ageing. Journal of Epidemiology and Community Health 2005, 59(7):574–579.

18. Glass TA, Mendes de Leon CF, Seeman TE, Berkman LF: Beyond single indicators of social networks: A LISREL analysis of social ties among the elderly. Social Science and Medicine 1997, 44:1503–1517.

19. Giles LC, Metcalf PA, Anderson CS, Andrews GR: Social networks among older Australians: A validation of Glass' model. Journal of Cancer Epidemiology and Prevention (formerly Journal of Epidemiology and Biostatistics) 2002, 7(4):195–204.

20. Rosenbaum PR, Rubin DR: The central role of the propensity score in observational studies for causal effects. Biometrika 1983, 70:41–55.

21. Joffe MM, Rosenbaum PR: Invited commentary: propensity scores. American Journal of Epidemiology 1999, 150(4):327–333.

22. D'Agostino RBJ: Propensity score methods for bias reduction in the comparison of a treatment to a non-randomized control group. Statistics in Medicine 1998, 17:2265–2281.

23. Imai K, van Dyk DA: Causal inference with general treatment regimes: generalizing the propensity score. Journal of the American Statistical Association 2004, 99(3):854–866.

24. Imbens G: The role of the propensity score in estimating dose-response functions. Biometrika 2000, 87(3):706–710.

25. Cochran WG: The effectiveness of adjustment by subclassification in removing bias in observational studies. Biometrics 1968, 24(2):295–313.

26. Huber PJ: The behavior of maximum likelihood estimation under non-standard conditions. Volume 1. Edited by: LeCam LM, Neyman J. University of California Press; 1967:221–233.

27. White H: Maximum likelihood estimation of misspecified models. Econometrica 1982, 50:1–25.

28. Kelman HR, Thomas C: Transitions between community and nursing home residence in an urban elderly population. Journal of Community Health 1990, 15(2):105–122.

29. Vicente L, Wiley JA, Carrington RA: The risk of institutionalization before death. The Gerontologist 1979, 19:361–367.

30. Gaugler JE, Edwards AB, Femia EE, Zarit SH, Parris Stephens MA, Townsend A, Greene R: Predictors of institutionalization of cognitively impaired elders: family help and the timing of placement. Journal of Gerontology: Psychological Sciences 2000, 55B(4):P247–P255.

31. Greenberg JN, Ginn A: A multivariate analysis of the predictors of long-term care placement. Home Health Care Services Quarterly 1979, 1:75–99.

32. Liu K, Coughlin T, McBride T: Predicting nursing-home admission and length of stay. Medical Care 1991, 29(2):125–141.

33. Palmore E: Total chance of institutionalization. The Gerontologist 1976, 16:504–507.

34. Lowenthal MF, Haven C: Interaction and adaptation: intimacy as a critical variable. American Sociological Review 1968, 33:20–30.

35. Bowling A: Measuring social networks and social support. In Measuring health: a review of quality of life measurement scales. 2nd edition. Philadelphia, Open University Press; 1997.

36. Buckwalter KC: Everybody needs a confidant (Editorial). Journal of Gerontological Nursing 2001, 27(5):4–5.

37. Wolinsky FD, Johnson RJ: The use of health services by older adults. Journal of Gerontology: Social Sciences 1991, 46:S345–S357.

38. Carstensen LL, Fung HH, Charles ST: Socioemotional selectivity theory and the regulation of emotion in the second half of life. Motivation and Emotion 2003, 27:103–123.

39. Anstey KJ, Luszcz MA: Mortality risk varies according to gender and change in depressive status in very old adults. Psychosomatic Medicine 2002, 64(6):880–888.

40. Luszcz MA, Bryan J, Kent P: Predicting episodic memory performance of very old men and women: contributions from age, depression, activity, cognitive ability, and speed. Psychology and Aging 1997, 12:340–351.

41. Van Tilburg T: Losing and gaining in old age: changes in personal network size and social support in a four-year longitudinal study. Journal of Gerontology: Social Sciences 1998, 53B:S313–S323.

42. Department of Health and Ageing: Aged Care Act 1997. Canberra, Department of Health and Ageing; 1997.

43. Sanchez L: Impairments of hearing and/or vision in the elderly and their relationship to health and functioning. In Speech Pathology. Adelaide, The Flinders University of South Australia; 1997.

44. Rosow I, Breslau N: A Guttman health scale for the aged. Journal of Gerontology 1966, 21:556–559.

45. Giles LC, Metcalf PA, Glonek GFV, Luszcz MA, Andrews GR: The effects of social networks upon disability in older Australians. Journal of Aging and Health 2004, 16(4):517–538.

46. Radloff LS: The CES-D scale: a self-report depression scale for research in the general population. Applied Psychological Measurement 1977, 1:385–401.

47. Folstein MF, Folstein SE, McHugh PR: Mini-Mental State: a practical method for grading the cognitive state of patients for the clinician. Journal of Psychiatric Research 1975, 12:189–198.

48. Barbor TF, de la Fuente JR, Saunders J, Grant M: AUDIT The alcohol use disorders identification test: Guidelines for use in primary health care. Geneva, World Health Organization; 1992.

49. Risk Factor Prevalence Study Management Committee: Risk Factor Prevalence Study: survey no. 3, 1989. Canberra, National Heart Foundation of Australia and the Australian Institute of Health; 1990.

CITATION

Originally published under the Creative Commons Attribution License. Giles LC, Glonek GFV, Luszcz MA, Andrews GR. "Do social networks affect the use of residential aged care among older Australians?," in BMC Geriatrics 2007, 7:24. © 2007 Giles et al; licensee BioMed Central Ltd. doi:10.1186/1471-2318-7-24.

Copyrights

Index

A

AAP. *See* Adelaide Activities Profile (AAP)
aBIC. *See* adjusted BIC (aBIC)
ACAT. *See* Aged Care Assessment Teams (ACAT)
ActiTrac° monitors, 145
active ageing, definition of, 121
activities of daily living (ADL), 19, 281
  predictor variables for functional deterioration in, 128
  relative operating characteristic analysis for, 129
  social participation and independence in, 129–33
    background of, 120–22
    data gathering, 123
    independence, measurement of, 125
    mental and physical health, measurement of, 124
    model A, 127–28
    model B, 128
    outcomes, 126–27
    personal finances, measurement of, 124
    population and sample, 122–23
    predictors of functional deterioration for, 128–29
    socio-demographic variables, 123–24
    statistical analysis, 125–26
AD. *See* Alzheimer's disease (AD)
Adelaide Activities Profile (AAP), 198, 202
Adelaide Statistical Division, 310
adjusted BIC (aBIC), 228
adjusted Lo–Mendell–Rubin likelihood ratio test (aLMR-LRT), 228
ADL. *See* activities of daily living (ADL)
Aged Care Assessment Teams (ACAT), 320, 321
age-related attenuation, of dominant hand superiority, 140–41
  age-related discrepancy, 151–53
  aging, general influence of, 153

cortical correlates, of hand dominance, 154–55

hand dominance with age, changes in, 153

methods, 141–42
  experiment 1, 142
  experiment 2, 145
  hand movements in everyday activities, assessment of, 145
  motor performance test-series, 142–45
  statistics, 145–46

outcomes
  experiment 1, 146–50
  experiment 2, 150–51
  questionnaires *vs.* practical performance, 155–56

aging, general influence of, 153. *See also* dominant hand superiority, age-related attenuation of

agitation, definition of, 278. *See also* older adults with dementia, discomfort and agitation in

AIC. *See* Akaike Information Criteria (AIC)

aiming, definition of, 143. *See also* dominant hand superiority, age-related attenuation of

Akaike Information Criteria (AIC), 228

aLMR-LRT. *See* adjusted Lo–Mendell–Rubin likelihood ratio test (aLMR-LRT)

alprazolam, 79, 165

ALSA. *See* Australian Longitudinal Study of Ageing (ALSA)

Alzheimer's disease (AD)
  and healthy aging, mental rotation of faces in, 174–77, 187–89
    face processing, AD effects on, 183–87
    face processing, aging effects on, 181–83
    materials and methods, 177–81

amitriptyline, 77, 79, 83

analysis of variance (ANOVA), 180, 183

usage of, 145

Anatomical Therapeutic Chemical (ATC) classification, 165

Annett handedness questionnaire, 141

ANOVA. *See* analysis of variance (ANOVA)

antidepressants, usage of, 163

antipsychotic medication, 166

ATC classification. *See* Anatomical Therapeutic Chemical (ATC) classification

Australian Longitudinal Study of Ageing (ALSA), 310

Australians, residential aged care in, 318–21
  background of, 308–10
  methods
    data collection and measures, 310–11
    propensity scores, 312–15
    residential care use, 311
    sample, 310
    social networks, 311–12
    statistical analysis, 315
  outcomes, 316–18

B

Backward Stepwise Wald Method, 125

BADLs. *See* basic or personal care activities (BADLs)

Barthel index, usage of, 267

Barthel scores, 36

baseline discomfort, 287

basic or personal care activities (BADLs), 125

Bayesian Information Criteria (BIC), 228

Beijing. *See also* China
  chronic disease prevalence and care in, 24–27
    analysis, 19–20
    background of, 14–16
    measures, 17–19
    outcomes, 20–24
    study design and catchment areas, 16–17

social-demographic characteristics in, 20
Beijing Longitudinal Ageing Study (BLAS), 15, 25
benzodiazepine, 77, 167
Beth Israel Deaconess Medical Center, 296
BIC. *See* Bayesian Information Criteria (BIC)
bifocal glasses, 195
bivariate correlation matrix
    usage of, 125
    for variables, 127
black box warning, 168
Blake shapes, 176
BLAS. *See* Beijing Longitudinal Ageing Study (BLAS)
Bonferroni correction, usage of, 180
bootstrapping technique, usage of, 95
buspirone, 77, 83, 165

C

Canadian Study of Health and Aging (CSHA), 91, 108
    composition of, 92
    secondary analysis of, 113
Capital District Health Authority, 108
captive audience, 35
Carstensen's socioemotional selectivity theory, 320
case-crossover, definition of, 210. *See also* hip/pelvic fractures in elderly people, in Stockholm
CCI. *See* Charlson Comorbidity Index (CCI)
Census Enumeration Areas, 91
ceramic biaxial piezoelectric accelerometer sensors, usage of, 145
change-in-support (CIS), 216
Charlson Comorbidity Index (CCI), 297
China. *See also* Beijing
    chronic disease prevalence and care in, 24–27

analysis, 19–20
background of, 14–16
measures, 17–19
outcomes, 20–24
study design and catchment areas, 16–17
China Health and Nutrition, 15
Chinese National Health Service Survey (CNHSS), 15
chi-square statistics, usage of, 297
chi-square tests, usage of, 80
chlorazepate, 79
chlordiazepoxide, 77, 79
chronic obstructive pulmonary disease, definition of, 17
CI. *See* confidence intervals (CI)
cigarette smoking, prevalence of, 249
CIS. *See* change-in-support (CIS)
CLESA Project, 132
Clinical Frailty Scale, 110, 240
clinical syndrome, 109
CMAI. *See* Cohen–Mansfield Agitation Inventory (CMAI)
CNHSS. *See* Chinese National Health Service Survey (CNHSS)
COGNITIVELY IMPAIRED (Cog-Imp), 233
Cognitively & physically impaired (Cog&Physic-Imp), 233
Cohen–Mansfield Agitation Inventory (CMAI), 279, 281
cohort, derivation of, 109
community-dwelling older adults, exercise impact in, 107–8, 113–14
    methods
        analysis, 110
        ethics statement, 108–9
        frailty index, 109–10
    outcomes, 110–12
community involvement, 122
community-living elderly, health status transitions in
    background of, 224–26

methods
   analysis strategy, 227–30
   data sources, 226–27
outcomes
   health state profiles, 232–36
   latent transition analyses, 236–37
   sample description, 230–31
confidence intervals (CI), 254
conventional fine motor test-series, usage
  of, 141
Cox proportional hazards regression, 94
CSHA. *See* Canadian Study of Health and
  Aging (CSHA)

D

DAT. *See* dementia of the Alzheimer's type
  (DAT)
Daxing, 17
  health status in, 21
  social-demographic characteristics in,
  20
DDD. *See* defined daily dosages (DDD)
Declaration of Helsinki, 142
defined daily dosages (DDD), 164
dementia, behavioral symptoms of, 277
dementia of the Alzheimer's type (DAT),
  277
dementia rating scale (DRS), 177
10/66 dementia research group cross-
  sectional survey, in China
  chronic disease prevalence and care in,
  24–27
    analysis, 19–20
    background of, 14–16
    measures, 17–19
    outcomes, 20–24
    study design and catchment areas,
    16–17
dementia with older adults, discomfort
  and agitation in, 286–88
  background of, 277–79
  methods
    measures, 280–83

participants, 279–80
outcomes of, 283–85
diazepam, 77–79
disability, definition of, 227. *See also*
  health status transitions, in older person
discomfort, definition of, 287. *See also*
  older adults with dementia, discomfort
  and agitation in
Discomfort Scale for patients with
  Dementia of the Alzheimer Type (DS-
  DAT), 281
Disease Analyzer database, 78, 79
dixyrazine, 165
dominant hand superiority, age-related
  attenuation of, 140–41
  age-related discrepancy, 151–53
  aging, general influence of, 153
  cortical correlates, of hand dominance,
  154–55
  hand dominance with age, changes in,
  153
  methods, 141–42
    experiment 1, 142
    experiment 2, 145
    hand movements in everyday activi-
    ties, assessment of, 145
    motor performance test-series,
    142–45
    statistics, 145–46
  outcomes
    experiment 1, 146–50
    experiment 2, 150–51
  questionnaires *vs.* practical performance, 155–56
doxepin, 77, 79, 83, 84
DRS. *See* dementia rating scale (DRS)
DS-DAT. *See* Discomfort Scale for patients with Dementia of the Alzheimer
  Type (DS-DAT)

E

Edinburgh Handedness Inventory (EHI),
  142

Edinburgh Handedness Questionnaire, 141
EHI. *See* Edinburgh Handedness Inventory (EHI)
elderly people, social vulnerability, frailty and mortality in, 90–91, 99–102
  materials and methods
    measures, 92–94
    statistical analysis, 94–95
    study samples, 91–92
  outcomes of
    descriptive analyses, 95–97
    mortality, 97–98
    re-sampling techniques, 98
elderly smokers, mobile smoking cessation service in, 258–60
  background of, 249–50
  methods
    data analysis, 254
    data collection, 252–53
    evaluation, 253–54
    follow up assessment, 253
    MSCP, 250–51
    target population and recruitment, 251–52
  outcomes of
    MSCP, costs of, 257–58
    outcome evaluation, 256–57
    quitting at 6 months, factors associated with, 257
    utilisation and process evaluation, 254–56
elderly, suicide risk in, 167–68
  background of, 163–65
  clinician, implications for, 169
  methods, 165
    ethics, 166
    statistics, 166
  outcomes of, 166–67
  strengths and limitations, 168–69
elderly women's mammography screening decisions, factors influencing, 301–3
  background of, 294–95

methods
  data collection, 296–97
  statistical analyses, 297
  study sample, 295–96
outcomes of, 298–301
emotional stress
  assessment of, 218
  definition of, 211
  in hip and pelvic fractures, 216
    background of, 209–10
    methodological considerations, 217–19
    methods, 210–14
    outcomes of, 214–16
Ethics Committee at the Regional Government of Aragon, 126
European Society, of Hypertension criteria, 17
exercise impact, in community-dwelling older adults, 107–8, 113–14
  methods
    analysis, 110
    ethics statement, 108–9
    frailty index, 109–10
  outcomes, 110–12

F

Falls Efficacy Scale—International (FES-I), 198
FAST. *See* Functional Assessment Staging (FAST)
FES-I. *See* Falls Efficacy Scale—International (FES-I)
FI. *See* Frailty Index (FI)
fine motor performance, laterality indices for, 148, 149
finite mixture models, 225
Fisher's exact tests, 80
flurazepam, 77, 79
Frailty Index (FI), 95, 109–10, 240. *See also* older adults, exercise impact in community-dwelling
frailty, social vulnerability and mortality, in elderly people, 90–91, 99–102

materials and methods
measures, 92–94
statistical analysis, 94–95
study samples, 91–92
outcomes of
descriptive analyses, 95–97
mortality, 97–98
re-sampling techniques, 98
Functional Assessment Staging (FAST), 280
Functional Autonomy Measurement System (SMAF), 226, 281
functional limitations, definition of, 227. *See also* health status transitions, in older person

G

GAD. *See* generalized anxiety disorder (GAD)
GDS. *See* Geriatric Depression Scale (GDS)
General Household Survey, 249
generalized anxiety disorder (GAD), 76
inappropriate prescribing in older patients with, 76, 83–85
background of, 76–78
IMS MediPlus, data from, 78
IMS MediPlus, information in, 79–80
outcomes of, 81–83
medications for treatment of, 80, 82
General Linear Modeling, usage of, 20
Geriatric Depression Scale (GDS), 227, 268, 273
Geriatric Mental State, usage of, 17
Germany, inappropriate prescribing in older patients with GAD, 83–85
background of, 76–78
IMS MediPlus
data from, 78
information in, 79–80
outcomes of, 81–83

GoM. *See* grade of membership (GoM)
Gothenburg
elderly, suicide risk in, 167–68
background of, 163–65
clinician, implications for, 169
ethics, 166
outcomes of, 166–67
statistics, 166
strengths and limitations, 168–69
Gothenburg Institute of Forensic Medicine, 165
GPower software, 283
grade of membership (GoM), 225

H

halazepam, 79
hand dominance, cortical correlates of, 154–55
handedness, definition of, 141
hand preference, determination of, 142
hand superiority, age-related attenuation of, 140–41
age-related discrepancy, 151–53
aging, general influence of, 153
cortical correlates, of hand dominance, 154–55
hand dominance with age, changes in, 153
methods, 141–42
experiment 1, 142
experiment 2, 145
hand movements in everyday activities, assessment of, 145
motor performance test-series, 142–45
statistics, 145–46
outcomes
experiment 1, 146–50
experiment 2, 150–51
questionnaires *vs.* practical performance, 155–56

HAROLD. *See* hemispheric asymmetry reduction in older adults (HAROLD)
health care utilisation and SMC, association between, 53–55
  background of, 46–47
  ethics, 50
  measurements, 47–49
  outcomes, 47, 50–52
  statistical analysis, 49–50
  study population, 47
health status transitions, in older person, 237–40
  background of, 224–26
  methods
    analysis strategy, 227–30
    data sources, 226–27
  outcomes
    health state profiles, 232–36
    latent transition analyses, 236–37
    sample description, 230–31
healthy aging and AD, mental rotation of faces in, 174–77, 187–89
  materials and methods
    face processing, AD effects on, 180–81
    face processing, aging effects on, 177–80
  outcomes
    face processing, AD effects on, 183–87
    face processing, aging effects on, 181–83
hemispheric asymmetry reduction in older adults (HAROLD), 154
hip/pelvic fractures in elderly people, in Stockholm, 216
  background of, 209–10
  methodological considerations, 217–19
  methods, 210–14
  outcomes of, 214–16
Hong Kong, mobile smoking cessation service in, 258–60
  background of, 249–50

  methods
    data analysis, 254
    data collection, 252–53
    evaluation, 253–54
    follow up assessment, 253
    MSCP, 250–51
    target population and recruitment, 251–52
  outcomes
    MSCP, costs of, 257–58
    outcome evaluation, 256–57
    quitting at 6 months, factors associated with, 257
    utilisation and process evaluation, 254–56
Hong Kong Smoking Cessation Health Center, 259
Hosmer–Lemeshow test, 125
HREC. *See* Human Research Ethics Committee (HREC)
Hubei Province, 25
Huber-White robust variance estimator, 315
Human Research Ethics Committee (HREC), 197
hypnotics
  classification of, 165
  usage of, 163–64

I

IADLs. *See* instrumental activities (IADLs)
ICP. *See* principal investigator (ICP)
IMS MediPlus. *See also* generalized anxiety disorder (GAD)
  data from, 78
  information in, 79–80
institutional review board (IRB), 79
instrumental activities (IADLs), 125
InterASIA survey, 26
inter-quartile range (IQR), 71
IRB. *See* institutional review board (IRB)

J

jackknifing technique, usage of, 95
Jebsen Test of Hand Function, 141

K

Kaplan Meier curves, usage of, 94
Kendall's Tau-b correlation coefficient, usage of, 125

L

latent class analysis (LCA), 225, 226
latent transition analysis (LTA), 225, 226
LCA. *See* latent class analysis (LCA)
levomepromazine, 165
LGA. *See* Local Government Area (LGA)
line tracing, definition of, 143. *See also* dominant hand superiority, age-related attenuation of
Local Government Area (LGA), 310
logistic regression model, 125
    usage of, 94
London
    research with older people living in nursing homes in
        background of, 34–35
        challenges, 36–38
        ensuring privacy and reassuring residents, 41–42
        methods, 35–36
        residents' feelings about taking part in research, 38–40
        seizing opportunities and staff involvement, 41
        taking informed consent, 40–41
lorazepam, 77, 79
lost to follow-up (LTF), 229, 230
LTA. *See* latent transition analysis (LTA)
LTF. *See* lost to follow-up (LTF)

M

Madrid
    older people, self-rated health in, 271–73
analysis, 268–69
background of, 265–66
chronic conditions, 267
cognitive status, 267–68
data collection, 266
depressive symptoms, 268
functional status, 267
outcomes of, 269–71
population and selection of participants, 266
self-rated health, 267
social interaction, 268
socio-demographic variables, 267
vision and hearing, 268
MAR. *See* missing at random mechanism (MAR)
Markov model, 111
    usage of, 226
Max-Planck face database, 178
Max-Planck Institute for Biological Cybernetics, 178
MDS-COGS. *See* Minimum Data Set Cognition Scale (MDS-COGS)
Melbourne edge test, 197
mental rotation of faces, in healthy aging and AD, 174–77, 187–89
    materials and methods
        face processing, AD effects on, 180–81
        face processing, aging effects on, 177–80
    outcomes of
        face processing, AD effects on, 183–87
        face processing, aging effects on, 181–83
mini mental state examination (MMSE), 142, 177, 197, 198
Minimum Data Set Cognition Scale (MDS-COGS), 267, 273
missing at random mechanism (MAR), 228
MLS. *See* motor test-series (MLS)

MMSE. *See* mini mental state examination (MMSE)

mobile clinical service, usage of, 250

Mobile Smoking Cessation Program (MSCP), 250–51
  costs of, 257–58
  subject recruitment in, 252

mobile smoking cessation service, in elderly smokers, 258–60
  background of, 249–50
  methods
    data analysis, 254
    data collection, 252–53
    evaluation, 253–54
    follow up assessment, 253
    MSCP, 250–51
    target population and recruitment, 251–52
  outcomes
    MSCP, costs of, 257–58
    outcome evaluation, 256–57
    quitting at 6 months, factors associated with, 257
    utilisation and process evaluation, 254–56

motor performance test-series, 142–45. *See also* dominant hand superiority, age-related attenuation of

motor test-series (MLS), 151
  device, usage of, 143, 144

Mplus, usage of, 227

MSCP. *See* Mobile Smoking Cessation Program (MSCP)

multivariate logistic regression, usage of, 166

N

Nagelkerke's R2 coefficient, 125

NAPB. *See* non-aggressive physical behavior (NAPB)

National Population Health Survey (NPHS), 91

Nicalert test, 256

non-aggressive physical behavior (NAPB), 278

NPHS. *See* National Population Health Survey (NPHS)

O

OA. *See* overall agitation (OA)

OARS-MAFQ. *See* Older Americans Resources and Services Program-Multidimensional Functional Assessment Questionnaire (OARS-MAFQ)

OARS questionnaire, 123

odds ratio (OR), 318

"Old Age and Dependency in Aragon", 122

older adults, exercise impact in community-dwelling, 107–8, 113–14
  methods
    analysis, 110
    ethics statement, 108–9
    frailty index, 109–10
  outcomes, 110–12

older adults with dementia, discomfort and agitation in, 286–88
  background of, 277–79
  methods
    measures, 280–83
    participants, 279–80
  outcomes of, 283–85

Older Americans Resources and Services Program-Multidimensional Functional Assessment Questionnaire (OARS-MAFQ), 123

older Australians, residential aged care in, 318–21
  background of, 308–10
  methods
    data collection and measures, 310–11
    propensity scores, 312–15
    residential care use, 311
    sample, 310
    social networks, 311–12

statistical analysis, 315
outcomes, 316–18
older multifocal glasses wearers, preventing falls in, 203–4
background of, 195–96
methods
initial assessment, 197–98
intervention, 198–201
participants, 196–97
randomization, 198
outcomes
compliance and adverse events, 202
primary outcome measure, 201–2
sample size calculation, 203
statistical analysis, 203
older patients with GAD, inappropriate prescribing in, 83–85
background of, 76–78
IMS MediPlus
data from, 78
information in, 79–80
outcomes of, 81–83
older people in Madrid, self-rated health in, 271–73
analysis, 268–69
background of, 265–66
chronic conditions, 267
cognitive status, 267–68
data collection, 266
depressive symptoms, 268
functional status, 267
outcomes of, 269–71
population and selection of participants, 266
self-rated health, 267
social interaction, 268
socio-demographic variables, 267
vision and hearing, 268
older person, health status transitions in
background of, 224–26
methods
analysis strategy, 227–30
data sources, 226–27

outcomes
health state profiles, 232–36
latent transition analyses, 236–37
sample description, 230–31
OR. *See* odds ratio (OR)
ordinal logistic models, 314
orientation agnosia, 176
overall agitation (OA), 281
oxazepam, 79, 165

P

Pearson correlation coefficient, 280, 283
Pearson's bivariate correlations, 186
pegboard test, 141
personal finances, measurement of, 124
PHYSICALLY IMPAIRED (Physic-Imp), 234
physiological profile assessment (PPA), 198
Poisson regression, usage of, 19
PPA. *See* physiological profile assessment (PPA)
presbyopia, 195
prevalence ratios (PRs), 19
primary outcome measure, 201–2. *See also* older multifocal glasses wearers, preventing falls in
principal investigator (ICP), 282
propensity score, 313–14
PRs. *See* prevalence ratios (PRs)
psychological autopsies, 168
psychotropic drugs, usage of, 163
PsyScope experimental software version 1.2.1, 178

Q

quitting smoking, definition of, 254

R

receiver operating characteristic (ROC), 128, 129
registered nurses (RNs), 279

RELATIVELY HEALTHY (R-Healthy), 234
Research Agenda on Ageing Project, 16
Research Ethics Committees, 108
Revised Waterloo Handedness Questionnaire, 141
R-Healthy. *See* RELATIVELY HEALTHY (R-Healthy)
RNs. *See* registered nurses (RNs)
ROC. *See* receiver operating characteristic (ROC)
Ruhr-University Bochum, 142

S

SAS software package version 8.02, 214
SAS statistical software, 297
SCT. *See* social cognitive theory (SCT)
secondary outcome measure, 202. *See also* older multifocal glasses wearers, preventing falls in
sedatives
    classification of, 165
    usage of, 163–64
selective optimisation with compensation (SOC), 121
selective serotonin reuptake inhibitors (SSRIs), 77
self-rated hand dominance and active performance, age-related discrepancy in, 151–53
self-reporting questionnaire, 19
sensors, usage of, 141
Serial Trial Intervention, 286
short portable mental status questionnaire (SPMSQ), 211, 227, 267
Short Psychiatry Evaluation Schedule, 124
single-lens distance glasses, for older multifocal glasses wearers, 203–4
    background of, 195–96
    methods
        initial assessment, 197–98
        intervention, 198–201
        participants, 196–97
        randomization, 198

outcomes
    compliance and adverse events, 202
    primary outcome measure, 201–2
    sample size calculation, 203
    statistical analysis, 203
SIPA program, 226, 227
SMAF. *See* Functional Autonomy Measurement System (SMAF)
SMC. *See* subjective memory complaints (SMC)
SOC. *See* selective optimisation with compensation (SOC)
social cognitive theory (SCT), 251
social engagement, 122
social networks, affects residential aged care
    background of, 308–10
    methods
        data collection and measures, 310–11
        propensity scores, 312–15
        residential care use, 311
        sample, 310
        social networks, 311–12
        statistical analysis, 315
    outcomes, 316–18
social participation and independence, in ADLs, 129–33. *See also* activities of daily living (ADL)
    background of, 120–22
    data gathering, 123
    independence, measurement of, 125
    mental and physical health, measurement of, 124
    model A, 127–28
    model B, 128
    outcomes, 126–27
    personal finances, measurement of, 124
    population and sample, 122–23
    predictors of functional deterioration for, 128–29
    socio-demographic variables, 123–24
    statistical analysis, 125–26

social vulnerability, frailty and mortality, in elderly people, 90–91, 99–102
   materials and methods
      measures, 92–94
      statistical analysis, 94–95
      study samples, 91–92
   outcomes
      descriptive analyses, 95–97
      mortality, 97–98
      re-sampling techniques, 98
social vulnerability index, usage of, 99
Spain
   older people, self-rated health in, 271–73
      analysis, 268–69
      background of, 265–66
      chronic conditions, 267
      cognitive status, 267–68
      data collection, 266
      depressive symptoms, 268
      functional status, 267
      outcomes of, 269–71
      population and selection of partici-pants, 266
      self-rated health, 267
      social interaction, 268
      socio-demographic variables, 267
      vision and hearing, 268
Spanish language version, 123
SPMSQ. *See* short portable mental status questionnaire (SPMSQ)
SSRIs. *See* selective serotonin reuptake inhibitors (SSRIs)
STATA 9.2, 20
STATA 8 software, 95
steadiness, definition of, 143. *See also* dominant hand superiority, age-related attenuation of
Stockholm
   hip or pelvic fractures in elderly people in, 216
      background of, 209–10
      methodological considerations, 217–19

methods, 210–14
outcomes, 214–16
subjective memory complaints (SMC)
   and health care utilisation, association between, 53–55
      background of, 46–47
      ethics, 50
      measurements, 47–49
      outcomes, 47, 50–52
      statistical analysis, 49–50
      study population, 47
successful ageing, 120
suicide risk, in elderly, 167–68
   background of, 163–65
   clinician, implications for, 169
   methods, 165
      ethics, 166
      statistics, 166
   outcomes of, 166–67
   strengths and limitations, 168–69
Sweden
   elderly, suicide risk in, 167–68
      background of, 163–65
      clinician, implications for, 169
      ethics, 166
      outcomes of, 166–67
      statistics, 166
      strengths and limitations, 168–69
   hip or pelvic fractures in elderly people in, 216
      background of, 209–10
      methodological considerations, 217–19
      methods, 210–14
      outcomes of, 214–16
   sales of psychotropic drugs in, 164

T

tapping, definition of, 143. *See also* dominant hand superiority, age-related attenuation of
tapping-tasks test, 141

TCAs. *See* tricyclic antidepressants (TCAs)
temazepam, 79
Third Chinese National Health Services Survey, 25
Tiananmen Square, 17
timed up and go test (TUG), 197
ToFa. *See* triggers of fall (ToFa)
triazolam, 79
tricyclic antidepressants (TCAs), 77
triggers of fall (ToFa), 210
    in elderly people, 216
        background of, 209–10
        methodological considerations, 217–19
        methods, 210–14
        outcomes of, 214–16
TUG. *See* timed up and go test (TUG)

U

University of Manitoba's Research Ethics Board, 177
urban and rural Beijing
    chronic disease prevalence and care in, 24–27
        analysis, 19–20
        background of, 14–16
        measures, 17–19
        outcomes, 20–24
        study design and catchment areas, 16–17
urine cotinine test, 256
U.S. Medical Expenditure Panel Survey, 78
U.S. national cohort study, of older adults, 69–71
    background of, 60–61
    outcomes of, 64–69
    statistical analysis, 64
    subjects, 61–63

U.S. National Medical Expenditure Survey, 78

V

variable inflation factor (VIF), 125
venlafaxine, 77, 82
Vienna-test-system software version 5.05, 143
VIF. *See* variable inflation factor (VIF)
Visual Intervention Strategy Incorporating Bifocal & Long-distance Eyewear (VISIBLE), 196

W

WatHand Box Test, 141
WHODAS. *See* World Health Organization Disability Assessment Schedule (WHODAS)
Wilcoxon Rank Sum Test, 297
women's mammography screening decisions, factors influencing, 301–3
    background of, 294–95
    methods
        data collection, 296–97
        statistical analyses, 297
        study sample, 295–96
    outcomes of, 298–301
World Health Organisation, active ageing introduction by, 121
World Health Organization Disability Assessment Schedule (WHODAS), 18

X

Xicheng, health status in, 21

Z

Zarit Burden Interview (ZBI), 18, 23
zolpidem, 79, 165